Global Technological Change

Global Technological Change
From Hard Technology to Soft Technology

Zhouying Jin

Translated by

Kelvin W. Willoughby
Professor of Management Graduate School of
Business, Curtin University of Technology

and

Ying Bai
Ph.D., Beijing Academy of Soft Technology

intellect Bristol, UK / Chicago, USA

First published in the UK in 2005 by
Intellect, The Mill, Parnall Road, Fishponds, Bristol, BS16 3JG, UK

Second edition published in the UK in 2011 by
Intellect, The Mill, Parnall Road, Fishponds, Bristol, BS16 3JG, UK

Second edition published in the USA in 2011
Intellect, The University of Chicago Press, 1427 E. 60th Street,
Chicago, IL 60637, USA

The second English edition translated from the second Chinese edition
Published by Peking University Press in Beijing, July 2010.
"JIN Zhouying, Global Technological Change From Hard Technology to
Soft Technology", ISBN 978-7-301-17548-4

Copyright © 2011 Intellect Ltd

All rights reserved. No part of this publication may be reproduced,
stored in a retrieval system, or transmitted, in any form or by
any means, electronic, mechanical, photocopying, recording, or
otherwise, without written permission.

A catalogue record for this book is available from the
British Library.

Cover designer: Holly Rose
Copy-editor: Michael Eckhardt
Typesetting: Mac Style, Beverley, E. Yorkshire

ISBN 978-1-84150-376-9

Contents

Preface to the English Edition: Theodore Gordon	11
Acknowledgements	13
Introduction	15
Chapter 1: What is Technology?	19
A. The Evaluation of Technology	22
B. A Fresh Understanding of Technology - To Understand Another Paradigm of Technology - Soft Technology	26
1. Many Hard Technologies Derived from Natural Science-based Knowledge are Softening and Becoming 'Soft' Technology	27
2. The Process of Transferring Hard Technologies: Commercialization and Industrialization are 'Process Technologies'	28
3. Various Commercial Technologies are the Direct Driving Forces of Socio-economic Development	29
4. There is Technology in the Innovation Process of Social Activity - Social Technology	32
5. The Innovation Process of Cultural and Artistic Activities is a Technology - Cultural Technology	33
6. Social Progress Requires the Integration of Technology and Art - Artistic Technology	34
7. The Diagnostic and Therapeutic Technology of Chinese Medicine: Another Type of Soft Technology - A Technology Derived from the Knowledge of Chinese Medicine and a Special Problem-Solving Approach	35
8. Technology Focused on the Human Mind and Relationship between Mind/Body - Psychogenic Technology and Psychosomatic Technology	40
Conclusion - The Need to Renew the Notion of Technology	43
C. What is Soft Technology?	44
1. Knowledge, Technology and Soft Technology	44
2. What is Soft Technology?	45
3. The Characteristics of Soft Technology	48

D. The Classifications of Soft Technology	54
1. Classification According to Sources of Knowledge	54
2. Classification Depends on Operational Field	55
3. Classification Depends on Operational Resources	55
E. Why Soft Technology Now?	59
1. Soft Technology Opens Up a New Discipline, a New Field of Research, Where the Value Creation Process is a Completely Different Paradigm	60
2. The Characteristics of Soft Technology are Suited for Coping with the Challenges in the Twenty-first Century	61
3. Grasping the Orientation of Technological Progress and Direction of Innovation - To Adhere to a Correct Soft-Tech Operation and Soft Environmental Design	63
4. To Identify 'Non-Technological Factors', Unveiling the 'Black Boxes' of Total Factor Productivity	65
5. Revealing the Essence of Creating Value for Intellectual Capital	68
F. Soft-Tech Trends and Challenges for the Future	71
1. Soft Technology is Transforming Hard Technology	71
2. More and More Soft Technologies are Creating New Industries	72
3. The Rapid Rise of New Soft Technology Fields	72
4. Increasing Soft Technology to Become High Technology	73
5. Intellectual Property of Soft Technology	74
6. Challenges for Soft-Tech Development	76
Chapter 2: Historical Antecedents of Soft Technology	**77**
A. The History of Commercial Technology: A Brief Analysis	79
1. Currency Technology	79
2. Accounting Technology	82
3. Patent Technology	84
4. Advertising Technology - Technology of Propagandizing and Inducement	88
5. Insurance System	90
6. Management Technology	92
7. Stock Technology and Securities Technology	97
8. Merger and Purchasing Technology	100
9. Venture Capital Technology	103
10. Logistics Technology	105
11. Supply Chain Technology	109
12. Incubator Technology	110
13. Tactics Technology	111
14. Business Model and Management Pattern	113
B. A Retrospective of Social Technology Development	117
1. Research about Relevant Social Problems in Industrial Countries	117
2. Innovation in Social Sciences and Social Technologies is Extremely Urgent	120
3. Distinguishing Social Science from Social Technology	121

4. Social Technology and Its Value		124
5. Social Resources		129
6. The Mechanism of Research Institutes		130
7. Virtual Technology and Organizational Innovation		131
8. Public Relations Technology		139

C. Soft Technology and Thrice Industrial Revolutions — 141
D. Soft Science, Soft Series of Science & Technology (SSST) and Soft Technology — 148
  1. Research Course of Soft Science — 148
  2. Soft Series of Science & Technology and Soft Science — 152
  3. From Soft Series of Science & Technology to Soft Technology — 157

Chapter 3: Soft Technology and Technological Competitiveness — 159

A. Knowledge and Technology are Merely Potential Sources of Competitiveness — 161
B. Where does Technological Competitiveness Come From? — 165
  1. Strong R&D Capacity is the Source for Creating Competitiveness — 165
  2. Soft Technology is the Tool for Creating Competitiveness — 166
  3. Hard Environments and Soft Environments are the Basic Conditions for Competitiveness - Beyond Technology: Institutions, Culture and Values — 170
  4. Exceeding the Power of Institutions - Cultural Innovation — 175
  5. The Three Key Factors of Competitiveness — 181
C. Comprehensive Competitiveness and Soft Power — 183
  1. How to Study Soft Power — 183
  2. From National Soft Power to Corporate Soft Power — 187
  3. Future 500: The Successful Case of Assisting Enterprises to Upgrade their Soft Power — 189
D. The Essence of the Gap Between Developed and Developing Countries — 190
  1. Keeping a Clear Head and Pinpointing of Their Own Strength and Weakness — 191
  2. To Actualize 'Leap Frog' Development and Boost Institutional Innovation — 191
  3. The Serious Shortage of Soft-Tech Experts is the Core of This Gap Regardless of the National Level, Industrial Level or Enterprise Level — 192

Chapter 4: Soft Technology and Innovation — 199

A. The Functions of Soft Technology — 201
  1. Soft Technology Provides Innovation Ability of Hard Technology — 201
  2. Soft Technology is to Act as the Core Technology to Create Independent Industries - Providing Core Technology for Soft Industries — 205
  3. Soft Technology Provides Content and a Basis for Institutional Innovation — 206
B. Soft Technology and Innovative Space — 207
C. Soft Technology and Institutional Innovation — 211
  1. The Essence of Institutional Innovation — 213
  2. Making Soft-Tech Institutionalization and 'Mechanism-ization' Keep Pace with Technological Innovation as well as Socio-Economic Development –

	The Enlightenment from Thousands of Years of Technology History and Hundreds of the History of Industrial Revolution	217

- The Enlightenment from Thousands of Years of Technology History and Hundreds of the History of Industrial Revolution — 217
- 3. The Relative Rigidity of Institutions and the Difficulty of Institutional Innovation — 221
- 4. Establishing an Institutional Audit and Evaluation System — 224
- 5. The Limitations of Institutions — 224
- 6. Technological Institutions — 228
- D. Soft Technology and the Innovation System Framework — 235
  - 1. Innovation Approach and Technology Innovation System Framework — 235
  - 2. The Main Body of Innovative Activities and the System of Technological Innovation — 239
  - 3. The Advantages and Disadvantages of the Technological Innovation System in China — 241
- E. Innovation and Corporate Competitiveness – Changing the Thinking Mode to Adjust to Corporate Strategy — 244
  - 1. Strategy Innovation – Opening Up and Adjusting the Business Divisions to Expand the Life Cycle of Enterprises — 244
  - 2. Strengthening the R&D for Soft Technology — 249
  - 3. Organizational Innovation — 251
  - 4. Talent Strategy — 252
  - 5. Product Strategy — 254
  - 6. Market Innovation — 255
  - 7. Changes in Corporate Culture and Values – Shaping 'Good' Enterprises of the Twenty-first Century — 256

Chapter 5: Soft Industries — 259
- A. Economic Softening and Soft Industry — 261
  - 1. Economic Softening in the Twenty-first Century — 261
  - 2. The Softening of Primary Industries and the Agriculture Service Industry — 266
  - 3. Soft Industry — 270
  - 4. Soft Industries and So-called Creative Industries — 274
  - 5. The Characteristics of the Soft Industry — 277
- B. Intellectual Service Industry — 279
  - 1. What is Service? — 279
  - 2. The Essence of Service and Service Innovation — 283
  - 3. Service Economy and Intellectualization of Service Industry — 286
  - 4. The Intellectual Service Industry in the Narrow Sense — 290
- C. Social Enterprise and Social Industry — 294
  - 1. More Attention is Being Given to Various Types of Social Industries — 294
  - 2. The Significance of Social Industry — 295
  - 3. The Social Market — 298
  - 4. The Characteristics of Social Capital and Social Industry — 299

5. The Types of Social Industry		302
6. Education Industry		304
D. Cultural Industries		304
1. Culture and Cultural Values		305
2. Understanding of Cultural Industry		307
3. The Classification of Cultural Industries		310
4. Culture Service Industry		315
5. A Reflection on the Commercialization of Culture and Arts		316
E. Think about the Future Industrial Structure		317
1. About High Technology and High-Tech Industry		320
2. Subdividing Service Industry		321
3. About Life Industry		322
Chapter 6: Soft Technology and the Fourth Generation of Technology Foresight		**329**
A. The Evolution and Development of Technology Foresight		331
1. The Three Climaxes of Technology Foresight		332
2. From Technology Forecasting to Technology Foresight		333
3. The Theory of the Four Stages of Technology Foresight		335
B. The Fourth Generation of Technology Foresight and Soft Technology		338
1. The Goals of Technology Foresight		338
2. Multiple Driving Forces of Technology and Technology Foresight		340
3. Soft Technology, Soft Environment and Technological Foresight		341
4. The Causal Analysis of the Failure of Technology Forecasting		343
5. Technology Foresight in Developing Countries		344
Postscript: The Principles for Development in the Twenty-first Century – Harmony, Balance, and Coexistence		**347**
About the Author		**353**
Bibliography		**355**

Preface to the Second English Edition

Eight years ago I wished Professor Jin good luck in implementing the new discipline she called 'soft technology'. The need for solutions to problems in this non-physical domain was clear enough, but starting new disciples is difficult and rarely successful. We read in the history of science about the cool reception given to new ideas that challenge old precepts,[1] the fate of most of the scientists that propose those new ideas, and finally, occasionally, the rallying around the new ideas that are destined to survive. Is it too early to consider whether the notion of 'soft technology' has taken root?

Professor Jin defines soft technology as the 'knowledge derived from the social sciences, non-natural sciences and non-scientific [traditional] knowledge to solve various practical problems'. It is that sphere of technology outside of the physical, beyond the machines and tools of physical technology. It is, she says, focused on human thought, not things, and is the realm of 'ideology, emotion, values, worldview, individual and organizational behaviours, as well as human society'.

Soft technology is probably older than hard technology, but hard technology is more systematically codified and understood. Hard technologies exist because of invention but the invention process itself and the uses made of the hard technologies come from the soft side. Moral and ethical considerations are not a part of hard technology. How often have we heard that 'technology cannot itself be evil, the evil lies in the way it's used'? Hard technology relies on laws of nature and information about how to do things; soft technology falls back on the inner self and ancient epistemologies.

Both hard and soft technologies involve knowledge systems, invention, and creativity; both are important because they affect the human condition but they operate on vastly different wavelengths.

So what has happened in the eight years since the first edition of this book?

- The Beijing Academy of Soft Technology was established in 2002 as a virtual institute; it now employs several full-time staff and several part-time members, who contribute depending on the project's need.
- Professor Jin and her colleagues have written several books on soft technology and a few of the issues it addresses.[2]

- The Institute has contracted with the Beijing Academy of Science and Technology, on future studies; with the National Academy of Soft Power at Peking University, and with the Chinese Academy of Social Sciences.
- Soft Technology as a new discipline was placed on the agenda of eleventh plan (2005–2010) for important research field of Institute of Quanti-Economics and Techno-Economics, Chinese Academy of Social Sciences.
- Over twenty papers dealing with soft technology topics have been published and delivered in symposia on topics that have included: the role of soft technology in the new economy, technology foresight, innovation, socio-economic development, trends in the service industry, technological progress and evolution, globalization, technological competitiveness of developing countries, strategic management and institutional innovation of China's coal-bed methane industry, the knowledge economy and the robot policy of China.
- Soft technology has become international. In many countries, some scholars are using soft-technology, and the US, Austria, Brazil and Malaysia have expressed interest in establishing a soft technology innovation system framework.
- Professor Jin has lectured in a dozen countries and thirty universities, corporations, and conferences about the theory and application of the soft technology; for instance she was invited as a speaker for the first 'Festival of Thinkers' with ten Nobel Laureates in 2005; and to the International Conference of Innovation Management in Brazil in 2009.

In the preface to the first edition, I wrote:

> One can think of dozens of necessary soft technology inventions. Consider inventing a way to protect intellectual property that rewards the inventor but does not withhold the fruits of the invention from people who need it but cannot afford it. Consider a soft technology for encouraging the use of futures research in decision-making or the development of a new decision science that goes beyond economic cost benefit and includes intuition, explicit risk-taking, artificial intelligence and neuropsychiatry. How can conflict resolution be improved? How can old ethnic hatreds be tamed? Or consider how, in this modern world, children, CEOs, clergy and politicians can learn values and moral behaviour. These are worthy soft technology research projects.
>
> There are a few other prospective soft technology inventions that deserve some thought. Science and the hard technologies that flow from it have contributed to our material world for better or worse; improved health, lengthened life and, despite the poverty gap, increased abundance for most people. But on the horizon are possible developments flowing from science that seem threatening and give us cause to pause. Further, science left to its own mechanisms seems unlikely to provide solutions to pressing global problems. In a nutshell, how can science help capture the best, and avoid the worst, that the future has to offer? Or to put it a different way, how can soft technology help shape science to make life better and less risky for all?

Not many of these problems and opportunities have yet been addressed fully by the fledgling discipline, but the road ahead is becoming clearer. More is possible, more is needed, but by any measure, there has been a good start.

Theodore Jay Gordon[3]
March 2010

Notes

1. See for example: Thomas Kuhn, *The Structure of Scientific Revolutions*.
2. Jin, Zhouying and Ren, Lin (2004), *Service Innovation and Social Resource*, Beijing: Financial and Economic Publishing House; Jin Zhouying; Jiang, Jinhe and Gong, Feihong (2006), *Long-term Strategy System Integration and Sustainable Development*, Beijing: Social Sciences Document Press, August.
3. Theodore Jay Gordon is a Senior Research Fellow for the Millennium Project. He started The Futures Group in 1971 and is the author of five books and hundreds of papers dealing with topics associated with the future, space, scientific and technological developments and global issues. He was a consultant to RAND, an early contributor to the use of the Delphi method and the inventor of several futures research techniques.

Acknowledgements

My sincere appreciation goes to those who helped so much during the preparation for the second edition of this book: Professors Theodore Jay Gordon, Hazel Henderson, Rinaldo S. Brutoco from the United States of America; Professor Karamjit S. Gill, Sir Geoffrey Lloyd from the UK; Alexander G. Welzl from Austria; Professors Yujiro Hayashi, Ryo Hirasawa and Shinichi Kobayashi from Japan; and my special thanks to Professor Kelvin W. Willoughby from Australia, Professor Guangbi Dong and Doctor Ying Bai from China.

Introduction

Technology has advanced greatly in recent years. The dream of people reaching the Moon has become a reality. It is now even possible to clone human beings. Knowledge and technology have had such a great impact that during the present age they have become an indispensable part of the vocabulary of international leaders, politicians, business executives and research personnel. In the meantime, the ruthless pursuit of economic profits and the unethical application of technology by immoral people, with the many tragedies that have ensued, force us to reconsider whether we have been warped in our understanding and application of 'technology'. They also force us to consider whether we need to revise and renew our understanding of 'technology' that underlies the concept of core competitiveness that is advocated by developed industrialized countries. In taking up this challenge we are forced to ask, what is the essence of the gap between the developed and developing or less developed countries?

A conceptual revolution – renewing the old concepts and thinking mode – is necessary for sustainable human development. In the twenty-first century, human beings have entered in to an era of the integrated development of the humanities, social sciences and natural sciences resulting in the elevation of the human awareness of knowledge. Therefore, the concept of technology needs to regain the full range of its original connotations, evolving from technology in the narrow sense to technology in the broad sense, and to express the comprehensive meaning it originally possessed. In other words, the notion of technology must undergo an expansion from hard technology to soft technology.

I discovered the alternative knowledge system and value-oriented paradigm described in this book – which are different from the orthodox system and paradigm – through long-term systematic thinking and rearrangement of ideas about the human knowledge system. Thereupon, I began research in the 1990s on the broad notion of technology and subsequently proposed the concepts of 'soft technology' and 'soft environment'.

For thousands of years human beings have been creating soft technology as well as using it and benefiting from it. However, soft technology has rarely, if ever, been generalized and categorized systematically and moreover developed consciously as a form of technology. It is because of the great impact of industrialization and the brilliant achievements of natural science and technology, as well as the excessive pursuit of material wealth, we have placed too much stress on the technology of 'remaking' and controlling nature, and have neglected to conduct enough research on the technologies that involve human beings themselves and human behaviours. Furthermore, to the extent that soft technology has been recognized, it tends to have been portrayed as a 'non-technological' factor or to have been relegated in recent decades to a 'black box' that covers everything.

This failure to appreciate the true nature of technology has prevented us from properly grasping and coping with the challenges facing human beings in modern society. As for hard technology, it has come into being through various disciplines, technical systems and industries, and through specialized research during several hundred years. Hundreds of works on the history of hard technology have now been published, while awareness of the alternative paradigm of technology – soft technology – has at best been minimal. Almost no organized research on the related technical systems and industries of soft technology has taken place as a formal discipline. Now is the time to lift the veil and to carry out detailed analysis, systematic research and application in the field of soft technology.

Advances in soft technology have opened up a new research field characterized by a completely different paradigm for the value creation process. Under this paradigm, interdisciplinary research is promoted along with a deeper and more comprehensive understanding of the human knowledge system. Accordingly, it is conducive to changing the thinking mode, the perspective on world affairs and the problem-solving orientation of people. This new paradigm reveals the essence of a great many of the core international and domestic issues of present society.

The study of soft technology – soft-tech – will enable us to assemble the missing half of the history of technology in the world, the absence of which has until now severely impeded our ability to make sense of the contemporary world.

From a practical point of view, so that each country may cope with its challenges during the twenty-first century, it is not sufficient to depend only on hard technologies, the hard environment and hard capital. We must study and integrate soft technologies, the soft environment and soft capital to design systematic solutions, and to implement the right soft-tech operation and soft environment design for addressing a multitude of problems. This requires accelerating research, development and application of soft science and soft technologies. Moreover, soft technology focalises technology in economic, social, cultural and environmental changes and practices. On the one hand, it will provide new ideas and new approaches for innovation in service industries and even for innovation in the soft industries as a whole; on the other hand, it will be conducive to improving our ability to understand, cope with and solve complex problems, and to make appropriate adjustments at various levels of strategic systems.

Fortunately, the concepts and contents of soft technology, the soft environment, and soft infrastructure, etc., have recently been widely accepted and quoted by scholars, government officials and the media in various countries. In particular, since the first English version of this book on soft technology was published by Intellect Books in January 2005, I have been invited to speak about the topic by more than thirty universities, academic institutions and international forums in countries such as the United States, Britain, Japan, France,

Introduction

Italy, Austria, Brazil, Belgium, Switzerland, Sweden, the United Arab Emirates, Iran, Spain, Hong Kong and Malaysia. The innovation centres of some countries and regions wish to adopt a broad framework for reconstructing their innovation strategy along the lines of soft technology, and a steadily increasing number of scholars have fully affirmed the academic and practical value of soft technology.

In China, my book *Soft Technology: Space of Innovation, Essence of Innovation*, published in January 2002 by Xinhua Publishing House, can be considered as Volume I in the 'Soft Technology series'; the book *Service Innovation and Social Resources*, published in 2004 by China Financial and Economic Publishing House, can be taken as Volume II; and the book *Long-term Strategy-system Integration and Sustainable Development*, published in 2006 by China Social Sciences Academic Press, constitutes Volume III in the series. They represent applications of soft technology at the level of service innovation and national strategy. Concurrently, by observing recent research on subjects such as the softening of the manufacturing sector, new agricultural business models, strategy for development of coal-bed methane, policy research on the development of robotics in China and 'tactics study' by researchers in military science, we can find many cases where the concepts of soft technology and soft environment have been put into practice. Entrepreneurs in Silicon Valley have also recommended the preparation of a version of a book on soft technology especially for the commercial environment.

The background and trends in soft-tech described above encouraged me to compose this second English edition of my first book in the 'Soft Technology series'. The second Chinese edition will also be published this year.

In Chapter 1, I have tried to prove, by citing a large number of facts and arguments, that human beings need a conceptual revolution in technology, and for that purpose I have proposed a concept of technology that is defined more broadly than is typical in the orthodox literature. Moreover, I elaborate on what is technology, what is soft technology, and also on the significance of studying soft technology.

In Chapter 2, through analyzing the history of typical soft technologies such as commercial technology and social technology, etc., as well as the contribution of soft technology to the three great Industrial Revolutions, I point out that there are two engines for promoting the development of human society: hard technology and soft technology. The history of the soft technology development is also the history of human creativity. Gaining an understanding of soft technology and the soft environment will guide people to rethink and reflect about many basic domestic and international issues related to sustainable development, globalization, the new economy and future social development.

The content of Chapter 3 through to Chapter 6 of the book is focused on the application of the concepts of soft technology and the soft environment.

In Chapter 3, I elaborate the three essential elements of technological competitiveness and probe into approaches to enhancing national soft power and corporate soft power from the perspective of soft technology, the soft environment and soft capital. In addition, I discuss the essence of various 'gaps' between developed and less developed countries or regions.

Chapter 4 addresses the essence of innovation and the broad sense of the innovation space, the relationship between technological innovation and institutional innovation, and a broad portrayal of the 'six plus one' innovation system framework, as well as the application of soft technology and the soft environment to enterprise management.

Chapter 5 reveals the essence of industrial softening and puts forward the concept of soft industries. Industrial softening and industrial innovation driven by soft technology have, overall, exceeded the service industry, or the so-called 'modern service industry', in scale and importance. This prompts the question of what actually constitutes 'service'. What are service innovation and industrial innovation? What are the creative industries and the creative economy? Gaining an understanding of the soft industries has not only helped us to clearly delineate the trend and essence of future industrial structure, but has also opened up new areas for entrepreneurial activities which are familiar but have been neglected for a long time. Moreover, it has expanded entrepreneurial space and the channels for entrepreneurship, has helped entrepreneurs to change their strategic thinking, and has provided a theoretical basis for creating a new business model. It also carries creative thinking through to focus on the future structure of industry.

In Chapter 6, I present and discuss the role and significance of soft technology in the fourth generation of technology foresight. Finally, I also put forward the overarching guiding principle for development in the twenty-first century – Harmony, Balance and Coexistence.

This book also calls for developing countries, including China, to shift away from the strategy of 'catching up and then surpassing' developed countries by investing excessively the most funds, human resources and energy into hard technology, towards a strategy of consciously developing soft technology. Developing countries need to understand and create new rules of the game on their own initiative, to remain sober-minded in the wave upon wave of hard-tech and soft-tech innovation which has been raised by developed countries, and avoid blindly following the thinking patterns and criteria of developed countries when choosing new development strategies, routes and industry structures. In order to catch up and ultimately surpass developed countries, developing countries need to bring their own new advantages into play and follow their own route to success.

Although this second edition of the foundational book on soft technology embodies important progress in theory and application, the subject of soft technology is nevertheless still a new research area, and many problems and issues addressed in this book will require further study. The weaknesses of this book, of which there are no doubt many, are open to the criticisms and comments of its readers. I sincerely hope that entrepreneurs, scholars, social activists, government officials, managers and administrators (no matter what rank) who are concerned with creativity, innovation and business start-ups, can participate jointly in both the research and the practice of soft technology.

Zhouying Jin
Beijing, January 2010

Chapter 1

What is Technology?

Technology has advanced greatly in recent years. The dream of people reaching the Moon has become a reality. It is now even possible to clone human beings. Knowledge and technology have had such a great impact upon social and economic development that we now routinely speak of the knowledge society.

In the meantime, the ruthless pursuit of economic profits and the unethical application of technology by immoral people, with the many tragedies that have ensued, are increasingly generating criticisms against the further development of technology. John Naisbitt, a renowned futurist and author of *High Tech and High Touch,* calls the public's attention to the meaning of humanity and suggests that the development of science and technology should be based on human needs and should positively benefit humanity. Bill Joy, co-founder and chief scientist of Sun Microsystems, has written an article in *Wired* magazine suggesting that '[o]ur most powerful twenty-first century technologies – robotics, genetic engineering, and nanotech – are threatening to make humans an endangered species' (Joy 2000). This statement warns people against the dangers brought upon the human race by unplanned and uncontrolled technological innovations. As a solution, he suggests developing a system of efficient control over the development of some technologies. Some domestic and foreign forums and monographs have also carried out discussion focused on the negative impact of high-tech. However, the development of technology as the engine of human society and its economic development appears to be irresistible. Its significance, including its position in the national competitiveness of each country, can never be over emphasized.

Then how should technologies be controlled and developed? Are there actually any 'good' or 'evil' technologies? Have we a comprehensive understanding of technology? What should be noted here is whether the high-tech that Naisbitt talked about or those powerful technologies that Joy mentioned refers to technology as traditionally understood, i.e. the 'hard' technology that is derived from natural science-based knowledge.

In the twenty-first century, with the rapid development of high technology and information, economic and technological globalization, human concepts of country, enterprise, government functions, knowledge, work and even science are undergoing change. Should we re-understand technology? For instance, could the knowledge derived from non-natural science, such as social sciences and the like, form technologies? If so, we will have to rethink the essence of technological competitiveness, rethink the meanings of technological innovation and may even need to adjust the national strategic systems.

In fact, following the experience of the Industrial Revolution, technology has also evolved. The concept of technology is shifting from technology in a narrow sense to that in a broad sense and from hard technology to soft technology. We have reached the era where the development of soft sciences and soft technology needs to be accelerated.

A. The Evaluation of Technology

From the Ancient Greek era until today human beings have studied what technology is and have investigated the essence of technology from a variety of perspectives. For example, it is now widely recognized that primitive men differed from anthropoid apes in at least the following four ways: they were able to walk on two legs; make, use and ameliorate tools; utilize fire; and communicate via language. In retrospect, the making and use of tools and the utilization of fire and language are technologies. It is obvious that technology existed in the primitive period. At that time, the so-called tools were produced with the intended purpose and played the role as extending or assisting parts of the human body (Gan & Hiromasa 1986).

During the past two thousand years human understanding of technology has varied a great deal. In Ancient Greece, the scope of technology in general was very wide. It included everything from farming techniques and ancient medical practices involving leeches, to political techniques, gymnastics and arts. The most representative view of technology in Ancient Greece is contained in the theoretical work of Plato, one the three great philosophers of ancient Greece. In his *Apology and Other Dialogues,* Plato points out that technology includes the technology of acquisition, as well as the technology of manufacture.

The technology of acquisition included the technologies of learning, acquiring knowledge, making profits, Agon[1] technique and hunting; while the technology of manufacture consists of practical manufacturing technology and image manufacturing technology, i.e. the technology (in a narrow sense) and art. Practical manufacturing technology included farming techniques, ancient medical practices involving leeches, construction techniques and tool technology, while the image manufacturing technology included the technology of imitation and idolization. Plato considered the creation of art to be a manufacturing activity, so he included the activities of art and tool creation in the technology of manufacture (Honda 1975).

The Ten Books of Architecture by Marcus Vitruvius, an architect of Ancient Rome (first century BC) is considered the technological encyclopaedia of Roman times (ibid.). In the first chapter, the author describes the qualities of an architect. Some of these qualities are as follows: fluent writing skills; drawing skill; knowledge of geometry, optics, mathematics, history, philosophy, music, ancient medical practices, law and astronomy, etc. Vitruvius regarded these abilities as organisms of the human body that must be integrated into an entire indivisible organic system.

China has a long tradition of thought about the nature of technology. The Chinese notion of technology, stemming from the ancient age, emphasizes technique, skill, feat and methodology (*Etymological Dictionary of Chinese Characters* 1998: 658).

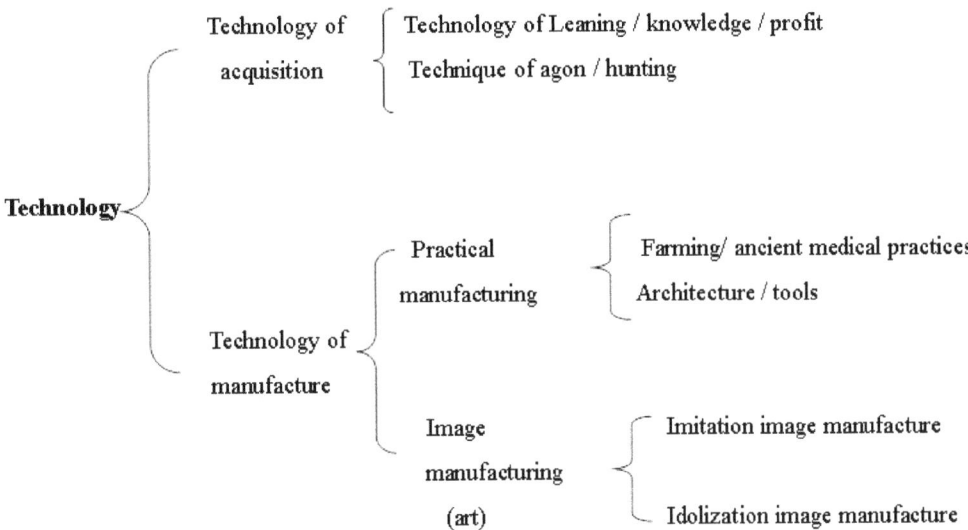

Figure 1: Plato's Concept of Technology.

Francis Bacon (1561–1626), the British philosopher famous for his aphorism 'knowledge is power', believed that if knowledge of the Greek period was used to seek kindness and beauty, and the knowledge of the medieval period was used to pursue beliefs, then the third period of knowledge is one which could ensure humans the power of domination over nature. Bacon thought:

> Human mastery over things depends upon science and technology. Only through submission to nature could a human being control it. That is to say, in the purpose of controlling nature, we should learn from nature and need to submit to and control it; the technological knowledge of modern human beings is how to deal with, understand and manipulate nature. (quoted in Honda 1975)

It is evident that Bacon's theory embraces the idea that 'technology enables mastery over nature'.

In the late eighteenth century, Denis Diderot (1713–1784), French philosopher and chief editor of *The Encyclopaedia*, stated that technology is the system of various tools and regulations organized for a common purpose (in Song 1984).

Friedrich Dassauer (1881–1963) of Germany identified three essential features of technology. He said that it should conform to natural laws, operate with a common purpose and also operate with a creative purpose (ibid.).

In the early twentieth century, the Japanese academic community began arguing over the concept of technology. Tosaka Jun, the leader of the Japanese Materialism Research

Association, which was founded in 1932, classified technology as notional technology and material technology. The former includes the means that constitute the subjective existence of technology, such as skill and intelligence; the latter includes means that constitute the objective existence of technology, such as machines and instruments. However, Jun also thought that notional technology was only the means for the subjective existence of material technology and not true technology in itself. Affected by the view that 'technology is embodied by the machine as the important means of labor [sic], or that of production, etc. in large industries; technology is infiltrated into the labor [sic] process', the majority of researchers in the Japanese Materialism Research Association think that technology is the means of production or even the system of the means of labour.

Many scholars at that time who opposed the theoretical notion of technology as the means of production believed that even Karl Marx himself did not define technology in such a manner, and it was actually very similar to that of people such as Bukharin who supported the notion of technology as mechanism. For example, Haruki Aikawa pointed out that the concept of technology should be divided into three categories: the concept derived from natural science; social sciences; and philosophy. From the perspective of social sciences, 'technology is the existing means of labor [sic] during the process of production' (Aikawa 1941: 131). Another Japanese scholar in this group, Mitsuo Takeya, saw technology in this way: 'technology is the conscious application of the objective laws during the production practice' (Takeya 1968: 139). Since then, many people of the Japanese philosophical community have begun serious research about technological philosophy under the general umbrella of this school of thought.

In the 1960s, an advisor to the OECD, Erich Jantsch, who was working in the general field of technology forecasting, made some important contributions to the ongoing debate about the definition of technology. Jantsch pointed out that technology included the conscious application of materials, life science and behavioural science, so that technology also included all the means of ancient medical practices, agriculture, business and other fields, including hardware and software (Jantsch [1967] 1968: 15).

F. R. Bradbury pointed out in his *Economics of Technology Development,* that 'technology is the way of doing' and 'the values of technological development are dependent upon the improvement of the ways we use resources to satisfy human needs' (in Williams 1979).

The Japanese *Yuhikaku Economic Dictionary,* published in 1979, defined technology as the following: 'a process of supplying better means of utilizing nature for the purpose of developing and improving human lives. There are roughly two explanations of technology: the conscious application of the objective laws and the system of labor [sic] means' (*Yuhikaku Economic Dictionary*: 88).

In the 1985 Chinese version of the *Concise Encyclopaedia Britannica*, technology is defined as 'the means or activities employed by human beings to change or operate the external environment' (*CEB* 1985c: 233).

In 1990, the Nomura Research Institute in Japan published an important document, *The Strategy for Technology in 2000*. This drew attention to 'the technological stream of human sciences' (*Japan Nomura Research Institute* 1990), and pointed out that the changes of the definition and coverage of technology was due to the orientation of technological developments in the 1990s. If hard technology was based and focused on natural sciences such as physics and chemistry, etc., then the trend of the twenty-first century should be directed towards soft technology based on human sciences. Hard technology, aimed at turning natural objects into artefacts, is the technology of controlling the 'object'. The high technology of the future will be the technology of controlling and commanding the 'human mind' and the management technology for group or organization, which is based on psychology.

British historian of science and technology Charles Joseph Singer defined technology in his monumental work, *A History of Technology*, as the following: 'technology is the sum of skill, art, craft, means and knowledge by which humans can take advantage of the large number of raw materials and energy stored in nature in accordance with the direction of their own desire' (quoted in Williams [1979] 1989).

The author of *Building a Win-Win World* and the founder of Ethical Markets Media, Hazel Henderson, used a broad definition of technology: 'Human knowledge applied to Human purposes' (Henderson 2002). She included design of political and economic systems, software and social security, etc. as technology.

From the preceding review of ideas about technology, covering an array of time periods and a diversity of cultures, we may draw the following conclusions:

- The concept of technology has been evolving.
- Throughout the long history of formal discussions of the concept of technology almost all serious commentators and analysts have included soft dimensions of technology (e.g. regulatory systems such as techniques, approaches, programmes and processes of activities, arts and the like) as part of the definition of 'technology' – in addition to the hard dimensions of technology (e.g. tools, machines and equipment and other means of labour). Plato's technology of acquisition is an example of 'soft' technology and so is the dimension of technology classified by Erich Jantsch as the application of behavioural science. As has been pointed out by the Nomura Research Institute, hard technology is technology for controlling material substance, while future high-tech will be technology for controlling and dominating the human mind based on psychology, etc., and management technology for groups or organizations.
- Nevertheless, since the Industrial Revolution, the concept of technology extant in the literature has gradually evolved to include such notions as 'the means of dominating and controlling nature', 'the initiative relation with nature', 'the system of labour means' and 'the means of changing or controlling the external environment' and so on. Hundreds of works on the history of technology have now been published in East and West and in a multiplicity of languages – with titles such as *Chronology of Science and Technology, History of Modern Technologies, History of Technology and Technological History,* etc. I have observed that almost all of these works tend to focus exclusively on research in the natural sciences and upon what I have labelled 'hard technology'.

- This focus has come about because, during the age of industrialism, material production played a vital role in the economy, and natural science and technology, taking substance as its carrier, made outstanding contributions to the improvement of material productivity. Especially in the past two centuries, the invention and extensive application of new technologies such as the steam engine, electrical technology, steel technology, chemical engineering, the telephone, wireless communications, the transmitter, the computer, large-scale integrated circuits and so on, have generated technology revolutions that, in turn, have promoted the development of productivity and changed human survival conditions and lifestyles. Human understanding of knowledge has therefore leaned towards the natural science-based knowledge; the rules, approaches and means, which were created during the process of problem solving in the material production by applying the knowledge of natural sciences, are usually called 'technology'; the focus of technological development has also leaned towards the use and transformation of nature.
- Contemporary dictionary definitions of technology typically define and connote technology in a manner that emphasizes utilizing and altering nature. For example, *The Modern Chinese Dictionary* defines the term 'technology' as, 'experience and knowledge in the process of utilizing and remaking nature', the term 'technology revolution' as 'thorough revolution of production technology' and the term 'technology innovation' as 'improvements in production technology, e.g. improvements in the process of machine components' (*Modern Chinese Dictionary [MCD]* 1991: 553). *The Economics Dictionary*, published by the Japanese Yuhikaku Press, defines technology as 'the means supplied to utilize nature' (*Yuhikaku Dictionary of Economics [YDE]* 1979: 88).

Many issues that human beings have faced since the late twentieth century, including the devious applications of technology and the disruption of the human living space, force us to reconsider whether we have overemphasized the 'remaking' of and mastery over nature in the understanding and application of 'technology', whilst ignoring the technologies that involve human beings and human behaviours. We have not conducted enough research regarding how technology changes and controls our subjective environment. Therefore, we need to revert to Plato's technological definition and gain a fresh understanding of technology.

B. A Fresh Understanding of Technology – To Understand Another Paradigm of Technology – Soft Technology

Now, we need to renew the understanding of the so-called 'technology' advocated by the industrialized developed countries as the core competitiveness.

1. Many Hard Technologies Derived from Natural Science-based Knowledge are Softening and Becoming 'Soft' Technology

Today we have discovered that some high technologies, which are the great driving forces for economic and social development, are different from traditional technologies. For example, computer software is becoming the strategic technology of economic development and national security. The total value of output of the information industry in the world was 587.71 billion US dollars in 2000, of which software made up 61% and hardware 39%, and was expected to reach 1022.23 billion US dollars by 2005.[2] By 2008, the total value of global output of the electronic information industry reached 243 trillion Yen,[3] which was about 2150.44 billion US dollars (1 US dollars was approximately 113 Japanese Yen at that time). China's software industry began to develop in the middle of the 1980s and achieved RMB 75 billion by 2001, compared with sales of only RMB 220 million in 1990. During recent years China's software industry has averaged an increase of two digits per year. According to MII (Ministry of Information Industry) statistics, the scale of China's software industry reached RMB 583.4 billion in 2007, which was about 82.2 billion US dollars, with an increase of 21% over the previous year, and it amounted to RMB 750 billion in 2008. It has become the fastest developing high-tech industry ever, and the scale of the industry ranked fourth in the world after the United States, the European Union and Japan.

However, it is not just with the contributions of 'new' technology that enables the software technology to become an impressive economic growth point. The successful design, implementation and application of software requires the integration of technology and different cultures, languages, arts, ways of thinking, working styles and procedures. This gives software the characteristics of humanity and locality. Software technology is thus no longer a technology in the traditional sense (i.e. hard technology). Furthermore, the greater the success of the software industry, the more that it will include 'non-technological' factors and the less that it may legitimately be seen as indifferent to human factors and values. Software itself has become a kind of service, and the development of the Linux operating system serves as one such example.

Moreover, it is now widely recognized that the majority of the value-added of many high-tech products mostly come from the service activities associated with the products and not from technology itself. Thus, high technology industries gradually evolve into service industries. Some technologies derived from natural science are softening because the more 'human' factors are involved in the technology, the more softened the technologies become.

Results of a recent investigation conducted by government authorities of computer users in the Asian-Pacific region concluded that hardware makes up 21 per cent of the overall expense of the computer, and the rest goes to software and service such as the use of products, maintenance, training and the upgrading of products. Accordingly, in successful software companies, more often than not, there are more 'soft-tech' experts rather than 'hard-tech' experts. The experience and practice of Computer Integrated Manufacture Systems (CIMS) in China also proved that the key to success is the integration of soft technology and hard technology; namely, the integration of various information, automation and manufacturing technologies with the organization, culture, regulation and abilities of the enterprise in question.

2. *The Process of Transferring Hard Technologies: Commercialization and Industrialization are 'Process Technologies'*

In order to get an upper hand on international competition, most governments have put a great deal of manpower, materials and funds into high technology. However, the same technology in different countries or regions transfers at different rates and brings about different 'results'. What functions, then, bring about the transfer of technology? What 'injects' technology into products and 'pastes' it to the market? What turns knowledge, technology and ideas into valuable products, services and results? Is the process of technological transformation from invention to production a kind of technology in itself?

Actually, we cannot expect technology to turn into a product or acquire market share if we rely simply on 'hard technology' itself. A series of other functions and facets are involved. For example, we have to set down a correct strategy, raise the necessary funds, design products that satisfy different customers, guarantee the quality of products, perform cost control and promote sales activities before enlarging the market share and making profits possible. All of these are part of a series of technologies that are used for inner management within enterprises.

In the meantime, for the sake of their survival, enterprises not only need to adjust their product and organizational structure by means of cooperation, alliance, purchasing, merging and foreign investment, they also need to seek help from the 'external brain'. They need to publicize their products and their images through advertisements and public relations. The regularization and formalization of these external activities may also be considered to be technologies, albeit technologies that differ from 'technology' as traditionally understood.

Another familiar example lies with the automobile industry. The automobile, the most common means of transportation, owes its success to continuous innovation of production and organization technologies. In the history of the automobile industry, the technology of the flow production line, developed by Ford, and the method of lean production, originating with Toyota (the new division of industrial labour and the strategy of sharing a common platform across models, etc.) are the critical soft technologies in the process of the batch production of automobiles, which has played a revolutionary role in moving the worldwide automobile industry towards the goal of large scale, low cost and high quality production. Just as was the case with hard technologies, involved in applications such as engine technology and new material technology, these soft technologies have contributed positively to the human race.

To sum up, in the process of technology transfer and competition, enterprises have developed a series of process technologies for value creation.

3. *Various Commercial Technologies are the Direct Driving Forces of Socio-economic Development*

A technology that has had a great influence upon the development of society and its economy – and which is familiar to us but which is not generally considered to be a kind of technology – is nothing other than what we might call 'market exchange technology' or 'commercial technology'. From the early days of double-entry book-keeping and shareholding systems, to the contemporary financial derivatives tools and e-commerce systems, various exchange technologies have been invented to facilitate human economic and social activities. These socialized exchange technologies have largely been developed in the manner of 'learning by doing': practitioners and observers summarize the perceived rules of their activities and accumulated experiences, following thousands of tests and experiments (in which the human production activities themselves serve as the laboratory), eventually adapting these results to different cultures, social institutions and technological levels.

For decades Hong Kong has not had any high technologies worthy of showing off, yet it has noted a remarkably fast economic growth. This indelible contribution is to due to Hong Kong's commercial technology.

Charles Jones, the Stanford University economist, has pointed out in his research about the changes in living conditions of the nineteenth and twentieth centuries, 'even by 1790, average per capita consumption in France was no greater than it had been during the days of the Roman Empire' (Jones [1999] 2000). It was only during the late nineteenth and twentieth centuries that unprecedented rapid growth produced a much higher living standard than the past several thousand years. It also produced electricity, cars and railways, and air flights were widely utilized. Population and educational levels also changed fundamentally. One of the main causes lay with the 'improvement in institutions that promote innovation, such as property rights'. These institutions also include the patent system; the establishment of limited liability companies; the development of stock markets and venture capital; the establishment and broadening of research institutes within companies and governments; as well as tax deduction and other preferential government policies toward research and development. He argued that encouraging innovation was the driving force of social development.

I have studied the relationship between the introduction of various commercial technologies and per capita growth of GDP over the last 200 years in the US. As is shown in Figure 2, the growth of GDP per capita during the 40 years from 1910 to 1950 (even taking into account the effects of the Great Depression that began in 1929, leading to five years of negative growth in the first half of the 1930s) far exceeds the growth of the 90 years from 1820 to 1910; the growth during the 40 years between 1950 and 1990 exceeds twice that of the previous 40 years. Growth has accelerated, especially since the 1940s. These two periods are characterized by the comprehensive application of various technological inventions of the past 200 years, such as wireless communications; automobile production line; electrification of railroads; thermal power stations; large capacity hydroelectric power stations; and airplanes. In the meantime, with social progress the average human life expectancy grew from 35 years in the Renaissance period and 36 years in the eighteenth century, to 45 years in the late nineteenth century. In addition, it became 55 years in 1920s, and over 65 years in 1990s (see Table 1).

It is generally considered that during these 200 years four technological revolutions took place: the first was centred around the scientific and technological principles of Newtonian mechanics and mechanical techniques and took place in the middle of the eighteenth century; the second was centred around electromagnetic theory and electrical technology and took place late in the nineteenth century; the third was based on the application of the new technologies of modern physics, computers, nuclear energy and space technology and took place in the middle of the twentieth century; and the fourth was based on the integration of microelectronics, computing and communications technologies, and the breakthrough in Internet and biotechnology, which took place during the 80s to 90s in the twentieth century. However, these revolutions do not seem to relate directly to changes in the American per capita GDP. This may partly be due to the 'delay time' lag effects on the

Table 1: Increase of Average Human Life Expectancy.

Historical Periods	Average Life Expectancy (years of age)
Bronze Age (2000 BC)	18
Ancient Roman Period (500 BC–AD 400)	29
Renaissance Period (Fifteenth century–sixteenth century)	35
Eighteenth century	36
Late nineteenth century	45
1920	55
1980	62.7
1985	61.74
1990	65.52
1992	65.9
1995	66.43
1996	66.71
1997	66.7
1999	66.25
2000	66.5
2003	66.8
2007	71

Sources: Statistics of the Bronze Age (1920) come from *A Handbook of Population* by Hongkang Liu and Zhongguan Wu from the Southeast Finance Economy University Press. The statistics of 1980–1997 came from the *International Yearbook of Statistics* (1998 edition) and the *International Yearbook of Statistics* (1999 edition) by Hong Liu, China, Statistics Press and *Wold Health statistics 2009*.

economy. What actually is the direct reason? How can we determine the differences in 'delay time' between different countries?

We found that the key factors closely related to above growth of American GDP per capita in the twentieth century were several waves of soft-tech development in the United States (see Figure 2).

A series of innovations encouraging commercial technologies were creatively developed and applied in the late nineteenth and early twentieth centuries in the United States. In addition, noticeable innovations in economic institutions were developed during this period. For instance, although patent technology (the patent system) used to protect the inventors and their intellectual property rights was first introduced in the fifteenth century, it was not applied to all industrialized countries until the end of the nineteenth century. After 1883, the Paris Treaty was issued with the aim of protecting industrial property, but by 1870 America had already performed radical reforms in the patent system (see the details in the patent part of commercial technology history, elsewhere in this book). The stock company system was originated in Europe in the early seventeenth century but the stock market was later successfully redeveloped in America during the early twentieth century. The research institute system was initiated primarily in Germany in the middle of the nineteenth century

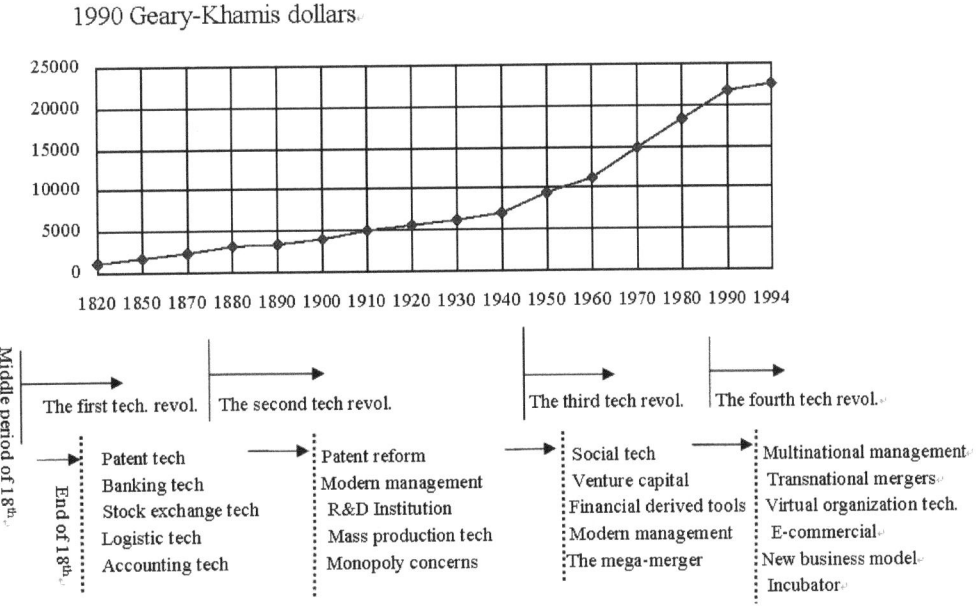

Figure 2: Soft Technology and American per Capita GDP.
* Sources: Zhouying Jin, Oct. 1997, statistics from:[4, 5]
** Geary-Khamis dollars is the standard currency measured with the Geary-Khamis conversion factor.

but many industrial R&D laboratories were established in America by the beginning of the twentieth century. Frederick Winslow Taylor developed modern Scientific Management Theory and Ford initiated production line in 1910, radio broadcasting was first introduced to America in the 1920s and the neon sign was introduced, as an advertising medium, in the 1930s.

In the 1950s, modern management technology, public relations, large-scale mergers, venture investments and military technologies such as logistics technology had their first big impacts in America. What must be noted is that the immigration policy and the intellectual personnel system of America greatly contributed to creating innovative environments, thereby attracting to the USA a great number of scientists and technology experts from other countries after World War II. In the third wave of commercial technology beginning from 1980s, the United States also played a leading role.

In the development of America's 'new economy' in the 1990s, we can place the role of the fourth wave of soft-tech development on a par with the activating forces of information technology (see details in the section 'Soft Technology and Thrice Industrial Revolutions' in Chapter 2). A series of soft technology innovations have prepared the way for the rapid application of information technology, the Internet and biotechnology in the world's markets, and they have enabled overall improvements in industrial efficiency.

4. There is Technology in the Innovation Process of Social Activity – Social Technology

Science and technology in the twentieth century have greatly contributed to the economic development and abundance of material possessions but have also left a great negative 'heritage' in its wake – such as people caring more about the economic benefits of modernity than the negative influences it has created in the human spirit, morals, education and psychology – that tend to be neglected. It is incarnated in some critical issues, such as public investment; land development; transportation; layout; protection of ecology systems and the environment; prevention and cure of certain diseases, etc. These problems have also appeared as side effects of the establishment of new enterprises, community development plans and urbanization. Namely, many 'social issues independent of the economy and industry' (Kusaka 1978) have appeared in our modern society. These problems are hard to solve using natural science and technology alone. Collaboration between social sciences, non-traditional sciences and the infusion of interdisciplinary learning are required in order to solve these problems.

From another perspective, we live in a new period of social changes. However, the effects of the rapid changes in our economy, culture and even in our institutions are unprecedented. Social problems resulting from these influences, ranging from the level of the individual person to the global strategic level, have developed, with the result that people are apparently turning to the fields of philosophy and social sciences for answers. Only after a thorough study of the application of the characteristics of contemporary social sciences – incorporating attention to comprehensiveness, realism, internationalism and compatibility with local and domestic considerations – could it be possible to find solutions to our complicated social problems. This situation raises philosophy and the social sciences to a higher level of responsibility in the development of the economy and technology. Meanwhile, the clear trend of contemporary social sciences is to provide services which are oriented to application, development and decision-making.

In view of the above understanding, the methods, means and procedures to solve the above-mentioned 'social problems' by applying the knowledge of social sciences are gradually reorganized and summarized as 'social technology'.

However, in reality, social technology as the innovative process of social activities is much more complicated and by no means is the applied technology of social sciences simple. Social technology carries much more resources, connotation and significance than is suggested by the above view. Its concept and signification will be further discussed in the next chapter.

5. *The Innovation Process of Cultural and Artistic Activities is a Technology – Cultural Technology*

Cultural knowledge and resources, like all natural science knowledge, do not naturally turn into valuable products or commodities. Only by the means of careful re-creation, development, processing and production could we convert cultural knowledge, cultural resources and values into products, commodities and services that can be 'appreciated, used or consumed' by final consumer, and then their true value can be achieved. The means and processes of converting cultural knowledge and resources into products, commodities and services – i.e. the means and approaches by which a culture can show its value, including economic value and social value – is cultural technology.

In this sense, cultural technology is a technology of culture-creation and cultural innovation. This so-called 'culture-creation' refers to the creation, renewal and development (according to the context of today) of new cultural resources and values that are conducive to social development; while, on the other hand, cultural innovation refers to the conversion of cultural contents and resources into products and service with social and economic values, i.e. improving the added-value of culture. Today, cultural resources include education, science, arts, morals, laws, customs, beliefs, natural environment and historical heritage. Globalization has enriched the content of cultural innovation and the key of cultural innovation will lie in discarding the dross and in selecting what is essential to the process of achieving coexistence and harmony between different cultures. In keeping with American futurist Daniel Bell's notion that the old concept of culture is based on the continuity whereas the modern concept of culture is based on the multiplicity, the old values are the tradition and the contemporary ideal is an integration of different cultures (Bell 1986: 211).

Details of the dualism of culture technology will be discussed below in Chapter 5 in the section devoted to 'culture industry'.

6. *Social Progress Requires the Integration of Technology and Art – Artistic Technology*

In modern society, people generally do not consider art to be a technology. The arts and technology are, however, essentially one indivisible whole. The term technology is derived from a combination of Greek words *techne* (techniques and skills) and *logos* (word and speech), which mean the exposition of modelling arts and applied technology (*CEB* 1985c: 233). In 400 BC, Plato placed artistic creation within technology. During the nineteenth century, Sir Edward Burnett Taylor, the British anthropologist, believed that 'we should not only study culture according to the achievements of arts and spiritual civilization, but also investigate culture according to technology and moral perfection in the development phase of various countries' (*CEB* 1986a: 644).

Since the middle of the nineteenth century, the dichotomies in the ways in which humanistic scholars and natural scientists look at the world have become wider. Humanistic scholars have tended to lack interest in natural science and technology and, therefore, have lacked knowledge of natural science and technology. Natural scientists have tended to focus on scientific and technical specialization and have often thereby become insensitive to the humanities, with some eventually even preferring to have nothing to do with the subjects of art, beauty, society, etc.: 'This inharmoniousness of these two groups of educated people, who are indispensable to the constructive criticism of the economic system and of social institutions, has caused great losses to the development of the times' (Bernal [1965] 1970). In other words, the industrialized society that was the crucible for the mechanization of

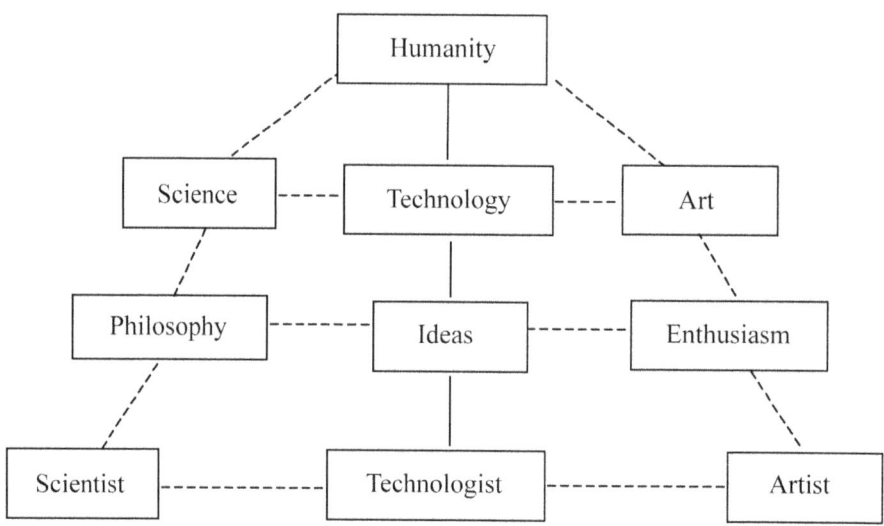

Figure 3: Science – Technology – Arts.
Sources: *Technological Anthropology* (Honda 1975)

technology has caused technology and human nature to increasingly drift apart, while the arts have become out of touch with reality.

The Japanese scholar Syuro Honda followed in the tradition of Plato's ideology by broadly advocating the integration of technology and art (Honda 1975). Honda employed the label 'profit technology' to describe technology as traditionally understood and the label 'technology of beauty' to describe the arts, which can not be stayed in a simple sense of the experience of beauty. 'Science is rational exploration and art is the creation of emotion; only the interaction of scientific reason and artistic passion and enthusiasm can drive the imagination to invent new technologies'. Therefore, technology is not simply subjective techniques or intelligence. Neither is it only an objective scientific and technological system. It is the human activity system that integrates subjectivity and objectivity.

Certainly, as the cultural 'products', the process technology or methods and means of re-design, development and processing, by which artistic knowledge and resources as well as aesthetic conception and values could be converted into products, commodities and services that can be 'appreciated, used or consumed' by a final consumer, is artistic technology.

Today's entertainment technology serves as an example. Entertainment technology combines music, movies, TV, video games and the Internet into common media, thus integrating technology and art. Furthermore, entertainment technology finally integrates the technology of beauty with economic profits. In 'The Future of Entertainment Technology', the editorial board of *American Science* stated that the integration of technology and arts will further new tendencies such as 'interaction technology without heroes', 'movies without professional actors' and 'the integration of the medias with the principles of families'. This requires a redefinition of technology to include soft technology involving cultural technology and artistic technology. The recombination of technology and arts will become an inexhaustible source of technology innovation (see the section of 'Artistic Industry' in Chapter 5).

7. *The Diagnostic and Therapeutic Technology of Chinese Medicine: Another Type of Soft Technology – A Technology Derived from the Knowledge of Chinese Medicine and a Special Problem-Solving Approach*

The methodology of traditional Chinese medicine is not only an important part of Chinese culture, but also is another example of soft technology that can be successfully applied to the human body to address human disease and health. The relationship between man and nature, and the concept of disease and health used in Chinese medicine are worthy of extensive additional research and development.

Since medical work began, as documented in the 'Yellow Emperor's Cannon of Medicine' (770 BC–221 BC), which is the first monograph on the classical theory of traditional Chinese medicine and came out more than two thousand years ago, Chinese medicine has formed and developed a systematic theory of unique treatments and therapies. It considers the human body to be whole, with the viscous and the 'meridian – the inner cycle-channel' as the core. There are 'Yin' and 'Yang', which obey the law of unity of opposites, in all human beings and in nature (Cheng et. al 1982: 68). Diseases are considered to be the process of maladjustment between 'Yin' and 'Yang', or in other words, the maladjustment between evil and good. With regard to the relationship between the whole and its parts, the former is emphasized. Traditional Chinese medicine focuses on recuperating the entire body, paying attention to control the 'QI',[6] 'Spirit', 'Viscera state', 'Meridian' and 'Symptom' only at the level of the entire human body (Ji 1999). This approach embodies a unique theory for approaching humans, nature, illness and health. Even more remarkable is that after several thousand years of practice, it has evolved unique and efficient means of solving problems, under the guidance of the traditional Chinese medicine theory, such as diagnostic technology, therapeutic technology, health care technology, health preservation technology and longevity technology.

The clinical treatment based on syndrome differentiation system of traditional Chinese medicine is a typical example of soft technology. Firstly, the diagnosis of the symptoms is based on collecting information on people's health status in four ways – looking, listening, questioning and feeling – and analysis by application of treatment based on syndrome differentiation, which includes: 'syndrome differentiation by eight principles in which Yin and Yang are the general principles'; 'syndrome differentiation of viscera'; 'syndrome differentiation of QI, blood and body fluid'; 'syndrome differentiation of six meridians'; and 'differentiation of syndromes according to defensive QI, nutrient and blood system'. The treatment principle and medicine prescription is then based upon the patient's symptoms. In other words, in Chinese traditional medicine diagnosis and treatment are based on an overall analysis of the illness and the patient's condition. Once the diagnosis is made, the process of treatment determination (based on syndrome differentiation) is to establish a detailed therapy programme under the guidance of the traditional Chinese medical theories (based upon perception and rational speculation) which can promote human 'ecological' coordination while the illness could be cured (soft characteristics). Excellent doctors that practice traditional Chinese medicine rely on their outstanding abilities with remarkable speculative ability, and LPFE (Learning through Personal Feeling and Experience) technology may compare favourably with the advanced equipment. The soul of treatment based on syndrome differentiation system, the diagnostic and therapy technology in traditional Chinese medicine is soft technology.

Secondly, from the perspective of operational object (see Table 3) and basic thinking mode, traditional Chinese medicine is derived from its fundamental theories and operating techniques, which include the theory of Yin and Yang as well as five elements: the viscera and meridian theory; the 'QI, blood and body fluid' theory; the pathogenic and pathogenesis theory; the drug property theory; and the preserving health and recovering theory. It is also comprised of the approach of treatment based on syndrome differentiation; rule and principles of treatment; diagnostic technology; therapy technology like acupuncture, moxibustion, massage, etc., which basic thinking mode is 'concrete thinking' and is different from abstract thinking and imagined thinking because it involves acquiring concrete sensation by sense organs (Tianjun Liu 1996). Here, subjective experiences – such as pain, danger, fear, stimuli, happiness and discomfort, etc. – are felt directly though sense, mood and action. During the past two thousand years of history, traditional Chinese medicine has been focusing on perfecting the inner and individualized insight – feeling technology and speculative technology. Thus, a type of technology that is different from technology in the traditional sense comes into effect.

From the perspective of technological parameters, in traditional Chinese medical theories human beings are considered to be an organic part of nature and society (Qui 2001). Factors regarding humanity, society, time and geography are all important parameters of traditional Chinese medicine. Almost all of the traditional Chinese medical theories and technological operations have embodied this different diagnostic result and prescribe, on the one hand, according to different human conditions (body constitution, heredity, acquired nourishment, personality, opportunity and the environment in one's life), and, on the other, with regard to the different time (year, season, time of day, prevailing period of diseases, etc.), and also with regard to place (geographical orientation, topography, temperature in the sun, humidity, water and other natural environment, etc.), as well as with regard to social context (interpersonal relations, cultural environment, etc.).

Analyzing these characteristics, like human status, standardization and regional features, in traditional Chinese medicine the human body is considered to be a being with consciousness, emotions, thinking and wisdom capabilities, rather than just a biological machine (as is sometimes the case in the western medical tradition). In other words, traditional Chinese medicine respects the integrity of human life and makes the entire human body and the living individual the focus of observation and thinking. Moreover, since human beings are a part of nature and society, people with different characteristics and conditions react and behave differently at different times, in different places and in different environments. This means that the problem-solving approach of traditional Chinese medical technology (including both diagnosis and therapy) does not readily accommodate standardization. Rather it allows individualized explanations for the same phenomenon, which are customized to take regional considerations into account.

It can be said that in the long history of the Chinese nation's evolution, the traditional Chinese medical tradition has not only become an indispensable part of Chinese cultural system but that its diagnostic and therapy technology has also formed an unique soft technology for China in the world medical field. For thousands of years, Chinese medicine and Chinese traditional medicine (drugs) have been integrated and constantly innovated and developed, showing the incomparable value and exuberant vitality of the soft-tech of traditional Chinese medicine through implementing the function of treating illness while nursing one's health and its unique efficacy.

China is a multi-national country and Chinese national medicine incorporates Mongolian medicine, Tibetan traditional medicine and many other minority medical traditions. Tibetan traditional medicine, for example, has a history of over two thousand years and the foundation of its approach involves concentrating on specific geography, climatic environment and culture. Its theory considers three factors, including *lung* (main breath), *chiba* (quantity of heat) and *peigen* (body fluid) that make up the material foundations of the body that are essential for its maintenance. If the three factors are balanced and harmonious, the human body is healthy; if not, the human body will become sick. According to this theory, Tibetan medicine mirrors dialectic prescribing. This theory expresses the wisdom and soul of the medical experience of this snow-covered nation. As a result, the Tibetan traditional medical industry is beginning to gain wider acknowledgement and acceptance.

In fact, Chinese medicine is the gem in Chinese culture which is most practical and which embodies the spirit of assisting one's generation. For over 5000 years it has made a tremendous contribution for Chinese people's propagation and health. However, after the entrance of western medicine into China, Chinese medicine has been declining, and has even been criticized as a kind of sorcery associated with certain extremist views. In the early days of the new China, the Chinese government even issued official documents disallowing Chinese medicine to be practiced in the hospitals, and doctors of traditional Chinese medicine were required to study western medicine. Attempts were made to reconstruct Chinese medicine by learning from western medicine, with the consequence that Chinese medicine has moved in to a subordinate position, even facing the brink of extinction. As a result, China has assembled neither the intellectual resources nor the financial resources to enable research into traditional Chinese medicine and medical technology to progress and develop. Even worse, Chinese traditional medicine and technologies are not considered to be part of advanced science and technology, and have sometimes even become associated with superstition, false beliefs and fetishes. A vacuum of its talent has been created by China's failure to adequately support traditional Chinese medicine with a scientific approach.

Let us compare the development of traditional Chinese medicine and western medicine in China during the past fifty years. There were 276,000 Chinese medical staff in 1949. The number dropped to less than 210,000 in 1972 and increased to 337,000 people by 1999. It increased by only 61,000 people (a gain of 22.1 per cent) over 50 years. The number of traditional Chinese medical staff per ten thousand people dropped by more than 50 per cent compared from the early 1950s. And for western medicine, there were only 87,000 staff throughout China in 1949, but the number reached 1.696 million by 1999 – namely, it increased 18 fold over 50 years. In 1999, there were 10,793 polyclinics in China, among which there were only 2449 Chinese medicine hospitals (not including the number of affiliated polyclinic and Chinese medicine hospitals of medical colleges). According to the 2009 China health statistical yearbook, among the 19,712 of polyclinics by 2008, there are 13,119 general hospitals and the hospitals that practiced traditional Chinese medicine only increased to 2,688 (*China's National Health Statistics Yearbook 2009*).

In fact, after the reform and opening up of China, the government's unsupportive policies towards traditional Chinese medicine have been corrected to a certain extent. In the early 1980s, a clause on developing traditional Chinese medicine was written in the Chinese Constitution. The State Administration Bureau for Traditional Chinese Medicine was set up in 1986 and based on this organization. The State Administration of Traditional Chinese Medicine of China was established two years later, which also took into account Chinese Traditional Pharmaceuticals. Nevertheless, due to the reason of awareness and notions, something is finally being done about this problem but the pace is very slow and only meagre resources have been allocated to its solution. Furthermore, research on subjects such as 'channels' and 'collaterals' (key topics within traditional Chinese medicine) were listed as significant state key research projects for basic research in the late 1980s. In keeping with this slow process of establishment and recognition, traditional Chinese medical research did not appear as a project within China's National High-Tech Research and Development Plan until the late 1990s. Although the content and goals of the project were not quite clear, at least the importance of the topic has finally been recognized.

An important reason for the obscurity of the content and goals of traditional Chinese medicine development is that China has neglected research on its scientific essence. In addition, its methodology has not been subject to research and development as a unique domain of technology. This means that even those R&D efforts that occurred have taken quite the wrong approach: they have tried to force the 'square peg' of hard technology frameworks into the 'round hole' of the soft-tech, traditional Chinese medicine. This has forced traditional Chinese medicine to face the fact that there was no scientific platform upon which it could develop, and even that it had made some serious mistakes in spite of its long history and rich cultural heritage. For example, a trend of thought against traditional Chinese medicine had emerged again in 2006, beginning with an article titled 'Farewell to Chinese medicine and Chinese Traditional Pharmaceuticals' which ignited contention. And then the 'Open letter to the national network readers' was published, following a series of activities, such as opening a special blog for anti-Chinese medicine, engaging in online signature event for clampdown on traditional Chinese medicine, resulting in a very adverse effects.

It cannot be denied that although traditional Chinese medicine has its distinctive superiorities as a discipline of medicine, traditional Chinese medicine is clearly not as good as western medicine in areas like emergency, trauma treatment and some diagnostic means, just as western medicine is inferior to traditional Chinese medicine in some areas. Therefore, the development of the soft-tech of traditional Chinese medicine must be integrated with hard-tech of western medicine, taking the path of integrating tightly traditional Chinese medicine and western medicine. However, it is noticeable that traditional Chinese medicine can in no way lose its own characteristics – the core soft technology due to the integration. In order to transform traditional Chinese medicine and Chinese Traditional Pharmaceuticals into a world-recognized domain of scientific knowledge and technology, and for it to form one of China's strategic industries, the study and the right positioning of traditional Chinese medicine in the field of science and technology will be crucial. How Chinese medicine takes its own road of development is the first challenge facing Chinese medicine in the twenty-first century, and is also a challenge facing Chinese culture.

Owing to its different cultural background and different way of thinking, it has been difficult for the western world to accept and understand the theory of traditional Chinese medicine and its diagnostic and therapeutic technologies. However, people around the world have gradually begun to realize that traditional Chinese medicine is different from traditional science and technology because it integrates science, Chinese culture and traditions in a unique way and that this technology is closely related to its culture.

From the perspective of soft technology, some exciting reasons to study traditional Chinese medicine and its various technologies are related to its thinking processes and its approach to understanding and analyzing problems. In addition, there is much to learn from the approaches developed by the traditional Chinese medicine to deal with contradictions. These features may assist in the comparative study of life and non-life systems and they may also help promote our understanding of the complex social systems in their entirety. In addition, they may help us to understand and deal with the relationship between the various factors that make up the society-economy-technology system, as well as to understand the individual relationships between each factor and the entire system (Jin 2001). To some extent, this may provide a more efficient platform for building harmony, integration and balance for mutual learning, infiltration and coexistence of the East and the West than is possible through politics.

As China succeeds in promoting its traditional Chinese medicines in the larger world, complete with the 'Chinese Brand' within international markets, it will have succeeded in making a great contribution to the twenty-first century. Improved Chinese medicine will constitute a new kind of industrial technology, and if developed wisely with appropriate investments, it will attract the world's attention.

Since China's reform and opening up, it has begun to place due importance on research and the application of traditional Chinese medical theory and its soft technology. Examples of this new attitude may be found in *Concrete Thinking is the fundamental thinking model of traditional Chinese medical* (Liu 1995); *Experimental Science and Experience Science – Contrast between Traditional Chinese Medicine and Western Medicine* (Liu 1996a); *Medicine and Human Culture* (Qiu 1993); and *Future Medical Thinking* (Ji 1999).

8. Technology Focused on the Human Mind and Relationship between Mind/Body – Psychogenic Technology and Psychosomatic Technology

As intimated above, technology started at the stage when humans first walked on two legs and used their two hands as tools. Then the technology of making tools and using artificial tools gradually developed and with it has come the increasing exclusion of the human body as a tool, namely so-called labour-saving technology, as popularized in economics (Honda 1975). Further, automation technology and robotics-based production has been pursued. Thus, human beings have become increasingly separated from 'human body technology' through the pursuit of mechanical technology. The primary emphasis of technological object in these developments has been placed on external (nature) matter and substance rather than the internal dimensions of the human 'heart' and 'mentality', such as 'spirit', 'psychology', 'sentiment' or 'feeling'.

From the point of view of problem-solving, technology may be seen as an extension of human abilities: the human body, sense and consciousness, etc. These all owe their externalization and expansion to the development of technology. Throughout most of its evolution, science and technology involving the human body – medical science and technology – have successfully treated the 'human body' as a complex organic system. The human body structure (static function) is refined to a digestive system, a circulatory system, a metabolism system, a genetic system, a nerve system and a motor system. Contemporary research in the life sciences has come to focus on the human memory system, cells, genes and the actual sources of life, etc. Namely, the majority of current 'high-tech' approaches to the life sciences assume the human body to be an organic 'system' or a special kind of 'material structure'. However, conventional science-based technology approaches have not been oriented towards the human mind (Kurihara 1987).

With improvements in the level of material civilization, people have come to care more about the sensate aspects of life – sight, sound, taste, smell and touch – and also the intangible aspects of life – intuition, moods, emotions, feelings and moral sentiments – and they expect these aspects of their experience to be treated with dignity and respect. To fulfil this requirement, our future efforts in technology should not simply focus on 'efficiency first' or 'benefits first' but rather should focus on the development of technological means that, even at some cost of efficiency (to sacrifice some efficiency), can make our living and working conditions easier, more comfortable and more convenient and that will respect human moods, feelings and morals. In turn, our understanding of the value and capabilities of human beings as well as 'psychosomatic relations' is not enough. These factors are the driving force behind the recent 'softening' of hard technology and they are also the main ingredients behind the rise of service-oriented manufacturing, which are important contents and topics for improving the value-added of soft technology.

Recently, a variety of in-depth research activities and applications focusing on the human mind have continued apace. More and more people realize that research on human beings is a unique discipline, and believe that all human activities, such as art, music, language, literature and architecture, are products organized by the human brain and in accordance with its unique rules. For this reason, many people are engaged in research on unveiling the secrets of human thinking. For instance, it is one of the objectives of Nerve Aesthetics Association.

Scientific research on exploring the mystery of human emotion is also very active. Happiness, anger, sorrow and joy are human emotions. Sometimes the human mind may be at ease, sometimes it may be excited and sometimes it may be depressed. Accordingly, developing the environment or products and creating comfortable living conditions which help put people's minds at ease have become the goals of many companies, and some companies have even developed emotion-oriented products – such as the shirt that helps people to relax. Through studying the way human beings accept flavour, an emerging taste science is concerned not only with research on dainty and delicious food, but also allows people to experience unusual tastes, and enables the adaptation of food products to different consumers tastes.

Various psychological methodologies have now been widely applied in clinics; and an operable knowledge system which may be called 'psycho-technology', derived from psychological science, has been developed and put to use. In 1991, the Chinese scholar Yang Xinhui founded China's first research institute in psycho-technology to foster masters and doctoral students in the application of psycho-technology. He founded the national research association of psycho-technology application in 2000, and is also the author of *Modern Psychotechnology Studies*, in which psychological evaluation techniques, social-psychology investigation techniques for reference groups, techniques of psychological counselling and psychotherapy, as well as economic psychological techniques, are elaborated. The treatise also introduces the relevant thinking of psycho-technology and psychotherapy of Chinese traditional medicine in Ancient China.

The philosopher René Descartes put forward a form of thinking subsequently known as 'Cartesian dualism', in which the body and mind are separated into two discrete domains. This form of thinking has become deeply rooted in science, education and medical research, and for a long time now has affected people's attitudes and prejudices towards the body.

A new concept of health suggests that human psychological activities should be in balance and harmony with the physiological system, such as the circulatory system and metabolic system, i.e. that psychological health should be harmonious with physiological health. It may be argued that in modern society people face more complicated and more profound social problems and their nerves are therefore more fragile. It is reported that there are over 400 million people with various mental diseases in the world. It makes both the use of the principle of psychosomatic medicine in health care and the generation of more psychosomatic physicians more urgent. Psychosomatic medicine will become a division of medicine.

It is noteworthy that Somatics originated in the Occident in the late eighteenth century and has subsequently gradually taken shape as a set of theories and methodologies adapted, contended and developed by a variety of schools of thought.

Kaiser Permanente, the largest health maintenance organization in the United States, has a large number of psychosomatic physicians and has launched a programme that uses mind-body interaction techniques to help people with chronic health conditions. 'I am surprised that it has taken medicine so long to recognize what is obvious; how single-minded and relentless we've been in reducing and separating mind and body in medicine,' said Dr. David S. Sobel, director of patient education at Kaiser Permanente (Roan 2003). Psychosomatic disorder is a kind of real disease caused or aggravated by human psychology or emotion.

Taiwan's Dafeng Lin and Meizhu Liu defined Somatics as follows:

> [...] [I]t is an empirical science to explore the relationship between body and mind as well as learn from the body itself and comprehend the physical wisdom. It is an art and 'ology' that attaches importance to realizing and reflecting internal experiences to explore an interactive relationship among the apperception of human body, biological function and external environment. (Lin & Liu 2003)

Somatics research aims to conduct a dialogue between body and mind through systematic theories and methods, so as to open up the body's perceptive ability and enhance the body's adaptability to the environment. The ten basic points of view include: the merging of body and mind; the body is not only a living organism, but also the source of feeling and thinking as well as wisdom; it is necessary to observe the human body from the first-person viewpoint; experience by the self-inherence; to emphasize on process-oriented; exploitation of the perception ability; to be conscious of changes and choices; pay attention to the power of touch and the re-patterning of habitual comportment; respect for the body wisdom; emphasizing body mind or body wisdom, as well as harmony between internal and the external environment.

Somatics has been applied in education, treatment, movement training, etc. For example, in education, followers of Somatics believe that gymnastic education can be the best place to implement directly body and mind education. In accordance with its views concerning physical fitness, the American Association of Health Physical Education Recreation Dance (AAHPERD) put forward the concept that health should include physical fitness, emotional fitness, social fitness, spiritual fitness and cultural fitness. These five fitness-oriented aspects can be integrated into the all-round education which contains three levels, namely, 'personal progress', 'interaction among human, society and culture', and 'the human and the nature, action facing matters'. The Somatics point of view provides people with opportunities for self-awareness, allows us to change our point of view on our body, and exploits or improves our perceptive abilities in relation to emotion and society as well as environment. It enables us to maintain physical and mental health by learning and enhancing relaxation and stress management, choosing a good posture and enhancing the efficiency of movement so as to encourage people to have even more respect for the body, experience and themselves. Somatics can also provide the method for communicating with the body to make the merging of body and mind present a high degree of skills and increase the demonstration of human vitality.

It is worth mentioning that despite Somatics being a research area proposed by western scholars, many fitness methods and psychosomatic techniques of traditional Oriental culture have been re-interpreted by western psychosomatic scholars and placed among the best techniques for physical and mental coordination including Qigong, Tai chi, sitting meditation, traditional health law; as well as Chinese physical and breathing exercises, martial arts, etc. of China; Aikido, Kendo, Zen-Meditation of Japan; Yoga, sitting meditation of India. This also reveals how the centuries-old Oriental culture is an endless treasure of psychosomatic soft technology, which needs to be reorganized, upgraded, exploited and applied systematically.

Conclusion – The Need to Renew the Notion of Technology

In the twenty-first century the explosion of knowledge, the softening of the economy, the changing of values, the integration of art and science and the human mission of sustainable development require us to renew the notion of technology, transforming the traditional understanding of technology from a narrowly defined concept to a broadly defined one. In addition, it will be necessary to enhance research, development and the application of soft technology. In other words, in the wake of several previous Industrial Revolutions, it is time for human beings to create a conceptual revolution in technology.

C. What is Soft Technology?

1. Knowledge, Technology and Soft Technology

Technology is the indispensable vocabulary for international leaders, politicians, business executives or research personnel in the present age. However, what is the technology? How does it affect the development of human society and the lives of each one of us? We need to put technology in the whole human knowledge system via systematic reflection and sifting of knowledge to clarify the meaning, classification and function of technology. It will help us to change our thinking mode and our vantage point for observing the world.

Ever since the beginning of the 1980s, international scholarly attention has been focused on a new phenomenon: the knowledge-based economy. Scholars from a variety of countries have conducted empirical research demonstrating that knowledge is the driving force behind social and economic development, allowing us (as shown in Table 2) to renew our understanding of the value of knowledge as an important productive force. Furthermore, this research has deepened our comprehensive understanding of knowledge from the perspective of its sources, its character, those who possess it, its functions, its levels and its feasibility of operation (Jin 1998c). Awareness of the sources and operability of knowledge is significant for soft-tech research.

Table 2: The Understanding of Knowledge – Knowledge in General.

	Category	Contents
1	Sources of Knowledge	*Scientific Knowledge and Non-scientific Knowledge *Knowledge of Natural Science and Non-natural Science *Innate Knowledge[7] and Acquired Knowledge
2	Character of Knowledge	External Knowledge and Implicated Knowledge
3	Holder of Knowledge	Organization Knowledge and Individual Knowledge
4	Form of Knowledge	'Knowledge of why', 'Knowledge of what', 'Knowledge of how to do it', 'Knowledge of who', 'Knowledge of who can help', 'Knowledge of when', 'Knowledge of which place is the best.'
5	Level of Knowledge	Local Knowledge and International Knowledge
6	Operability of knowledge	Operable knowledge and inoperable knowledge (scientific knowledge and technological knowledge)

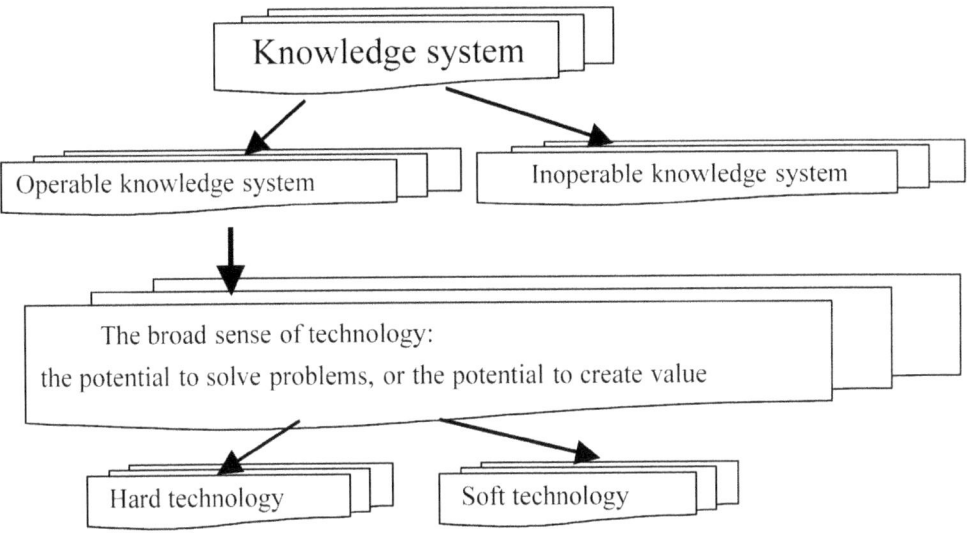

Figure 4: Knowledge – Technology – Soft Technology.

The sources of knowledge are not only natural sciences but also non-natural sciences such as philosophy, the social sciences, cognitive sciences and thinking science, etc. Knowledge of an unscientific nature, as found in art and religion, is also included. Through the analysis conducted earlier in this chapter, it became evident that knowledge derived from the non-natural sciences and from non-scientific sources also has a direct and powerful effect on social and economic development.

From the perspective of operability, knowledge can be divided into operable and inoperable knowledge. In fact, the operable knowledge system is the broad sense of technology and the so-called 'operability' refers to the potential to solve problems or the potential to create value. This kind of classification has always been neglected but it has a great significance for the understanding of scientific knowledge and technological knowledge, as well as institutional designs pertinent to scientific and technological development. Moreover, it is also conducive to understanding the deeper meaning of technology.

2. What is Soft Technology?

For a long time, people have created an air of mystification around technology. Simply, technology is the means and tools of 'problem solving'. There are two types of 'problem solving': one is concerned with tangible things such as products, and the other is concerned with intangible phenomena such as processes. We can call the former 'hard technology' and the later 'soft technology'.

```
┌─────────────────────────────────────────────────────┐
│              The broad sense technology              │
│  * Tools, means, method, rules for the solution of problems │
│  * Operable knowledge system                         │
└─────────────────────────────────────────────────────┘
           ⇓                              ⇓
┌──────────────────────────┐  ┌──────────────────────────────┐
│      Hard Tech           │  │      Soft Tech               │
│ * Knowledge of natural science │ *Knowledge of non-natural & non-science │
│ * To operate in the physical world │ * To operate in the spiritual world │
└──────────────────────────┘  └──────────────────────────────┘
```

Table 5: The broad sense technology.

Therefore, in general, technology is composed of hard and soft technology. From the perspective of problem solving, technology refers to the skills, tools, rules methods and procedures that are employed by humans to improve, alter, accommodate (humans can only accommodate nature, not control it) and manage or control the subjective and objective world for human survival and development.

As economies develop and as technology changes, the boundary between hard and soft technology blurs. However, in general, we can say that hard technology is manifested mainly through material forms, and its knowledge system is derived mainly from the knowledge of natural science. Soft technology is manifested mainly through human psychology and behaviour, and its knowledge system is derived mainly from non-natural sciences and non-scientific (traditional) knowledge.

Here, 'hard' refers to the physical entities through which operations are conducted. 'Soft' refers to entities without physical form. In other words, 'hard' refers to tangible phenomena, while 'soft' refers to intangible phenomena.

Soft technology must, by attribute, exhibit two general sets of characteristics: it must be *technology* and it must also be *soft*.

From the vantage point of its *technological attribute* we may say the following things about soft technology:

- It should be an operable knowledge system with tools, procedures, system of rules for the solution of problems (practicality or 'realizability').
- It should be directed towards practices for providing service for social change and economic development (services). The services provided by technology here must take account of multi-dimension of objectives – to improve the economic, social, ecological, resources and environmental benefits.

From the vantage point of its *softness attribute* we may say the following things about soft technology:

- Its operational knowledge is deeply rooted in internal, human, conscious activities.
- Its operating objects include the processes of human psychological activities, processes of thinking, and those human behaviours controlled by and embodying human psychological activities related to perceptions, emotions and values, as well as the field of relationship between body and mind. The various psychological, social and cultural factors are therefore the distinguishing parameters of soft technology.
- The way in which soft technology provides service, besides tangible products, is mostly through intangible modes such as services, procedures, rules, etc.
- The 'plasticity' of soft technology. The meaning, functions and characteristics of soft technologies can be presented, formed, modified and expressed according to the changes of natural and social environments as well as our capacity and understanding level of the subjective and objective world around us.

The reasons I have chosen to use the phrase 'soft technology' are that, firstly, the above *softness attribute* distinguishes the primary subject matter of soft technology from that of traditional hard technology; secondly, the term 'soft science' has entered into international usage, perhaps making it easier for people to accept the term 'soft technology' rather than some other term that I may have chosen to coin; and, thirdly, I wish to emphasize the distinction between what I call soft technology and the concept of 'humanistic technology' that is discussed in some literature (L. Gao 1996).

In short:

- Soft technology is an operable knowledge system which is realized from the practice of economic, social and humanistic activities, and which then shapes the rules, mechanisms, means, processes and institutions of problem solving through summarizing and consciously using its common laws and experiences, which have guiding significance of implementing the amelioration, adaptation and control to the objective and subjective world.
- Soft technology is the intellectual technology or 'intangible' technology of creation and innovation centred in human thought, ideology, emotion, values, world views, individual and organizational behaviours, as well as in human society.
- Soft technology is technology which is drawn from multiple perspectives, such as the economic, social, philosophical, cultural and artistic angles. Its objective is to find an alternative paradigm that will be differentiated from traditional concepts of knowledge, technology and value-orientation in a variety of knowledge systems, and make known how to use these tools to analyze and deal with the complex challenges, and gain access to the implementation capacity of problem-solving. Therefore, soft-tech research will be advantageous for rethinking and reflecting on the multitudinous challenges with which human kind is faced. It can also be embedded in different cultural systems to promote sustainable development and interdisciplinary research, so as to obtain broad insights and effective solutions for solving global, national or corporate level problems as well as individual problems.

3. The Characteristics of Soft Technology

In order to gain a deeper understanding of soft technology, we will now discuss its basic characteristics.

Hard technologies and soft technologies, since they are both technologies, have many attributes in common. From the perspective of their intrinsic nature as 'technology' they both exhibit the following features:

- They take the form of skills, tool, rules, mechanisms, means, methods or procedures for the solution of problems.
- Their purpose is to practice and they are intended to provide 'service' for social progress and economic development.

However, as we can see from Table 3, when compared with hard technology, there are fundamental differences between them, and soft technology evokes a completely new technological paradigm.

In general, soft technology contains the following characteristics when compared with hard technology:

1. **Soft technology exhibits a closer relationship with humanity and culture.** Soft technology takes the internal psychological activities and the external behaviours of human beings as its operational object, and its content and levels are determined by its focus on the ways of thinking, levels of understanding and action modes of human beings. Its application and popularization are directly related to local morality for particular times, cultural backgrounds, habits, knowledge levels, etc. At the same time, during the process of the operation of soft technology, different features are formed according to different operational objects. We can say that human factors and the social environment control the formation, presentation, moulding and innovation of soft technology. Therefore, soft technology is a technology that contains human thoughts, viewpoints and strong individuality and it controls the direction of hard technology applications.
2. **Soft technology embodies distinctive concepts of humanity and human factors.** The concept of Man in soft technology is different from that in hard technology where the object is 'outside the human body'. Although the human body is the operational object in western medical science and life science, it is treated as a 'physical' object or a complex of cells, even a crowd of gene, and can be duplicated and cloned in the scientific sense; whereas in soft technology, the human body is regarded as an organic whole with consciousness, feelings, purpose, thoughts and values. Human growth and the formation of personality are not solely determined by genes, but result from multiple genes and a complex interplay of nature, society and environment in the long term.

Hard technology also stresses these human factors but is more concerned with human reaction and capacity towards the 'outside' and 'substance', while soft technology pays more attention to human psychological activities such as sense, feeling, mood, thought, thinking patterns, values, traditions, habit, personality and the ability to control psychological activities. Therefore, the main parameters of soft technology are various psychological, social and cultural factors.

3. **Soft technology is rooted in the spiritual world.** The so-called spiritual world includes the abstract world (the object that is conceptualized through the processes of internal psychological activities), the visual world (the reappearance of images of events through memory and the mind's eye) and the concrete world (the reflection of sensory experience, emotion/mood and action, e.g. heartache, dread, enjoyment, etc.) (Tianjun Liu 1994; 1996). Whereas abstract thinking operates concepts, visual thinking operates images, and concrete thinking operates the consciousness itself. The concept of the concrete thinking has already been applied in arts circles, which can be further divided into three branches, namely, emotional thinking, sense thinking and action thinking.

The physical world, in which hard technological operation is conducted, includes natural and artificial fields. The spiritual world, in which soft technological operation is conducted, includes an 'inner orchestration-action system' and an 'outer behaviour system'. The former includes the abstract field (concepts, modes, systems, etc.) and the consciousness field (sensation, emotion and mood/feeling). The latter includes the social behaviour field (performance is dominated by inner orchestration action and value view, world view, ethics/morals, sentiments, etc.).

4. **Soft technology is not 'neutral'.** Soft technology is fundamentally dualistic. The dualism of soft technology stems from its dualistic functions, in that it simultaneously manifests both productive forces and the relations of production.

On one hand, because of its technological attributes (natural attributes), soft technology not only generates added-value and stimulates the formation of new industries as core technology, but also provides directions, tools and means for innovation in hard technology to improve the efficiency of innovation belonging to the *productive forces*. On the other hand, owing to its 'soft' attributes (social attribute), soft technology plays a critical role in determining the substance and content of new institutions. Hence it may also form the basis of institutional reforms, thereby assisting with decisions associated with adjusting the benefits distributional system and with changing, restraining and regulating various relationships in economic and social activities including the interest relationship (belonging to the *relations of production*). These functions of soft technology manifest distinctly the dualism of soft technology. Using the language of political-economy we may say that soft technology involves both the productive forces and the relations of production.

For a long time, profit-oriented technology development and its unethical application have caused numerous disasters for humanity. Besides facing ecological damage to the external ecological environment, human beings have also faced technology-induced threats to their interior nature and to the social world. Examples include the issues associated with the Y2K threat, which brought about a 'global earthquake' during 1999. Countless observations may be found in the scholarly literature and in the popular news media about the negative effects of almost every kind of technology, such as atomic technology, genetic technology, nanometer technology and the technology of high-energy physics.

In the information society created by information technology, there exists the 'chiphead', the 'network aficionado' and the Internet criminal, and moreover 'hackers' infiltrating information networks; and the negative influence of the modern alchemy of finance and its worshippers results in a 'stock aficionado' and 'financial crimes'. In the biotechnology era of the twenty-first century, if left unchecked, human beings may create countless 'new species', and 'transgenic animals' and 'cloned animals' may be mass produced in biological factories; mass-scale 'human cloning' may very take place clandestinely as a secret weapon; and the risk of 'cloning crime' and 'clone terror' will grow.

However, it is unfair to blame technology for all these disasters and potential disasters. The responsibility for the defects of technology, the so-called menace of technology, and for the direction in which the application of these hard technologies has moved, lies completely in the hands of the operator of the technology – people. Moreover, we need to continuously invent and develop new technologies to meet human demands, to deepen our understanding of nature and to serve the needs of society for solving many problems that are brought about by hard technology industrialization.

Only by dint of soft technology can human beings provide effective control and the right direction for hard technology, e.g. soft technology is a tool of innovation, which has, as one of its purposes, the manipulation of innovation in hard technology. In this sense, soft technology is imbued with 'thought' and 'opinion', and hence soft technology is influenced by morality, values and world-views, like the dualism manifested by cultural technology and social technology themselves. Therefore, soft technology has both positive benefits and negative effects.

That is why governmental and social interventions are necessary in the application and promotion of soft technology, and its activities must be encouraged and regulated and sometimes even prohibited or censured by institutions, laws, standards and policies. Namely, it is necessary to carry out the institutionalization of soft technology and institutional innovation.

5. **Thinking mode.** The phenomena, problems and subjects which soft technology needs to address are invariably complex and interdependent. Therefore, for designing the scheme of problem solving, a systematic and holistic approach is needed. And it is necessary to first find out the background, various causes and potential impact of the 'problem', and only by grasping the whole can we identify the key issues and look for breakthrough points. This is why the thinking mode of soft technology to solve the problem or design the scenario must be from the whole to the part and from the macro to micro, and not the other way around.

6. **Soft technology is resistant to standardization.** The fact that soft technology embodies psychological, social and cultural factors creates severe obstacles to its standardization. On the other hand, soft technology includes explicit and tacit technologies (Nonaka & Takeuchi 1997). The former can be presented by words, data, standardized procedures and general principles that are disseminated and shared by way of books, lectures and training; whereas the latter cannot be properly presented in the form of written documents and formal languages, so it becomes difficult to communicate and share in a standardized manner, e.g. LPFE technology. This is due to what Nonaka refers to as 'implicit knowledge' embodied in technological phenomena.
7. **Soft technology has imprecise boundaries.** Since all soft technologies are closely related to human factors, and although soft technology can be systematically categorized, as we shall see below, the boundaries between different types of soft technologies are very vague and each influences and infiltrates the other.
8. **It may not be necessary to convert all soft technologies into products and services.** A great number of soft technologies, just like hard technologies, can form the basis of industries and can be used to provide products and services, e.g. financial derivative tools, various kinds of incubators, cultural products and so on. However, most soft technologies do not necessarily provide the tangible product, but a kind of processes, efficiency, procedures, results, mechanisms or institutions.
9. **Innovation in soft technology has distinctive causes and characteristics**. The reason for the obsolescence of hard technologies is usually the emergence of new inventions to replace the old technology. Innovation and renewal of soft technology, however, is more often caused by the changes in peoples' lifestyles, values, levels of demands and thinking patterns.

    Innovation in soft technology is differentiated by the fact that it is more strongly limited by relevant institutions, systems, laws, regulations and policies than innovation in hard technology. Hence, the creative destruction of old institutions and laws should take place to enable soft technology innovation. This process usually follows this pattern: designing the system/method → implementing the operating system/method → providing concrete services (the process of operating or implementing is the process of providing customers with services) → developing criteria to form institutions and organizations to facilitate promotion, popularization and application → creative destruction of institutions → operation in the new institutional environment → the next round of innovation.
10. **Soft technologies have to be combined or integrated in practice.** The implementation and success of soft technology requires a comprehensive and holistic approach to its application. As was mentioned above, hard technologies have to be integrated with soft technologies so that their social and economic value may be realized (for more details see the section of 'Function of Soft Technology' in Chapter 4). The primary criterion for judging the success of the commercialization and industrialization of hard technologies is whether they integrate well with continuously advancing soft technologies. On the

Table 3: Differences between Soft Technology and Hard Technology.

	Standard	Hard technology	Soft technology
1	Source	Knowledge based on natural science	Knowledge of non-natural science and non-traditional science
2	Operational object	'Substance'	'Human' psychological action and social behaviour
3	Operational field	Physical world	Spiritual world
4	Operational Goal	To change and control the nature and substance of material	To master, orchestrate and manage human ideology, emotion, thinking mode, values, as well as the behaviour mode of individuals, groups and organizations
5	Carrier	Tangible substance	Intangible human factors
6	Technological parameter	Physical factors	Psychological, social and cultural factors
7	Meaning of Human factors	Influence of extrinsic behaviour	1. Influence of extrinsic behaviour – the performance of psychological action 2. Influence of intrinsic behaviour, i.e. psychological action, such as feeling, sensation, emotion, ideology, culture, value view, world view, tradition, individuality, etc.
8	Position of human body	An organism; in the final analysis, a substance and a cellular combination	A life that contains consciousness, sensation and a spiritual dimensions
9	Source of innovation	New inventions and discoveries	(New inventions and discoveries) +(Result of human notions, life styles, values and points of view)
10	Characteristic of innovation	Not necessary to destroy and can coexist with old system	Need a creative new system to displace old broken system
11	Process of innovation	Materials-processing products; product design, manufacturing and marketing	Dreams/originality → forms systems/ modes/methodology → exercise/ regularization; Design system and methodology → run/implement → cultivate the process from which the new institution grows → displace the old system → create and build new system
12	Relationship with institutions	Institutions are the environment of hard-tech innovation and creation	Institutions are the innovation environment of soft technology, on the contrary, soft-tech innovation

			is the content and basic of the new institutional innovation
13	Thinking mode	From part to whole	From whole to part and holistic approach
14	Relationship with subjective purpose	Independent of human will; no subjectivity	Involves subjectivity; can be moulded, developed and affected by humans intellect, thinking modes and behaviours
15	Product	Physical form	Without independent physical form
16	The Method of resolving problems	Products and services	Processes, rules, institutions, products and services
17	Ontology	Neutral	Dualistic
18	Standardization	Tends towards standardization	Involves strong individuality and is difficult to standardize
19	Regional feature	Region-neutral	Regionally specific

other hand, soft technology will only succeed if it is combined comprehensively with other relevant soft technologies, modified according to the different conditions of local geographic and socio-political circumstances, and in accordance to the demands that stem from ostensible design goals. Similarly, integration with continuously advancing hard technologies is necessary for the promotion of soft-tech innovation and to ensure 'high-technologization' of soft technology.

11. **The characteristics of soft technology must accommodate distinctive regional conditions.** Differences in soft technology will be invented or produced by differences in cultures, economic levels, lifestyles, habits and thinking modes of communities. Hence, a soft technology developed in one region will always have to be redeveloped and innovated to suit the conditions in another if it is to function optimally in that other region.

12. **The relationship between soft technology and institutions is close.** The dualism of soft technology means that it is, by nature, entwined and infused with institutions. In the case of hard technology, institutions are part of the environment and conditions for innovation. In the case of soft technology, however, institutions are not the only part of the environment of technological innovation. Soft technology itself forms the foundation and content of innovation in relevant institutions, systems, law, regulations and policies. On the other hand, the invention, dissemination and application of many soft technologies are restricted intensively by relevant institutions, systems, laws, regulations and policies (for more details see the section of 'The essence of Institutional Innovation' in Chapter 4).

13. **Soft technology requires special talents.** Hard technology requires specialists in relevant technical fields, but soft technology requires people with talents derived from interdisciplinary and cross-sector knowledge and practical experience.

D. The Classifications of Soft Technology

There are a number of ways, using a wide variety of criteria, for classifying hard technologies. For example, in terms of operational matter, they may be classified as information technology, material technology, biotechnology, energy technology, ocean technology and space technology, etc.; in terms of function, they may be classified as automatic technology, environmental protection technology, sense technology, remote sense technology and so on; and, in terms of industry, they may be classified with such categories as oil exploitation technology, textile technology, coal technology, etc.

It is more difficult, however, to classify soft technologies. First, analyzing the sources of soft technologies is difficult. In the case of natural science and hard technology, the physical world is the object of cognition and operation. For non-natural science and soft technology, however, what needs to be 'cognized' and operated are cognition and operation themselves. In other words, hard technology can be regarded as a game object, while soft technology can be regarded as both the game object and game content (see the section of 'The Essence of Institutional Innovation' in Chapter 4).

Second, since all soft technologies have human factors as their carrier, they bear profound relationships with each other. For example, when we analyze the applicable characteristics of soft technology, we see that it is difficult to draw a strict line between business technology and other kinds of soft technology operating in the social or human resources; education technology is not only an intellectual technology but also a social technology.

For the sake of further study on soft technology and the conscious creation, development and application of soft technologies, it is nevertheless necessary to develop reasonably good categories for classifying soft technology. This book provides a first attempt to take up this challenge. We will begin by classifying soft technology according to the sources from which its constituent knowledge is derived.

1. Classification According to Sources of Knowledge

In light of our understanding of technology as an operable knowledge system, we may say that different types of knowledge form the basis for different kinds of soft technologies.

Category 1 – Technology originating from the knowledge of social sciences. Examples include various social technology and commercial technology (see details in Chapter 2).

Category 2 – Technology originating from the knowledge of the natural sciences with 'soft' features. Examples include network technology, software technology, biotechnology, environmental protection technology, artificial intelligence technology, etc. All of these technologies are rooted in natural science knowledge but the most added value will be from their 'soft' features as they are transformed into products and services.

Category 3 – Technology originating from the knowledge of Oriental culture and medical science. Examples include telepathy technology of humans for environmental change, health preservation technology, Chinese traditional diagnostic and therapeutic technology, psychosomatic technologies, and the three regulating technologies of breathing, mind and body in Qigong (Tianjun Liu 1999), longevity extension technology and the diagnostic technology of ethnic minorities, e.g. Tibetan medical therapeutic technology.

Category 4 – Technology originating from the knowledge of thinking science. Examples include psycho-technology, spiritual health technology, decision-making technology, systems technology and thinking technology.

Category 5 – Technology originating from non-traditional scientific knowledge. Technologies come from the knowledge of language, literature, philosophy, law, art, religion and special environment, including culture technology, indigenous technology and some social technology.

Category 6 – Technology originating from the intersection of the above fields of knowledge. These technologies are the most abundant examples of soft technology with the greatest potential.

2. *Classification Depends on Operational Field*

The most fundamental difference between soft technology and hard technology is the contrast between their operational fields. The operation of the former is mainly rooted in human spiritual world, which is human-centred; while the operation of the latter is conducted in the physical world, which is material-centred.

As shown in Figure 6 on the next page, the broad concept of technology includes the hard technology with which we are more familiar that operates Nature, Natural Matter, Artificial Matter, and the Human Body as Matter. Soft technology operates fields such as psychological activities, human behaviours controlled by and embodying human psychological activities, as well as the relationship between body and mind.

3. *Classification Depends on Operational Resources*

As shown in Figure 7, operational resources of soft technology can be divided into economic resources, social resources, cultural resources, natural resources, the operational resources focusing on psychology, life and psychosomatic relations, as well as the hard-tech resources that can be 'softened' – artificial systems.

The so-called operational field of soft-tech is actually the innovative resources of soft-tech, thus this classification can also be regarded according to the innovative resources of soft technology.

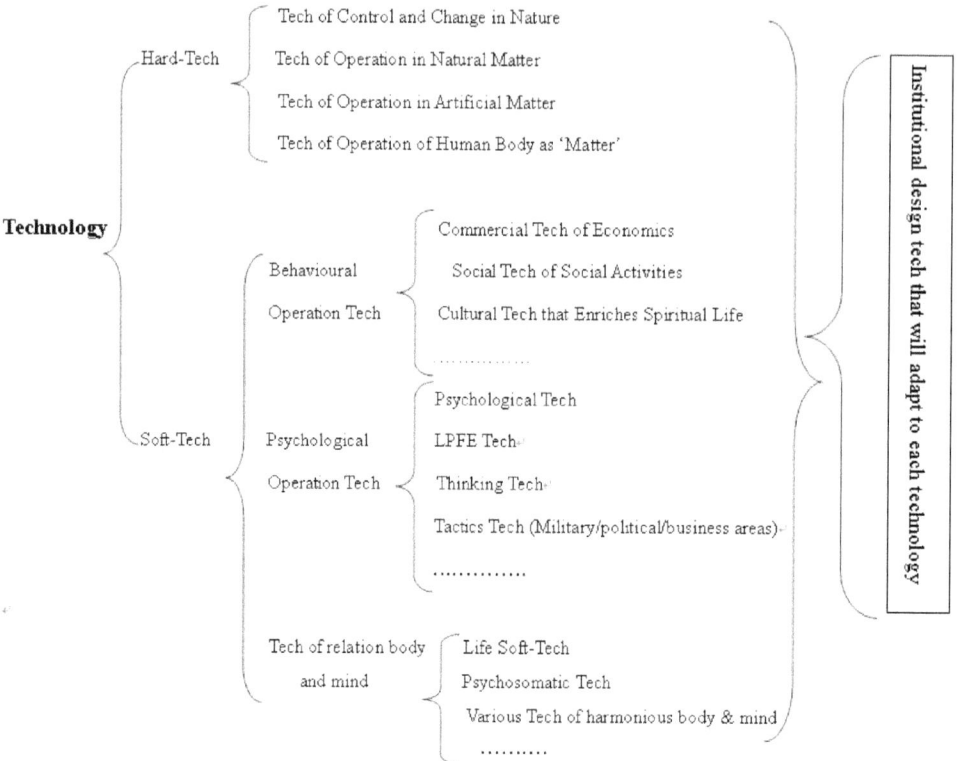

Figure 6: A General Classification Map of Technology.

1. **Soft technology that takes economic activities as its operational sources – Commercial technology.** Commercial technology is the process of technology addressing creative human economic activity for improving the efficiency of economic activities. Various technologies, such as exchange technology, currency technology, patent technology, accounting technology, stock technology, advertising technology, management technology, marketing technology, financial derivatives and incubator technology, supply chain technology, etc., are cases in point. Commercial technology can also be regarded as the 'game' in economic field that economists are often talking about.
2. **Soft technology that takes social activities and problem-solving of social issues as its operational sources – Social technology.** Social technology is technology directed towards social activities and social relations of human beings, who create value by exploiting and making use of social resources. The technologies for exploiting social resources include meeting technology, discussion technology, education technology, training technology, learning technology, coordination technology, alliance and cooperation technology, social programming technology, public relations technology,

What is Technology?

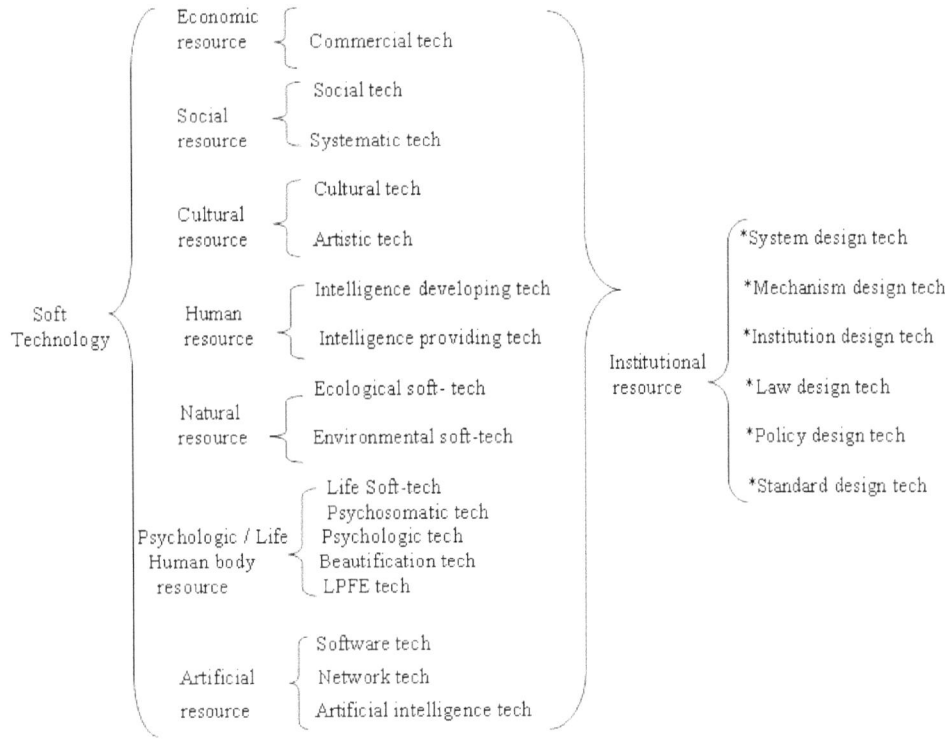

Figure 7: Categories by Resources.

human relation technology, organization technology, service exchange technology, etc. The technologies for dealing with social issues include all types of systems technology, planning technology, diagnosis and evaluation technology, foresight technology, decision-making technology, strategy-choosing technology, city technology, social simulation technology, monitoring technology for social crises, institutional innovation techniques, etc.

Political technology belongs also to the category of social technology. The problem-solutions or means used by governments, political parties, social groups and individuals involved in the management of foreign and domestic social affairs, or in the activities of dealing with international relations – such as political system design technology, diplomatic technology and canvasser techniques – may be called 'political technology'. Military soft technology that consists of military strategies and tactics may also be classified under 'social technology'.

3. **Soft technology that takes cultural resources as its operational sources – Cultural technology.** Cultural technology refers to the process of innovative activity associated with culture, which aims to enrich human spiritual life. There are distinctive technologies in the process of creating, producing, managing and marketing cultures, and also in the process of providing services to customers. Cultural technology can be divided into designing technology, producing technology, marketing technology and management technology. All kind of cultural technology, such as beautification technology (see details in the section of 'Cultural Industry' in Chapter 5), artistic technology, fashion technology, media technology, entertainment technology, movie and television technology, sports performance, game technology, amusement technology, drawing technology, performance-art technology and even cooking techniques are the process of embodiment for the market value and social value of various cultural resources. This process will enable culture value to be 'materialized' or 'solidified' through the exploration, design and manufacture of cultural resource.
4. **Soft technology that takes psychological activities, human life and human body as the operational sources.** It includes psycho-technology, longevity technology, diagnostic and therapy technology of traditional Chinese medicine, psychosomatic technology, the three regulating technologies of breathing, mind and body in Qigong, LPFE technology, and soft-life technology (see the section of 'Life Industry' in Chapter 5).
5. **Soft technology that takes artificial systems as the operational sources – Soft-engineering technology.** Now, more and more the framework of artificial systems can be used for simulating and manipulating human thinking mode, sensibility and contents of social systems, making the artificial systems – hard technologies – soften gradually. Examples include software technology, network technology, 'Internet of Things', artificial intelligence technology, system engineering and social engineering.
6. **Soft technology that takes human resources as the operational sources – Narrow sense of intellectual technology.** Soft technology mentioned here, refers in particular to the technology which focuses on human capital and takes developing, creating and applying the value of human capital as the main objective. In order to facilitate the research, we have labelled it as 'intellectual technology in the narrow sense', which may divide approximately into the intelligence-developing technology and the intelligence-providing technology:

- **Intelligence-developing technology**
  It is used to improve and develop human intelligence and abilities, for example, education technology, learning technology, training technology, and R&D technology.
- **Intelligence-providing technology**
  It is used to help people improve their problem-solving abilities and judgement by providing knowledge, wisdom, judgement, insight and experience. The prevalent intelligence-providing technologies include consulting technology, diagnostic technology, design technology, asset accumulation technology, etc.

7. **Soft technology that takes natural resources as the operational source.** There is both hard capital and soft capital in natural resources (see the section of 'Comprehensive Competitiveness and Soft Power' in Chapter 3). Soft technology taking natural resources as the operational resources mentioned here refers in particular to the technology operating on ecosystem, environment etc., i.e. environmental soft technologies, ecological soft technologies, etc.
8. **Institutional design technology.** Institutional design technology refers to the design technology of the institutional environment in addition to its system design at the macro level, mechanism design, institutional design, law design, design of regulations and policies, as well as standard design; and it also includes the institutional design focusing on a variety of types of soft and hard technology innovation (See Figure 6 and Figure 7). Rules, mechanisms, institutions, laws, regulations and policies themselves are the products of soft technology, which in turn promote or constrain technological development and technological innovation. Therefore, the key to success in institutional design technology is how to create an environment where we could not only promote innovation but also restrain those perverse or unjust innovations. Capability in institutional design technology can be obtained through continuously tracking and finding trends in a variety of soft technologies so as to summarize, forecast and analyze them, and gain foresight about them.

It is predictable that the existing fields of soft technology will further expand, divide and be further elaborated over time, in accordance with social and economic development.

E. Why Soft Technology Now?

Some observers aver that the concept of soft technology is so broad that it is all-encompassing except the natural science and technology and that it seems as if it is just another term for 'management'. As a consequence they question the feasibility of studying it.

First of all, management is the most mature field of soft technology, since it was developed relatively early. Secondly, because soft technology is so extensive and ubiquitous, we have neglected to study it 'consciously' and systematically (analogous to the manner in which water was for a long time not regarded as an important resource). Additionally, soft technology has been treated as a 'non-technological' factor in recent decades; and this has led to it being talked about as an all-encompassing 'black box'. Now it is time to lift its veil, analyze it in detail and carry on systematic research and application of soft technology.

Moreover, we have observed above that there are two basic approaches for solving problems. One of them (hard technology) has been studied extensively over the centuries and, as a result, various kinds of disciplines, technical systems and industries have been built on the foundation of hard technology. The alternative technological paradigm – soft technology – has missed out on such attention and research and, as a consequence, has not led to a concomitant suite of disciplines, technical systems and industries.

1. Soft Technology Opens Up a New Discipline, a New Field of Research, Where the Value Creation Process is a Completely Different Paradigm

Because soft-tech and hard-tech are different in their knowledge sources, carriers, operational objects and technological parameters, the value creation processes for each of them are based on entirely different paradigms. Graham R. Mitchell said:

> It is necessary to define technology formally as the two-part notion. Setting policy objectives for economic growth requires that we do far more than identify needed hard technologies. It seems to me that soft technology serves as an opening to the discussion of what constitutes much of the rest of the useful knowledge we need in order to create and maintain sustained economic growth. Meanwhile, the two-part definition for technology is particularly useful in demystifying technology in the service sector. It is necessary to note that the explosive growth of technology in the US service sector where core technologies often consist of hard technologies applied to soft objectives. (Mitchell 2001)

Research on soft-tech has provided a theoretical basis and means to identify, develop and innovate the core technology of the service sector. Mitchell believes that the discussion of soft-tech is often more of a problem for the policy analyst or planner than it is for the practicing engineer or scientist. The planner is often seeking to organize and allocate resources to a range of hard and soft objectives through a pre-designed, consistent theoretical framework, and often runs into the problem that value creation in manufacturing follows a different paradigm from that in the service sector.

In manufacturing we often start with materials, then design and manufacture products, and finally sell them for enough to cover all our costs. R&D and technology portfolios for manufacturing industries thus usually focus attention in three key areas – materials, design and manufacturing – as improvements here usually lead directly to the bottom line. For large service industries such as finance, transportation, wholesale and retail trade, communications and electric power, however, value is created by building a system or network first, then efficiently and competitively operating that network, and finally providing services to the customer who is very often part of the general public. Hence, R&D and technology portfolios in the large service industries frequently focus on system or network design, operations, design and delivery of customer services. In other words, the application process of hard technology is from part to whole, while that of soft technology is from whole to part, i.e. entire system design → operation/implementation → standardization → creative destruction → new system design.

To sum up, it is apparent that the value creation process applied to non-material 'production' areas is certainly different from that applied to material production areas, and we need to further encourage and protect such technological innovation.

2. The Characteristics of Soft Technology are Suited for Coping with the Challenges in the Twenty-first Century

The challenges facing humanity in the twenty-first century are different from those of the past, such as the global energy crisis, environmental crisis, population crisis, agricultural crisis, food crisis, etc.

First of all, as typified especially by the movement for 'sustainable development', which emerged into the limelight during the second half of the twenty-first century, humanity has begun to move away from the crude development mode of 'simplistically pursuing material wealth'. The language of 'sustainable development' has become an integral part of the world's common language. However, the question of how to implement it remains.

Secondly, the trend towards globalization has extended gradually from that of globalized production to that of economic, technological and information globalization, and this trend shows no sign of weakening. The challenges now are how we might fully utilize the benefits of globalization, such as the sharing and reasonable allocation of global resources (including hard resources like mineral resources, currency resources, etc., and soft resources like cultural resources, social resources, human resources, etc.), while avoiding a variety of cross-border risks, such as financial risks, transnational crime and the globalization of disaster. In addition, we need to consider how we might maintain equality, fairness and reasonable principles, along with protecting regional economic development and cultural diversity in the context of different degrees of development and increasingly widening gap between the rich and the poor.

Thirdly, we need to consider how it might be possible to transform the 'old economy' into the 'new economy' suited to sustainable development. Through centuries of practice, the world has become familiar with and has adapted to the so-called 'old economy', namely, the industrial economy making profits through conquering and controlling nature; the product economy based on the manufacture and sale of products as its core; the competitive economy with the main operating objective being the simplistic pursuit of economic profits; the economy based on a linear mechanical model of 'raw materials → manufacture → abandon' as well as 'mass production → mass consumption → mass abandonment'. How can we develop an economy with a completely different profit pattern, namely a harmonious economy in which profits derive from the harmony of humanity and nature; a cooperative economy which has shifted from absolute competition to a win-win mode; an economy consistent with a recycling society, featuring saving/moderate consumption/repeated use? Ultimately, it is necessary to gradually transform the national or regional development mode from the mode of simplistically focusing on GDP (pursuing economic profit first, with its inherent contradictions including periodic financial crises) to that of coordinated development of the economy, society, environment and resources (the quadruple bottom line of national development) (Jin 2007).

Fourthly, enterprises are faced with comprehensive challenges. For example, how might a corporate business model adapt to the new economy and to the notion of a harmonious society in which enterprises should not only create economic benefits, but also embrace social and environmental responsibility? Additionally, how might corporate culture and values adapt to the changes of a corporation's position – from that of a simple economic entity to that of a global citizen – and how might the core competitiveness of enterprises be reconstructed?

Fifthly, there are serious challenges in the sphere of culture and values. In order to meet the above-mentioned challenges it is necessary for new culture and new values to be cultivated. For example, in the relationship between humanity and nature, we should understand that we have only one Earth, that the supply of its natural resources is limited, and that people must live in harmony with nature; and that in the relationship between humans and humans, human beings must coexist and survive together, sharing and protecting the Earth together. From the perspective of intergenerational issues, our generation must learn to deal with relationships between the past, present and future, and consider what can and should be left for the next generation. From the perspective of awareness of capital, sunshine, blue sky, clean water and edaphic quality belong to natural capital, which should be valued as integral glue to GDP, social capital and human capital, and thus an important part of national wealth. We must mitigate the negative effects of rapid industrial development and lags in social development. From the vantage point of social lifestyle, it is necessary to promote an abstemious consumption pattern and to establish a culture conducive to the recycling and re-use of resources.

Human beings have never encountered such a severe challenge. Many of these problems can not be expected to be resolved through science and technology. In meeting these complex challenges, only some of the problems can be mitigated or resolved through the tangible solutions of hard technology. Taking environmental pollution issue as the example, we may observe that green technology can not eradicate the problem of pollution because the root causes of various forms of pollutions do not lie in the technology itself, and sometimes green technology even becomes a shield against systematic environmental protection solutions. In fact, the rapid industrialization that has progressed for more than 200 years has created many severe problems. Its main reason is that through industrialization human beings have simplistically pursued material civilization, and have thereby undermined the affinity of human communities for peaceful coexistence and sharing. On the other hand, the direction of innovation has been misguided, due to a one-sided worship of hard technology as well as a strong desire to conquer nature, thereby making us more effective in destroying homes and tramping on the Earth and leading more quickly to environmental catastrophe. In the present era, all social behaviours and psychological activities are influenced by the soft environment, composed of institutions and culture, both locally and internationally.

The characteristics of soft-tech make it suitable for dealing with the challenges mentioned above. For instance, for solving problems soft technology adopts the thinking mode that is from whole to part, which is a way of adapting to systematic design; its operational field is focused on the psycho-social world (hard technology is rooted in the physical world), the operational objects are human psychological activity and social behaviour; the tools of resolving problems are processes, governance or manipulation, rules and institutions; and its technological parameters are social, cultural, psychological factors, etc. Moreover, the cross-cutting and interdisciplinary knowledge sources of soft-tech are conducive to gaining insight, creative thinking and enlightened ideas. In brief, by using the route which makes soft technologies, soft environment and soft capital (see the section of 'Soft Technology and Innovative Space' in Chapter 4) integrated with hard technologies, hard environment and hard capital with which we are familiar, there may be a deep understanding of the difficulties we face, so as to design the systematic solutions which can integrate a variety of knowledge with technologies to 'resolve or cope with' the challenges we face in the twenty-first century.

3. Grasping the Orientation of Technological Progress and Direction of Innovation – To Adhere to a Correct Soft-Tech Operation and Soft Environmental Design

Bill Joy's admonition mentioned earlier in this chapter raised the argument about whether technological advancements would threaten to make humans an endangered species (Joy 2000). In fact, this sort of argument has recurred over the centuries. Bill Joy's worry is justified. For instance, life sciences and biotechnology have a great potential for benefiting mankind and have brought unprecedented new hopes for resolving crucial issues such as human survival and health, food shortages, resource and energy shortages, environmental pollution and so on. At the same time, because its object of study is a living being, if the technological applications are abused or directed towards inappropriate purposes they will lead to a variety of unpredictable hazards including the potential dangers of genetically modified crops, the enormous impact on the ethics of modern society by human cloning, the threat of biological weapons, etc. It is not excessive to talk even about the potential destruction of mankind.

However, it is just about the continuous in-depth research on nature and society that allowed humans to come to resolve the large numbers of 'problems' they encountered, enabling us to be always in a process of the social progress and the amelioration of the quality of life. Compared with human requirements, there are still too many problems we need to solve and we have to accelerate the pace of hard-tech progress. In fact, any direction of technological application depends on human beings themselves, and many technologies have both positive and negative effects – just as a coin has two sides, the same as nuclear technology can be used for the atomic bomb or nuclear electric power. Computer, genetic engineering and robotics can also be a two-way application, which may bring benefits to human or cause harm based on the aim of application: hard technology is 'neutral', and soft technology is the core to decide the direction for technology application.

Advocates of technology tend to believe that if human beings can develop a new technology, then they must surely also be capable of controlling it. However, it is highly questionable whether depending only on new technology (higher-level, hard technology) to control technology is really a viable solution.

Take environmental issues as an example. The Chinese scholar Hao Tang has raised some profound questions in his 'roadmap' of technical solutions to 'environmental problems' (H. Tang 2007b). Can technological advances solve our environmental crisis? Are they adequate to overcome these threats to our existence? If expectations of green technology become an excuse for ruining the environment today we are faced with absurdity. The inability of green technology to fully play the role in this world that is expected of it indicates that the root of the problem does not lie in technology itself. Ironically, the survival crisis we face is in large part caused by extraordinarily advanced technologies that give us mastery over nature. Otherwise we would not have brought devastation to our planet with such efficiency, thus bringing ourselves close to the brink of destruction. For instance, extensive use of fossil raw materials has greatly increased greenhouse gas emissions; agricultural mechanization has accelerated desertification in northern areas; marine and fishing technology has pushed the whale to the brink of extinction; and large dams constructed by high-tech have completely altered ecological environments over areas of tens of thousands of square kilometres. The more polluted the environment is the wider the market for technology that cleans it up. Moreover, imbalanced development of technology is also threatening the human ethical 'bottom line'. In the case of biofuels, an environmental expert has criticized how inhuman it is to burn massive quantities of food as fuel when hundreds of millions of people are malnourished. We need a global system of technological cooperation, with proper organization, planning and regulations, to evaluate the potential threats of new technologies and to deal jointly with the problem of losing control of technology. Certainly, the technology mentioned here actually refers to hard technology.

During the global financial crisis that erupted in late 2008, either those deep-seated reasons or the so-called innovation in financial derivatives which became a trigger for the crisis may be considered as part of the calamity of soft technology operation on different levels.

In short, soft technologies have also involved bad games and innovation is not always good. The direction of technological progress and innovation should be guided by fine values and subject to the constraints of society, and it ought not to be driven only by business interests. Albert Einstein once raised the question of how could a culture-loving era become so corrupted and depraved: 'Now I increasingly put kindness and love above everything else [...] all of our technological progress on which people lavish praise – our only civilization – seems to be a sharp ax [sic] in the hands of a criminals with a morbid mentality' (quoted in Dukas & Hoffman [1981] 1984: 78).

On of the important missions for soft technology in hard-tech innovation is to design solutions for different technologies according to sound ethics, thereby avoiding negative technological development and avoiding turning the 'innocent' hard technologies into evil things, and also in preventing their development in directions humans do not expect. It is necessary to publicize the positive side of technological applications, which will be conducive to appropriately controlling technology, and be advantageous for social progress; and, moreover, designing the soft environment to ensure that the 'solutions' are successfully implemented. That is, soft-tech innovation must consist of institutional design which will restrain those disadvantageous in humanly-gentle, sustainable development.

4. To Identify 'Non-Technological Factors', Unveiling the 'Black Boxes' of Total Factor Productivity

People have known for a long time that in addition to the science and technology with which we are familiar and which we worship, there are various 'other' factors playing a role in social and economic development, which are sometimes beyond the role of science and technology and which may therefore be considered to play the role of the so-called 'first productivity'. They are widespread, with strong impact, but do not belong to any category of traditional technologies. Thus, they need to be incorporated and collectively labelled as 'non-technological factors'.

In the late 1970s, after investigating problems occurring during software development projects, the American Department of Defense discovered that 70 per cent of all failed projects were due to insufficient management and not the lack of technical knowledge. Another example may be found in an international study by Kelvin Willoughby of the role of technology in local economic development initiatives from the 1960s onwards. Willoughby found that non-technological factors (which he labelled 'technology practice') surrounding technologies were key determinants of the success or failure of projects in fields ranging from agriculture and water supply to energy and small-scale manufacturing (Willoughby 1990).

In fact, improving the comprehensive economic power of a nation, a city or an enterprise has tended to depend mostly on the orientation of its development strategy and resource allocation and on innovation in existing technology. However, as the gap of competitiveness in science and technology between countries and regions continues to expand, it is important to adopt a new perspective to seek the growth factor or the driving force beyond science and technology. Some people refer to these non-technological factors as the second driving force. Non-technological factors are normally critical in determining the success or failure of technological projects. What are we referring to when we say 'non-technological factors' or 'the second driving force of productivity'?

It is noteworthy that with the advent of post-industrial society or the new economic era in the twenty-first century the function of the 'non-technological factors' has increasingly been seen as important, causing some senior economists to change their opinions.

In 2005, *The New York Times* published an article entitled 'What Makes a Nation More Productive? It's Not Just Technology', which pointed out the following: in the late 1990s, growth in labour productivity kicked into a higher gear, and economists generally believed that the higher rate was a by-product of the new economy. Much of the growth was spurred by the highly productive businesses that made information technology products – companies like Dell, Intel and Microsoft – and by their customers (Gross 2005). 'About half of the growth resurgence from 1995 to 2000 was due to IT,' said Dale Jorgenson, university professor at Harvard. As the technology investment boom of the 1990s gave way to a bust in 2000, many analysts feared that the productivity gains would dissipate. Instead, productivity since 2000 has grown at a substantially higher pace than it did in the late 1990s. And productivity growth is still strong. Therefore, some economists have changed their views and now believe that a major breakthrough in information technology is not required to fuel further productivity growth. Professor Jorgenson said, 'It's not research and development that cause the big gains in productivity. The real drivers are things like competition, deregulation, the opening of markets and globalization' (ibid.).

A report by the McKinsey Global Institute found that sectors other than technology drive productivity growth during the years after a sudden drop in technology investment. From 2000 to 2003, five of the top contributors to productivity growth were service industries including retail trade, wholesale trade and financial services. That is surprising since economists have generally believed that it is much harder for service industries to reap sharp productivity gains than it is for manufacturers to do so.

Ms. Diana Farrell, the director of McKinsey Global said, 'IT is a particularly effective enabling tool. But without the competitive intensity that drives people to adopt innovation, we wouldn't see these kinds of gains'. To compete with Wal-Mart, for example, retailers of all stripes have been working furiously to gain scale, to manage supply chains and logistics more effectively, and to negotiate better terms with suppliers and workers. A similar dynamic has played out in the finance sector where there has also been a huge gain in productivity. It is likely that competition and structural changes are responsible for those gains. One mystery of recent years has been the enduring gap in productivity growth between the United States and Europe. In this case, another structural force, regulation, may be at work.

Judge Randall Rader stressed that 'a country's future has an important relationship with its innovation and decision-makings on technology, the world economy is driven by new products and progressive solutions' (Rader 2007).

Recently, there are an increasing number of articles that pay attention to non-technological factors. Some experts believed that Chinese enterprises devote insufficient attention towards non-technological factors in the process of 'informationization', thereby forming a management gap in non-technological factors, resulting in the low success rate of 'informationization' projects (Min & Tang 2003). They have taken four factors – leadership, organizational change, change management and project management – as well as application management, as major non-technological factors; some people took the factors like corporate culture, brand, international image, marketing strategy, etc., as non-technological factors affecting the international competitiveness of Chinese enterprises. A research report looked at the technological innovation of the Long March launch vehicle series as a case study, and elaborated that the ability of technological innovation and technology diffusion depends not only on the knowledge accumulation and talent structure, but to a large extent also on the policy environment and the organizational structure of technological innovation (Q. Zhang 2002). There are other relevant articles like 'Non-Technical Factors in Tasks of Using Science and Technology to Defeat Terrorism' (Xu & Ma 2007); 'Non-Technological Factors Affecting the Hospital Network Security' (Shen 2002); 'Technical and non-technical factors influencing the process of Teleconsultation services development carried out in Krakow Centre of Telemedicine' (Duplaga et. al.), and so on.

Let's observe non-technological factors affecting economic growth from a view point of economics.

Research in the growth of total factor productivity (TFP) of China's large and medium-sized industrial enterprises attempted to answer what is the real driving force behind cutting-edge technological advancements (Tu & Xiao 2005). The article pointed out that technology has a very wide connotation including not only the direct technological factors like advanced technics, patents, technological innovation, high-tech equipment and talent, but also some non-technical factors like economic cycles, as well as the system shift in the area of economy, society and legal, such as the development of private enterprise, the restructuring of state-owned enterprises, tax reform, foreign investment, and the entrance of WTO, etc., which have far-reaching impact on output and affect the increase in productivity. This study incorporated various factors behind cutting-edge technological advancements into four major factors: firstly, the industry's fierce competition between enterprises brings market pressure to bear on cutting-edge technological advancements; secondly, globalization and foreign direct investment are sources of cutting-edge technological advancements; thirdly, changes in ownership structure is the internal driving force; and fourthly, the economic expansion period has created the external environment for cutting-edge technological advancements.

A study by the Chinese Academy of Social Sciences interpreted the growth rate of TFP (the so-called 'technological progress') in the productivity of research from a new angle (Jin, Jiang & Gong 2006), pointing out that besides technological progress factors as traditionally understood, it is the soft technology and the soft environment (usually called 'other' elements in conventional studies) which actually exert a larger influence than the factors that have conventionally garnered attention! In economics, the growth rate of TFP refers to the difference between economic growths and contribution of the input factors. For example, in the case of factor inputs of capital and labour, output growth can be decomposed into the contribution of labour, capital contributions and the contribution of technological progress (in an economic sense), and the latter is also known as the growth rate of TFP (total factor productivity). Because its contents are rather ambiguous, the growth rate of TFP is generally considered as the 'other' all contributions besides the contribution of quantifiable inputs like capital, labour, etc., which is even called 'black box' and also known as the 'Solow residual'.

A study of the World Bank concluded that 'residual' gathered social capital and human capital together (Dixon et. al. [1997] 1998: 160). According to our study, the 'black-box' or 'residual' consists of not only contributions (including quantifiable social capital, human capital) brought by the investment in quantifiable (tangible) technological progress such as R&D, education and ecological environment, but also contributions of soft factors such as the innovative ability of soft technology and that of soft environment, etc., which are intangible and difficult to be quantify. Our study further revealed that, although in the traditional sense the contribution rate of various investments in technological progress will gradually rise along with the improvement of its investment scale and environment, in China, the soft factors (in the 'black box') will always be over 60 per cent in the next 50 years.

Obviously, technological progress in economics is by no means the type of technological advancement as conventionally understood, and the broad sense of technological progress. Because our future study used scenario analysis, the data may not be so accurate, but this study illustrated sufficiently the essence of so-called 'contributions of technology to improve the productivity' as well as the role of 'non-technological factors' in economic growth, which will be conducive to adjusting economic strategies and investment priorities.

In conclusion, non-technological factors are actually the innovative abilities of soft technology and that of soft environment. If we generally only incorporate a large amount of factors which have impact on economic, social, environmental as well as corporate development into 'non-technological factors', and are only aware of their importance without studying their essence and the pattern of influences, it will not be conducive to their development and innovation.

5. Revealing the Essence of Creating Value for Intellectual Capital

The society of the twenty-first century is known as the knowledge-based society. In recent decades, intellectual capital research has gradually developed to become a discipline through the tireless efforts of a lot of scholars and entrepreneurs, and it has accumulated a great deal of theoretical and practical experiences from the different levels, such as the philosophy, measurement, management and the creation of intellectual assets, etc.

The concept of intellectual capital was brought forward in 1969 for the first time by the famous economist John Kenneth Galbraith (in a letter to Polish economists), and it has become a systematic theory thanks to the contributions of Itami Hiroyuki ('Mobilizing Invisible Assets' [1980]), David J. Teece ('Profiting from Technological Innovation' [1986]), Karl Eric Sveiby ('The Know-How Company' [1986]), Thomas A. Stewart, ('Intellectual Capital' [1994]), Patrick Sullivan, ('Value Driven Intellectual Capital: How to Convert Intangible Corporate Assets Into Market Value' [1998]), Leif Edvinsson (the world's first corporate director of Intellectual Capital), Ikujiro Nonaka ('The Knowledge-Creating Company' [1995]). In the practice domain, the experts who are currently active in the measurement of intellectual capital include Ludo Pyis, Alexander Welzl, etc.; in the field of intellectual capital management, there are many outstanding experts like Yasunito Hannado, Serafin Talisayon, etc.; and in the creation of intellectual assets, we can find Gordon McConnachie, Darrell Mann and so on.

Now that intellectual capital management has become the core content of the competitiveness of outstanding companies, many enterprises have set up knowledge management executive positions, and every country has their relevant regulatory agencies for intellectual property rights and, furthermore, there is the World Intellectual Property Organization in Geneva. However, the development of intellectual capital has still a long way to go.

How does intellectual capital create value? What is the connotation of its value? How can it create greater value? The study of soft technology provides a new perspective for addressing the above issues.

There are different definitions on Intellectual Capital (IC). For example, the OECD has defined it as the resource utilized in future value creation without a physical embodiment; the founding Board Chairman of the National Intellectual Assets Centre of Scotland, Dr. Gordon McConnachie, believes that intellectual capital is all knowledge that can be used to create value. However, while most of experts retain different views and understandings of IC, their classification of IC is almost identical (with only minor differences), which includes human capital (at individual level and throughout organization), relational capital (at external/networking level) and organizational/structural capital (at internal/company level).

From the perspective of soft technology, intellectual capital as mentioned above belongs to the category of soft capital (see the section on 'Comprehensive Competitiveness and Soft Power' in Chapter 3) in the enterprises dimension.

However, with either soft capital or hard capital, there is only potential value in capital; namely, capital is just one of the innovation resources which can not create value without being translated into practice or application. This 'translation' is the process of technological operation to the various types of capital. Intellectual capital needs to be translated into intellectual assets or intellectual property. Therefore:

1. **Capital is not equal to the value**. I proposed that in the stage of establishing the concept, it is necessary to separate 'capital' from the operational 'process' – technology which is aimed at creating value. Only by the understanding and in-depth study of value creation process of the capital can these types of capital be developed synthetically and systematically, thereby managing intellectual capital and creating intellectual assets more effectively and, moreover, bringing a large number of business opportunities and employment opportunities.
2. **To make clear the tools of creating value**. The process technology to create intellectual capital value or the means for creating value includes soft technology and hard technology. Because soft technology is the tool for hard-tech innovation (see the section of 'Functions of Soft Technology' in Chapter 4), the key means for creating IC value is soft-tech innovation as well as the integrated innovation of soft and hard technology (see Figure 8).
3. **To expand the contents of IC value**. The value that is pursued by excellent corporations must go beyond the mere pursuit of economic profit and beyond the level of the single enterprise. It should consider sustainability and the development of social industry, as well as the enterprise position ('corporate citizenship') in the twenty-first century. Therefore, intellectual capital should not be applied simply for the pursuit of economic value. Rather, it is necessary to pay attention to economic, social and eco-environmental value. This kind of understanding not only expands the space for intellectual capital to create value, but also provides another kind of basis to identify and determine value for measuring the IC value.
4. **Innovation of soft environment is the base for creating more IC value**. The successful creation process of IC value should be a process of innovation in the broad sense (see the section of 'Soft Technology and Innovative Space' in Chapter 4). The maximum value can

be created only when the process of value creation is integrated with the innovation of soft environment, especially supported by the innovation of soft infrastructure.

5. **To attach importance to soft-tech property rights**. Since soft technology is another paradigm of technology, the intellectual assets and intellectual property should include both the tangible part (the hard-tech property) and intangible part (the soft-tech property). The latter has typically been ignored in the past, thereby leading to the loss of great potential to create intellectual capital.

In short, awareness of soft technology will be conducive not only to clarifying the process of IC value creation and to revealing the essence of intellectual capital management – to improve the efficiency of soft-tech innovation – but also to extending the value creation chain of IC. Meanwhile, the management of intellectual capital, especially the progress in measuring intellectual capital, provides a guideline for further study of the measurement of soft-tech innovation.

Study of soft-tech is not about cultivating some kind of abstract principle or concept but rather it is directed towards understanding the main trends and the inherent demands

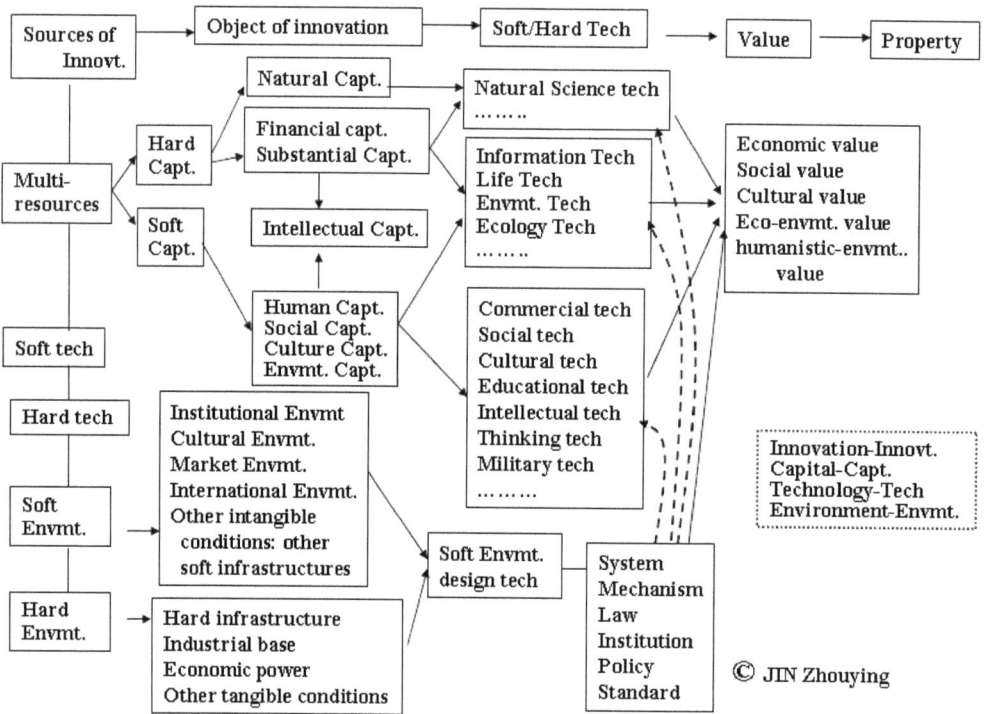

Figure 8: The Process of IC Value Creation.

of contemporary social and economic development. The study on soft-tech and soft environment can promote an idealistic revolution of technology and full-scale understanding of the human knowledge system. Decision-makers at all levels should understand that soft-tech will be a leading factor and force of the twenty-first century, leading to a shift in the traditional tendency to focus most manpower, resources and inputs on policies for hard technology and hard environment, so as to adjust the entire strategic system and to promote the balance between soft and hard.

The history of the soft technology development is also the history of human creativity. If we can incorporate the soft factors which were not regarded as 'technology' in the past into the new paradigm of technology, assort them systematically and consciously develop in-depth research to promote their innovation, so that we can not only help people change their mode of thinking and cultivate their creativity, but also guide them to rethink and reflect about many basic domestic and international issues related to sustainable development, globalization and new economy (including re-evaluating innovation, institutions, competitiveness and the essential features of various 'gaps'), we will be able to put together the half of the history of technology in the world that has been missing until now.

F. Soft-Tech Trends and Challenges for the Future

The development of soft technology is one of the most important trends in technology development worldwide. At present, its development exhibits the following characteristics:

1. Soft Technology is Transforming Hard Technology

The rapid development of technologies such as software technology, Internet technology, 'Internet of Things', bio-information-technology, gene technology and artificial intelligence and their industries can be attributed to their softening. These technologies have all contained 'soft' features from their origins. As these technologies mature and generate products, their 'soft' side is subject to constant innovation, which enables the products to adapt to the ever-changing marketplace, and becomes the main source for providing added-value. Take the mobile phone industry as an example. In the late 1990s, a recent model of a Panasonic mobile telephone had a listed sales price as a stand-alone product in France of 1400 Francs (one Euro is about 6.56 Francs); but its price at a French supermarket was set at only one Franc. The purpose of this artificially low sales price was to attract new customers, lock customers into a particular telecom network and, above all, to induce customers to purchase a complementary service through the supermarket, i.e. a subscription to a particular wireless telecommunications service. Accordingly, as noted by the French weekly magazine *New Observer*, the first of the ten keys to success of the new economy is that the best way to sell service is to give away equipment free of charge. In the year 2000, for example, many mobile phones can be bought for only one hundred Yen or are even free of charge in Japan, while the different fees will be charged according to the communications network and associated services. The aim here is to attract customers of cellular phones to eventually become permanent customers. The selling of mobile phones is no longer the core feature of mobile phone transactions. The profits of the mobile phone industry derive mainly from services, so the mobile phone industry has actually become a service industry rather than a manufacturing one. Now, China's mobile phone industry has generally adopted the above pattern.

Recently, service-oriented business strategies of established enterprises (including those in the equipment industry) – including increasing the service component of products, emphasizing services in the branding of firms and adding humorous, entertaining and human-performance dimensions to products – can become the primary means for attracting customers, creating markets and increasing added-value. At the same time, firms in a variety of industries, such as education and training, physical circulation, business, intelligent services, culture and art have begun to make heavy use of Internet services as a vehicle to promote and extend their businesses.

2. *More and More Soft Technologies are Creating New Industries*

The most familiar soft-tech industries are the specialization and industrialization of management technology. In addition to the specialization of various management functions, that of managerial approaches is also evolving quickly. Specialized companies that take certain management technology as their main product or service content are becoming widespread, and the programmes and curricula of education and training (based on some managerial approaches) are having importance increasingly attached to them. Examples include various finance companies, specialized management consulting companies and assessment companies.

A current trend within the intelligent services industries is towards specialization. Almost any of the soft technologies, including various cultural technologies and social technologies, are forming the basis of some kind of industry. Soft technologies are therefore rapidly becoming core technologies for established industries and are facilitating the innovation of new industries at a rapid pace. It can be evidenced by the so-called 'creative industries' which are very hot recently (see the relevant part of 'Creative Industries' in Chapter 5).

3. *The Rapid Rise of New Soft Technology Fields*

Since the 1980s new soft technologies have emerged in an endless stream along with the trend of economic globalization and the economic 'informatization'. For example, in the field of business technology, new technologies such as e-commerce technology, incubator technology, virtual technology and modern finance technology have emerged. Furthermore, this process by which intellectual service industries grow rapidly propels the upgrading of traditional soft technologies and contributes to the development of new stages in the evolution of soft technology. The emergence of the Nasdaq, multinational mergers, venture capital mergers and new business models are examples.

And nowadays, in order to meet a good many challenges of the twenty-first century, the fifth wave of commercial technology raised issues such as addressing globalization, sustainability and the overall softening of industries (see the section of 'Soft Technology and Thrice Industrial Revolutions' in Chapter 2).

As business technology has leapt forward, other types of soft technologies, such as social technologies, have begun to gain or regain attention. Innovations in culture technology including entertainment technology also continue. Furthermore, humanity is finding new theoretical perspectives through which to look at itself, such as being part of the ecological systems of nature, information-processing systems, information sources, information receptors, energy sources and life sources. Psychological science and psycho-technology are assuming a more important position in fields such as decision-making and health. Research into traditional Chinese medical science and technologies is progressing not only in the field of health care but even in agriculture, with the humanistic ecological view of man-nature harmony being more widely embraced to develop new technologies, which will stimulate the development and application of more new types of soft technology

4. Increasing Soft Technology to Become High Technology

The examples of the integration of soft and hard technology are too numerous to enumerate, and the common ones include contemporary music technology, Internet, software, modern logistics, 'Internet of Things', etc. The extensive integration of soft technology and hard technology is creating immense business opportunity, which at the same time increases the intelligence content of soft technology. In short, soft technologies are increasingly becoming exemplars of high technology. For example, in the field of product design, especially in applications such as computer software, advertising, movies, television, fashion and environmental management, etc., high technology is gradually merging with culture, art and aesthetics. Thus we can say that high technology does not belong solely to the domain of hard technology, and high technology can also be found in soft technology.

Since its birth fifty years ago, the credit card industry has progressed from being one kind of simple paper-based credit issued on trust by a barbecue restaurant, through being no more than 200 cards issued on trust by the Diner's Club for use in a group of restaurants, to its present status as a substitute currency embodied in more than a hundred types of credit card that are widely used throughout the world. The credit card industry is now big business. It is prevalent that everyone has credit cards and some people even possess seven or eight different cards. From the perspective of knowledge connotation, credit cards began as paper cards with a personal signature on them and have now developed into smart cards that incorporate a large amount of information and a plethora of functions. Today, many credit cards contain large-scale IC chips; new applications of the IC chips are infiltrating all aspects of modern life. Nobody could say that credit card technology is not high technology and indeed it is.

E-commerce is an electronic mode for conducting commercial transactions through a network, where the processes of advertising, ordering, making payments, sending out goods and delivery are all conducted through the exchange of digital information (or 'e-data'). E-commerce aims to reduce the cost of the transactions, achieve paperless trade, improve efficiency and make the whole transactional process transparent. On the other hand, the application of electronic transaction tools may bring revolutionary changes to many traditional transaction modes and their associated technologies. For example, paper receipts for financial transactions may disappear. It also creates new modes of commerce, such as attracting new customers through digital networks, transforming traditional transaction technology into intelligence-intensive transaction high technology. E-commerce technology can be regarded as the high-tech of trade.

5. *Intellectual Property of Soft Technology*

How might more incentives be generated for innovation in soft technology? From this perspective, there is no doubt that intellectual property rights, including copyright of soft technology, are needed to protect and to help innovators to appropriate the returns of their activities, thereby encouraging the continuing innovation in those firms, individuals and organizations engaged in soft-tech activities.

However, the task now facing those active in soft technology is how to identify and define the most appropriate forms of intellectual property protection, including patents, for soft technology inventions. In fact, a large number of financial products and new business models in the financial sector have already been protected as intellectual property in the US. Certainly, as has been the case with intellectual property protection for hard technology, debates will arise as to whether some categories of soft technology ought not to be allowed intellectual property protection while other categories should; and issues will arise concerning how the exceptions to IP protection for some categories of soft technology ought to be decided.

A particularly instructive example of the revolution that is taking place in the institutionalization of soft technology may be found the development in the patent system in the United States. In the US, 'anything under the sun developed by the hand of man [...] as long as it produces a [...] useful, concrete, and tangible result' (Morris 2001) is patentable, while laws of nature, natural phenomena, and abstract ideas are not patentable. In 1998, the *State Street* decision triggered an awareness of the 'business method claim' as eligible for patent protection. Ever since the US Supreme Court in January 1999 affirmed the court decision (State Street Bank v. Signature Financial Group), software-enabled business processes have indeed been patentable so long as they are novel, non-obvious and produce tangible results. As a consequence, the patenting of business models, marketing methods and other intangible innovations has mushroomed (Rivette & Kline 2000: 16).

The United States Patent and Trademark Office (USPTO) has set up a working group responsible for examining patent applications in automated business data processing technologies. On 29 March 2000, the USPTO announced a plan to improve the quality of the examination process in technologies related to electronic commerce and business methods. This white paper, 'Automated Financial or Management Data Processing Methods (business methods)', discusses the patent history of business data processing, and points out that we are at the beginning of a change in patent technology. The USPTO is attempting to improve the quality of the review process for business method patents, and has created a special patent class in to which various soft technologies in fields such as e-commerce or 'modern business data processing' are formally incorporated (Class 705, Modern Business Data Processing).

Once the USPTO began granting patents for business methods they received 7800 business methods patent applications between 2000 and 2001 alone, and during that same period it issued 899 business methods patents. The promulgations of law on relevant fields, like Patent Improve Act (PIA) in March 2001 and US Inventor Protection Law on 27 November 2000, have supported USPTO.

In other words, for at least the last decade the US government has been issuing patents for soft technology! Certainly, the action of the USPTO has generated quite a lot of controversy. Much of the controversy in the US concerning patent law now focuses on technology-enabled methods of doing business rather than on the devices or tangible inventions that used to be seen as representative of technology.

We can say that this development within the US patent system is a kind of revolution in a soft-tech institution. However, many problems currently still exit. For example, the current business technology patents are mainly concentrated in financial services and most of them refer to the application of computing technology. The question arises as to whether it really is business technology. What is a business method or a business technology exactly? What is the appropriate legal boundary for protecting business method inventions? Could business innovation be treated the same way that innovations in conventional technology have tended to be treated? Whatever answers to these questions eventually arise, it is important to consider the fact until recently patents have been largely limited to the field of hard technology. The recent developments within the patent system of the United Sates should still be viewed as a landmark event in the history of soft technology.

In fact, a revolution in the field of managing patents and copyrights is required in today's knowledge society. In theory, the challenge is to find ways in which knowledge, including the intellectual property that is the common wealth of humanity, may be shared with the general public while, in practice, maintaining the incentives for innovators, especially in fields of soft technology. Namely, both the social contributions and economic benefits of innovators should be recognized. From this point of view, traditional patent systems and institutions are inadequate, or even inappropriate, for the field of soft technology. However, soft technology innovation should also be protected. Addressing this issue is one of the biggest challenges for the development of soft technology.

Some activists in the world of software have been developing and promoting the GPL (General Public License) and focusing on the so-called 'Copyleft' (a special kind of open copyright license) as a potential solution to the problems of intellectual property and innovation in the area of computer software. However, at this stage it does not appear that the GPL has been able to address the issue of economic compensation for the innovators.

6. Challenges for Soft-Tech Development

The development of soft technology is facing serious challenges.

1. The very concept of soft technology itself faces challenges within academia. Such cross-disciplinary technology has not yet been recognized by the traditional disciplines of science and technology, and by the administrations of universities and academies.
2. It is not clear how soft technology innovation should be measured. What should be measured, what can be measured and where do the difficulties of measurement lie?
3. How should protection, motivation and incentive for soft-tech innovation be provided? Since the characteristics of soft-tech are non-neutral and since they are sometimes difficult to define precisely, challenges are created for deciding how to set standards for encouraging or controlling soft technology.
4. The issues of identification and management of intellectual property rights in the field of soft-tech remain salient. What kind of soft technologies should be granted intellectual property rights? As has been the case with intellectual property for hard technologies, the intellectual property management of soft technology faces the dual pressures of the theory and practice (and problems are perhaps even more incisive in the domain of soft technology).
5. Research and development in new fields of soft technology raises immense intellectual and practical challenges.

Notes

1. The Agon was an Ancient Greek festival featuring sports, musical and theatrical competitions.
2. Gui, Shou (2000), *Chinese Macro-economy Information Web*, 3 November, http://www.macrochina.com.cn/info.shtml.
3. Japan Electronics and Information Technology Industries Association (JEITA). December 19, 2007.
4. Maddison, Angus (1995), *Monitoring the World Economy, 1820–1992*, Paris: Development Centre of the Organisation for Economic Co-operation and Development, 4 September.
5. *The volume of historical economy statistical data for Soviet Union and the main capitalism countries (1800–1982)*. Beijing: People Press, October 1989.
6. 'QI' in Chinese Medicine refers to the microscopic substances moving continuously and with strong vigour. It refers to the nutritional substances in human body as well as the functional activities produced by these substances through the organs.
7. Innate knowledge refers to the 'collective, unconscious' knowledge that is endowment by nature (the inborn one).

Chapter 2

Historical Antecedents of Soft Technology

The process by which each soft technology emerges is part of the history of the enablement of human creativity.

Soft technology has existed for as long as human beings have been engaged in production activities. Examples range from Sun Tzu's the *Art of War* (from Ancient China) and the diagnostic and therapeutic technology of Chinese traditional medicine; through accounting technology, insurance technology and stock technology of the sixteenth and seventeenth centuries; and then the R&D-institute mechanisms of the nineteenth century; to the contemporary financial derivatives, virtual technology and incubator technology. The history of soft technology is just as long as the history of hard technology.

This chapter will take some familiar commercial technologies and social technologies as examples to provide a retrospective of how soft technology has driven the progress of human society, even though people have tended not to see soft technology as a unified field of technology in its own right. In the meantime, the history of each technology will reveal how soft technology was invented, how we human beings explore creative ideas from ordinary life of every day and receive enlightenment after solving difficult problems and how the new 'game' (soft technology) was invented, along with an explanation of the game rules (relevant institutions).

A. The History of Commercial Technology: A Brief Analysis

1. Currency Technology

Exchange technology is the most ancient technology in human development history and it has continuously developed, even up to the present day. It can be said that the history of exchange technology is the process of economic and social development. In the words of Adam Smith, 'Each individual supplies what the other one needs, exchanging article by article and these mutual transactions are the fundamental elements owned by and unique to human beings' (Smith [1880] 1997: 1–12). Human beings have developed innumerable exchange technologies. We will now discuss currency exchange mechanisms, the commercial means with which we are most familiar, as an example of the historical evolution of exchange technology.

So-called 'currency technology' is the means by which value flows and it is the most fundamental aspect of exchange technology. The most primitive currency technology –

bartering – is also very direct and can be highly efficient (within certain circumstances). However, if one lacked what the other needed or if the available quantity was inadequate, it was difficult for a deal to be struck.

In order to overcome this inconvenience, another medium of bartering was invented that has been labelled 'necessary currency'. Necessities differ according to culture, but as the medium of exchange, each necessity must possess universal value. For example, some tribes in South Africa accept cows as their necessary currency; Guatemala, in South America, accepts corn; Melanesian tribes in the South Pacific, typified by those in Papua New Guinea, accept pigs as a medium of exchange (Kato 1983: 248). In Ancient China, textiles were accepted as a necessary currency, and up to the Middle Ages, boiled-off silk was accepted as a medium for tributes to the palace in Japan.

Another necessary currency is common salt for it possesses both use value and exchange value. Salt can be long preserved and is necessary in life. In Ethiopia, Egypt and Ancient Rome, salt was used as the means of exchange. A vestige of this tradition is found even today in the English word 'salary', which is derived etymologically from 'salt'. This derives partly from the fact that Roman soldiers were paid with salt. In the slavery trade of Ancient Greece, as a further example, a slave was exchanged for his same weight in salt. In the language of economics, necessary currency must possess the duality of both use value and exchange value.

Decoration currency eventually evolved from necessary currency. Decoration currency possesses a rarity value; it is beautiful to look at and may be a piece of artwork. The stone currency used on the island of Yap of the Micronesian Archipelago in the West Pacific and cowry currency, once extensively used in China, Japan, Africa and some European countries, are some examples. In Chinese characters the words for business (買賣), goods (貨), currency (貨幣) and property (財産), all contain the character of cowry (from the ocean) '貝'. This reflects the fact that the cowry was the main decoration currency at that time in China. As a matter of fact, besides knife currency and textile currency, all kinds of material currencies and measurement currencies such as pearl, jade, tortoise shell, cowry, gold, silver and tin, existed in various places in China before Qin Shi Huang, the first emperor of China, united the currency system. The government of the Qin Dynasty decided that, in terms of measurement currencies, gold was the superior currency and the round copper coin with a square hole in the middle that was coined monopolistically by the government was regarded as the inferior currency. All other currencies were prohibited (*Encyclopaedia of Finance* 1990: 2097).

Before paper currency was invented, gold had been the worldwide means of exchange. Today, the gold reserve and the price of gold are still retained as important international financial indices.

China was the first country in the world to use paper currency. During the Song Dynasty and the Jin Dynasty (at around the eleventh century AD), military expenditure was high, and gold and silver flowed outward because of trade, and this promoted the invention of paper currency called, at that time, '*Jiao Zi*' (Hongsheng Wang 2001). During the thirteenth century, Italian and French craftsmen used paper with words on it to show that they had received a certain quantity of gold, i.e. a note of deposit was used to make value flow both convenient and safe (ibid.). This was the earliest paper currency in Europe.

Later, banks appeared. Paper currency guaranteed that you could make an exchange at any time necessary so long as you had sufficient gold deposited in the bank to cover the exchange. Paper currency was issued in the form of a coin certificate. This tradition lasted until the end of World War I. After World War I, paper currency was no longer restricted to the form of a convertible note and its use and popularization spread worldwide. In China, Sun Yat-sen suggested switching to the sole use of paper currency in 1912. He pointed out that the currency evolution from textiles, silk, knives and cowry to gold and silver and then to paper currency was natural evolution. This process was called 'the revolution of currency' in China.

Currency, as exchange value separated from use value, came into independent existence with the emergence of paper currency. The paper, which no longer possessed the features of the convertible note and did not guarantee a corresponding gold reserve, could represent different values mainly because of the trust in the country that issued the currency. In other words, paper currency is a product of trust.

Nobel economics prize laureate Robert A. Mondale believed that to adapt to the globalized economy, a globalized currency is needed. He had been keen on the idea of unified world currency. It was Robert Mundell who proposed the idea of the Euro in 1969 and who promoted the birth of the Euro in 1999.

In the information age in which we now live, e-commerce is becoming widespread. As a consequence, traditional ways of doing business and the conventional means of exchange with which we are familiar are undergoing fundamental change. The use of e-money is now emerging as a new kind of exchange mechanism.

Bernard Lietaer published a book in 1999 entitled *The Future of Money: Creating New Wealth, Work and a Wiser World*. In this book he summarizes and analyses complementary currency systems that can make up for the defects of the national currency systems (Lietaer 1999). Although the rudiment of complementary currency can be dated back to the ancient colonial era, the pioneering example of a complementary currency system in the contemporary era was the WIR, established in Switzerland in 1934. WIR (which, in German, is also the equivalent of the English pronoun 'we') is an abbreviation for *Wirtschaftsring-Genossenschaft* and may be translated roughly as 'economic mutual support circle' (see the section of 'More Attention is Being Given to Various Types of Social Industries' in Chapter 5). More recently, in 1982, LETS (Local Exchange Trading System) was established in Canada. By 1999 there were over 2000 complementary currency systems running in twelve developed countries alone.

With the accelerated advent of aging society, the Ministry of Health, Labour and Welfare of Japan has been implementing very early a special health care currency called *Fureai Kippu* (literally, 'caring relationship tickets'). It was designed specifically for nursing the elderly, the sick and the disabled, and the unit of account does not equal traditional Japanese Yen, but an hour of service. The working time of the volunteers is recorded in a 'time deposit book', which calibrates the time and value of a variety of services rendered. In addition, the volunteers may keep these credits for themselves, family members or someone else when they need help, then use the credits in exchange for services. Kato Toshiharu has applied the label 'eco-money' to the various currencies of this kind, reflecting the fact that they have been extensively applied to welfare services, medical treatment, education, disaster prevention, environmental protection, nursing for the aged, cultural communication, nursing for the sick and related purposes and, furthermore, the experimentation of 'Smart Card' that contains the medical health records, and can use eco-money and the usual Japanese Yen simultaneously, has been carried out in some areas of Japan. 'The future community network programme of Japan' integrates gradually various services into the supplement currency system, such as shopping cards, phone cards, preferential benefit service for long distance telephone, discount services of gasoline and other items, tram cards, bus cards, airline mileage cards, etc.

While complementary currencies have garnered less public attention than other new forms of exchange, such as the Smart Card or the 'Internet Purse', they represent a more profound currency revolution than these. Thus, it is evident that the time-honoured example soft technology – currency – is still in the process of innovation and development.

Obviously, the currency technology is the soft technology that is the most widely applied in the world as the incarnation of both trust and imagination. As a means of exchange and measure of value, it has played a significant role in establishing and improving the product concept, the market mechanism and the value mechanism.

Moreover, in addition to the evolution, development and innovation of the monetary form as a financial tool, the unceasing innovation in monetary applications, as well as national monetary policies (the operations of interest rate and exchange rate), have had great impact on the contemporary economy and society, as well as international relationships. Unceasing innovation in currency technology without adequate supervision sometimes becomes uncontrollable, like a wild horse without a bridle, and may become the cause of financial bubbles. Currency operations played an especially important role in the financial crisis of 2008 and, as a result, the reform of the monetary system worldwide has become imperative.

2. Accounting Technology

Accounting is a commercial language that no business or serious practical operation can do without. It is an approach to 'confirming, measuring and passing on economic information and a process for enabling information users to judge and make strategic decisions on the basis of it' (American Accountants Association 1996: 1).

In the Sumerian culture, originating from Mesopotamia in about 3200 BC, some symbols of accounting records are found on pieces of pottery; officials in charge of the accounting department had to complete something equivalent to a training course in accounting in Ancient Babylonia and Egypt in 3000 BC.

In Greece around 1400 BC, the work of accounting was normally the responsibility of slaves. In Ancient Greece, slaves were usually forced to conduct audits. Careless slaves could be punished at any time while the law protected free people from such complications. However, in later Greece, accounting gradually became an honourable job and various accounting records of the costs of governmental constructions were inscribed on the buildings.

In Persia, under the rule of Darius (King of Ancient Persia, 521 BC–485 BC), there were special officials who audited the accounts within the areas of their jurisdiction; similar officials existed in the Hebrew culture where the main audit officials held the second most important positions in government.

Constantine established a public administration school during the period of the Byzantine Empire in the early fourth century, and accounting was among the courses offered. The Holy Roman Empire, under the government of Charlemagne (742–814 AD), followed the accounting and auditing system of Ancient Rome and Persia. After Charlemagne's death, the system was abolished and soon after, the empire collapsed (Anthony et. al. [1995] 2000: 7–8).

The accounting system in Europe began to decay in the Middle Ages, and was not revived again until the Italian religious wars.

Double-entry book-keeping appeared in the early fourteenth century: the first buds of this technique were evident in Genoese accounting records in 1340. At the same time, financial departments were set up in Britain.

In the fifteenth century, branches of the Medici Bank had to submit an annual balance sheet to the headquarters in Florence.

In 1631, the investors in Plymouth and Massachusetts, in the American colonies, sent an accountant from the Netherlands to investigate the cause of increasing liabilities there, thus giving the people of the New World the experience of their first audit.

Over the past several thousand years, large-scale accounting activities have been mainly related to governmental activities, especially taxation, but the Industrial Revolution brought about other demands in accounting. The size of enterprises and the funds required to maintain them increased gradually while management services became increasingly complicated. This was accelerated by the functional separation of investors and managers. Once the function of management became the responsibility of people other than investors, the investors themselves needed to know the operating status of their capital, i.e. they required their managers to report accounting information to them (similar to an annual financial statement). At the same time, the managers of the enterprise needed to analyze the internal accounting information, which stimulated the development of management accounting.

In the early twentieth century, Frederick Taylor's 'scientific management' system was applied extensively. In order for the system of scientific management to work properly it was necessary for existing accounting practices to be improved. Accordingly, special techniques in cost accounting – such as the standard costing system, budget control and differential analysis – were first introduced, and this in turn stimulated the development of formalized management accounting. After the 1950s, modern management accounting took shape and modern accounting divided into financial accounting and management accounting. During the wide application of accounting technology, not only were relevant standards set up in individual countries and internationally, but also accounting and auditing were developed as a discipline (Wei et. al. 2000: 3–8).

Since the latter half of the twentieth century, with a better understanding of soft capital, especially for the role of intellectual capital, quantification and accounting calculation of intangible capital has become a tremendous challenge for accounting technology development.

3. *Patent Technology*

It is commonly believed that owning a patent for a valuable technology may be a pathway to making a fortune. In his thesis of 1985, *Japanese Technology or American Wisdom?*, Kimindo Kusaka reiterated the observations of a professor from Harvard University that, during the 37 years following World War II, America transferred to Japan all the inventions of the past 200 years and charged Japan only about 40 billion US dollars for them (Kusaka 1985). The GDP of Japan at that time was 1200 billion US dollars. Hence, *if* it would pay the 'royalty fee' by 3 per cent of the sales income in the light of the standard of 'patent royalty payment', Japan would have paid 40 billion US dollars every year. It is evident that the many American inventions that were supplied almost free of charge contributed much to the economic development of Japan after World War II.

Patent system is employed by governments to encourage invention and to protect inventors and their intellectual property rights. A patent is essentially a government-enforced right given to an inventor for a specified period of time to prevent others from commercially exploiting his or her *bona fide* invention without formal permission. In other words, a patent is a state-granted monopoly right given to an inventor for specified time. If imitators enter the market rapidly and gain a substantial market share, although inventors or patentees receive compensation for their stolen ideas, such reparations generally cannot make up for the costs and risks of losing proprietary control of their inventions. The patent system strengthened protection against competitors who would copy an invention and apply it in the market during the specified monopoly period. The reduced competition during the life of the patent helps inventors to avail themselves of the opportunity to earn monopoly rents, within proscribed time limits, as a reward for their inventiveness. The hope is that more people will thereby be encouraged to engage in invention and creation, activities that would not be accomplished through market forces alone. Therefore, patents function to 'extend the interval between invention and imitation to encourage people to create' (Petersen & Lewis 2000: 431). In addition to patent protection for technological invention, most governments also provide exclusive protection for copyrights, trademarks, trade names, business model and design rights for artistic conceptions of fine arts and industry (Machlup [1958] 1975).

In short, the term 'patent' refers to the rights that a government confers on inventors or their assignees to make, use or sell their inventions. The duration of a patent may, in principle, vary from country to country but almost all countries have adopted the WTO standard of a minimum of twenty years.

A king of Athens in the fifth century BC conferred the first patent on a cook for his innovative cookery; in 1236 King Henry III of England conferred on a citizen a 15-year monopoly to produce coloured fabric; and Edward III (1327–1377) gave a patent for weaving to John Kemp of Flanders. These were all primitive patents (Wang 1985: 2–5).

The first recorded industrial patent was the one conferred by the Republic of Florence in 1421 on an architect and engineer by the name of Filippo Brunelleschi. The patent in question was applied to the manufacture of lifter-wheeled barges in the transportation of marble (*CEB* 1986b: 541).

The first patent law was passed in the Republic of Venice in 1474. Later, in 1624, Britain passed the Statute of Monopolies, granting owners of patents monopoly rights on the manufacture and use of inventions for a defined period. American and French patent laws were passed over a hundred years later, i.e. in 1790 and 1791 respectively (Machlup 1975). Ever since then, patent technology has been institutionalized in much of the world.

In Europe, periodic resistance has arisen to patent laws. For example, the free trade movement in Britain and Germany from 1850 to 1873 promoted an anti-patent law campaign, probably in an effort to counter the overly coercive enforcement of patent laws. The government of the Netherlands even repealed its patent laws in 1869 (ibid.). However, stimulated by the industrialization waves in various European countries in the nineteenth century, more countries introduced patent laws. For example, Austria adopted the patent system in 1810, Russia in 1812, Sweden in 1819, Spain in 1826, Brazil in 1859, Italy in 1859, Argentina in 1864, Canada in 1869, Germany in 1877 and Japan in 1885 (Uchida 1987).

Thanks to the tenacious efforts of advocates of the patent system, the Industrial Ownership Protection Regulation (Treaty of Paris) was adopted in 1883, a remarkable event that strengthened patent protection throughout the world and moved the international patent system substantially in the direction of standardization. Patent systems have continued to develop throughout the world ever since. In 1900 there were only 45 countries that had established a patent system. By 1925 this had risen to 73, to 99 by 1958, to 120 by 1973, to 158 by 1980 and to more than 180 countries and regions at present (Jiqun Li 1999: 11–13).

What follows below is to analyze the history of how to protect and respect patent technology in the United Kingdom, and the case of constant reforms of the patent system in the United States, illustrating that the patent technology plays a significant role for hard-tech innovation as well as economic development.

British industries, which depended mainly on the export of wool, lagged far behind continental countries during the mid-sixteenth century. This was partly due to the fact that Spain, Portugal and the Netherlands controlled marine transportation. The British royal family at the time did not have sufficient financial resources to address the problem, and did not have an apparent suitable strategy for Britannic revitalization. Queen Elizabeth, the monarch of the day, adopted several economic policies aimed at improving Britain's position, including a policy that introduced technology and protected industries. Queen Elizabeth was in favour of this policy because by the end of the sixteenth century the European continent was weary from the chaos of more than thirty years (1562–1598) of war; the Protestants had been persecuted and deported in France; and other countries, especially Germany, had excluded Jews from all sectors; whilst Britain had relaxed restrictions, including religious ones. Such policy attracted many foreign craftsmen, merchants and technician's technology: for instance, a large number of glass-manufacturing experts from Italy and Protestant skilled craftsman from France fled to Britain; weavers from the south of Netherlands had come to Britain seeking asylum from religious persecution, thereby bringing the secret of weaving 'a new fold decoration'; Dutch farmers had brought the drainage and intensive arable cultivation techniques; and the Jew descendant who had been expelled from Spain and Portugal brought the methods and experience of business and financial management to Britain, etc. These technologists contributed much to the prosperity of economy as well as science and technology in Britain. In order to retain the skilled foreigners the Queen suggested granting monopoly rights to inventors and requested patentees to apply their patents in Britain within a certain period. In spite of strong resistance to the plan, the Queen's absolute authority over-ruled the resistance and the policy was put in place. Among the 55 charters granted, twenty-one were conferred on foreign dwellers or inventors from abroad who were willing to transplant their technologies. This policy was repealed in 1599 (Uchida 1987).

The patent system is certainly far from perfect, especially when considering the special role that patented technology plays in our economic development. In 1950, the United States Congress required that the patent, trademark and copyright subcommittee of the judicial committee of the United States Senate review the patent system. Under the leadership of John Stedman, the committee turned out fifteen research reports, each from a different angle, between 1956 and 1958. Fritz Machlup issued the fifteenth report, 'The Economics of the Patent System', in 1958. In his report, Machlup pointed out that the patent system contained profound economic contradictions from its very beginnings. In particular he was concerned about what he saw as the detached island of monopoly rights that is granted along with various forms of the patent. The preface by John C. Stedman read, 'In the free economy that is backed by competition, we should not only overcome the detached island of the monopoly that is restricted by various patent forms, but also encourage the patents. There are three million patents of this kind in American industrial history' (Machlup [1958] 1975). The American publication *Business Week* published an article entitled 'Intellectual Property: A New Approach to a New Problem' that stated:

> For more than a century, our system for protecting intellectual property rights has fostered the creativity of artists and inventors. But now it is being challenged by two major global

forces: the Internet, which can act as a gigantic international copying machine; and a global health crisis in which patented miracle drugs are too expensive for developing countries. It is time to re-examine how the intellectual property rights regime works. (Garten 2001)

Much of America's industrial competitiveness was lost during the 1970s. Many research funds did not earn efficient capital and this caused the discordant relationship between academic institutes and industries in the 1980s. Senators Robert Dole and Birch Bayh believed that the main problem was that the patent system at that time impeded competitiveness and obstructed the overall application of research achievements. The patent system's main problem was that because many products were developed with government funding or as the result of research conducted in government laboratories, the government owned the titles to the patents. Although research projects funded by the government did indeed produce inventions, it was rare to find innovations that were applied successfully. Enterprises had to deal with the bureaucracy when using patents because the government owned them. Therefore, they were not active and most innovations owned by the government were put on the shelf. The new patent system of Bayh and Dole changed the technology-sharing relationship between academic institutes and industrial corporations.

The United States Congress adopted the Bayh-Dole Act in December 1980 and amended the Federal patent policy, thereby laying the foundation for successful technology transfer from universities (the recipients of much of the government research funding). The main points of the Bayh-Dole Act are listed below:

- Universities and small-sized enterprises can select their own research subjects and make innovations that will be funded by the federal government.
- Universities must apply for a patent for all their innovations.
- Researchers of universities must publicize their inventions immediately and apply for patents.
- Universities are encouraged to cooperate with the industrial sectors to accelerate the utilization of inventions funded by the federal government.
- Universities must share with other inventors the use of the charter charge.
- Universities are responsible for reporting to the financing institution the serving conditions of their inventions.
- Small-sized enterprises and the manufacturing industry are first to enjoy the use the new inventions.
- The Government reserves the right to use all non-exclusive patents developed by US Government funding in the world.
- The Government reserves the right to 'march-in'.

The complicated history of the patent mentioned above involves the process of the development of patent technology (institutionalization), the acceleration of patent-assisted technological innovation and industrial innovation (the reform of patent system). In the contemporary era of 'globalization' and 'informatization', the negative effects of patent technology have come more strongly into view. This has created some challenges for the patent technology. The future patent system must protect not only the interests of artists and inventors but also the interests of public and the users of the inventions.

China's Patent Law came into effect in 1985, and China has more than 170,000 patent applications in 2005, becoming the world's third largest patent applicant after Japan and the United States. China's international patent applications reached 5456 in 2007, representing a year-on-year increase of 38.1 per cent, and ranking seventh in the world, and leaped to fifth place in the world by 2009. The report of WIPO (World Intellectual Property Organization) published in 2007, indicated that in the past ten years the number of patent applications in China has grown eightfold. The Director of the State Intellectual Property Office (SIPO), Tian Lipu, said at an annual meeting of SIPO that domestic patent applications for invention accounted for 62.4 per cent of total invention patent applications accepted by SIPO in 2007, while it was only 40 per cent in the early 1990s and 50 per cent in 2003. It indicates that China's independent innovation capability is increasing, and the quality of patent applications has been further strengthened.

4. Advertising Technology – Technology of Propagandizing and Inducement

The technologies of propagandizing and inducement are technologies that are implemented with the expectation of influencing and manipulating the mode of conduct of human beings (involving decision-making at all levels) by promulgating new information, new ways of thinking and new lifestyles.

Advertising technology is a most prevalent technology of propagandizing and inducement. It is employed with the expectation of inducing and changing the behaviour of consumers or catching the public's attention by promoting or publicizing certain products, services and concepts through media such as newspapers, broadcasts, mail and television. The term 'advertisement' comes from the Latin word *advertere*, meaning attention and temptation.

For a long time advertising has existed as a tool for spreading information. Advertisements written on papyrus appeared in Egypt in 1550 BC and in the ruins of Thebes in Ancient Greece archaeologists have found an advertisement letter dating back 3000 years. This advertisement offered a pure gold coin as reward for catching an escaped slave.

Among the earliest printed commercial advertisements in China are the steel plates of the Liu Furniture Canteen in Jinan (960–1127 AD). These plates are housed in the Shanghai Museum. In the centre of each plate is a painted depiction of a white rabbit pounding medicine in a mortar.

In 1450, Germany's Gutenberg invented the letterpress, making it possible to print numerous verbal advertisements quickly and at a low cost. British printer William Caxton printed many advertisements on religious books and posted them on London streets, thereby initiating the use of printed advertising in the West.

The first newspaper advertisement appeared in Germany in 1525, and up to the 1800s the British newspaper *The Thames* carried a hundred advertisements every day (R. Yang 1987).

The earliest advertising agency was established in 1610 by two knights, courtesy of James I, the King of England.

Owing to the Industrial Revolution and batch production, advertising became an important sales technique in the nineteenth century. This period was considered the expansion period of commercial advertising. The first professionally managed advertising company was set up in London in 1812, followed by the first American advertising company, established in Philadelphia in 1841. In 1865, wholesale advertising agency appeared. Following that, it could be said that advertising had become a new industry. In 1868, the first modern advertising company, Ayer & Son, was established. Advertising companies receive commissions from newspapers and other organizations by acting as agents for their advertisements, while simultaneously establishing a clientele of producers wanting to advertise their products.

In the twentieth century, broadcasting became a powerful means to occupy and control the market that provided extensive, rapid and early service. The United States was first to apply broadcasting to advertising. In 1902, the first licensed broadcasting station, the commercial broadcasting station of the Pittsburgh Westinghouse Electric Company, began broadcasting. Following success in America, broadcasting stations were soon established in other countries. Advertisements were broadcast in commercial programmes. In 1903, the American psychologist Walter Dill Scott wrote *Advertisement Theories* in which advertisements were first studied as a discipline of science (Y. Zhao 1987: 6–7).

A Frenchman decorated the Paris Basilica with the first neon sign. Neon signs were introduced to America in 1923 and became popular by 1930 (*CEB* 1985b: 524).

Because television commercial advertisements integrate speech, music and visuals, television has become the ideal way to advertise and television commercial ads have furthered the advancement of advertising technology to a new level of development. The first television station was set up in Britain in 1936. Television stations were set up across America in 1939, and in 1941 commercial broadcasting on television began.

The history of advertising proves that the integration of and interaction between soft technology and hard technology were key factors in the rapid development of advertising technology. Every new kind of hard technology, such as the letterpress, corporate systems, agent systems, broadcasting technology, neon technology, television technology and the Internet technology, appear, without exception, to promote advertising technology. We can assume that there is still great potential for its future development.

In the modern society, the level of advertising is one of the important symbols to measure a country or region's economic development level. During previous years, the top ten countries in the world for annual advertising turnover in the world were generally developed countries (*Reference News* 2008). The revenues of US advertising industry ranked No. 1 in the world with 106.7 billion US dollars in 1997 and 163.036 billion US dollars in 2006, increasing by 53 per cent in ten years. China's advertising industry ranks the second in size in the world. China's advertising revenues in 1997 were only 3.68 billion US dollars but soared to 48.52 billion US dollars in 2006, an increase of 1218 per cent over ten years. According to the data from TNS Media Intelligence, China's advertising expenditure amounted to 287.5 billion in 2006, ranking the third (together with the UK) after the US and Japan in the global advertising market. Japan was squeezed into third-place, increasing from 33.426 billion US dollars to 34.34 billion US dollars, an increase of only 2 per cent in ten years. These countries were followed next by the United Kingdom, Germany, Brazil, Mexico, France, Italy and Canada.

5. Insurance System

Insurance technology is a complicated mechanism, the establishment and maintenance of which requires great skill. Since the beginning of time, human beings have continuously been seeking ways to prevent disasters from occurring. Although it is impossible to prevent all catastrophes, human beings have found ways to make up for the losses they have to endure. Insurance is an established system that charges premiums, provides mathematical forecasting for unplanned calamities and transfers these risks from individuals in the form of a contract to an organization in which the members of the organization will fairly share the loss. The mechanism by which the majority will share the loss of the minority is not only an economic institution but also a form of legal regulation (Sun 2000: 20–36).

The earliest form of insurance is the ship mortgage contract that was found in sphenogram documents from the third century BC. It provided the ship owner with a loan that would be repaid if the voyage was safely completed. Later, the shipping mortgage developed into marine insurance. At the end of the Middle Ages, this kind of insurance was spread to the field of land transportation (*Encyclopaedia Britannica* 1999: 389).

On 23 October 1347, Georgius Lecavellum, a Genoese merchant, created the earliest certificate of insurance known to man for the voyage of the Santa Clara ship. Marine loans and loss deposits developed into two exclusive businesses in 1400, and merchants specializing in marine insurance began to appear. Business insurance contracts became common with the rapid development of overseas trade in the sixteenth century (*Encyclopaedia China-Finance, Revenue, Banking and Price* 1993: 8).

The Barcelona Royal Decree, the earliest law relevant to insurance, was issued in 1435. Later, the Venice Decree appeared. A relatively complete marine insurance regulation act was established in Florence in 1523, along with a standard insurance policy. The establishment of the insurance broker system originated from the insurance law that was enacted by King Philip II of Spain in 1556. British Queen Elizabeth I enacted the first marine insurance law in 1601 (ibid.).

The inventive process of fire insurance added a great deal to soft technology innovation. Owing to a baking oven that overheated, the royal bakery caught fire on 2 September 1666.

After spreading throughout London and burning for five days and nights, the fire caused great damage; 13,000 houses were burnt down leaving 200,000 citizens homeless. This fire made people aware of the importance of fire insurance. In 1667, Nicholas Barbon, a dentist, provided funds to establish the first fire insurer in the world. In 1710, Britain's Charles Povey established the earliest fire insurance joint stock company, the Sun Insurance Corporation.

The development of demographics and statistics marked the beginning of fire and life insurance joint stock companies. The first life insurance company to embody the elements of the modern life insurance system was established in Britain in 1762 (Yao & Kun 1999: 35–37). The insurance premium was based on the theory of an average insurance premium determined according to a table of mortality and insured individuals who did not measure up to the policy's standards were charged a different special premium.

During the nineteenth century, insurance technology entered the modern period and the scope of the industry expanded to include property loss, death insurance, survival insurance, liability insurance, credit insurance and reinsurance. Now a complete body of insurance theory and a complete insurance technology system is in place.

According to statistics, the total profit of insurance premiums around the world in 1950 was 21 billion US dollars; in 1976 this was 250 billion US dollars; in 1985 it was 630.5 billion US dollars (the USSR, Eastern European countries and China were not included), 2000 billion US dollars in 1995; and 2940.67 billion US dollars in 2003 (D. Cheng 1997: 15). By that time, the industrialized countries accounted for 89.32% of the world market share; the emerging market countries and regions accounted for 10.68%, and China accounted for 1.6% (Swiss Re-insurance 2004a). After 1986 the average rate of growth of insurance companies was 4.7% (the actual growth rate after adjusting for inflation and fluctuation in exchange rates), which surpassed the worldwide rate of GNP growth during the same period (Fang 1989: 39). The ranking list according to insurance premium income in 2002 was as follows: the United States, Japan, Britain, Germany, France, Italy, Korea, Canada, the Netherlands and Spain. The premium income of the US segment reached 1097.8 billion US dollars in 2004, and its rate of international market possession was up to 33.84%, while Japan ranked second place with a 15.18% market share.

With regard to insurance penetration (proportion of the premium income in GDP), the global average insurance penetration was 8.06% in 2003, and countries or regions ranking in the top ten in the world were as follows: South Africa (15.88%), United Kingdom (13.37%), Switzerland (12.74%), Japan (10.81%), Belgium (11.61%), Taiwan of China (11.31%), Barbados (11.29%), the Netherlands (9.77%), Korea (9.63%), United States (9.61%). China was ranked the 44th in 88 countries and regions, with an insurance penetration was 3.33%.

With regard to insurance density (per capita insurance premiums), the global insurance density was 469.6 US dollars in 2003. Switzerland ranked the top of the list with 5660.3 US dollars, followed by the United Kingdom with 4058.5 US dollars. The average insurance density of the industrialized countries was 2763.5 US dollars; while in the emerging markets it was 58.7 US dollars. China was ranked in the 71st place with 36.3 US dollars (Swiss Re-insurance 2004b).

China's insurance business was restored in 1980, and its national insurance premium income was only 460 million Yuan. The insurance premium income reached 159.6 billion Yuan in 2000, with an average annual increase of 34 per cent, but the level of the insurance industry was also quite low. By 2005, there were 93 insurance companies and their premium incomes totalled RMB 492.7 billion. According to the '11th Five-Year Plan', China's insurance penetration will reach around 4 per cent by 2010, while the insurance density will amount to 750 Yuan.

Thus, it is evident that the insurance industry is representative of a country's economic and social development. The development of insurance technology and its institutions suggests that as human beings expand the range of their fields of activity, as complicated social relations increase and as economic intercourse rises, people's worries about risk and security also tend to rise, leading to the rapid development of the insurance industry. The complete range of human wishes and problems pertaining to social and economic life is a rich source of soft technology.

6. *Management Technology*

Management technology aims at improving people's ability to coordinate, balance and control the things for which they are responsible. Management technology covers a wide range of topics and a wide scope of organizational domains, from state affairs and international relations, to industries, enterprises, families and individuals. Hence, management technologies and all the soft technologies closely related to management can be classified at the micro, meso and macro levels; and in terms of management domains, there are individual, enterprise, industry and state domains.

Management technology is one of the areas of soft technology that was developed earlier and is a special kind of soft technology.

First, management technology infiltrates and covers all fields of the economy, society, organization and technology – similar to the way in which behavioural science has reached into all fields of the social sciences. In the final analysis, the object of management is humanity. Management is therefore not only a technology but is also an art, closely related to personal factors such as judgment, insight and charm, moral character and human endowments.

Secondly, management technology is a kind of soft technology but not all soft technologies belong to the category of management technology. For example, education technology, design technology, culture technology, social technology and many of the business technologies are not a part of management technology. Even if some soft technologies closely related to management, because of many influential environmental factors they cannot be controlled or 'managed' by enterprises or industrial sectors, hence they should not be simply considered to fall within management technology, and moreover it will not be conducive to the further development of soft technology.

Management has been treated as a discipline of science and technology since early in the twentieth century. In the early stages of commercial and industrial history, when ownership was not yet separated from management, enterprise management relied mainly on the individual experiences of practicing managers, namely, the so-called epoch of experience management. For a long time managers had no sources of formal knowledge to which they could turn for insight; learning took place mainly through informal communication with friends, colleagues and associates in similar or related fields of business when they faced with the problems which is now considered as the managerial issues.

However, as industrial development and the division of labour became more specific, managerial problems became more complex. Gradually, managers came to realize that managers in quite different industries, such as textiles and railroads, confront many similar issues. Charles Babbage's work on 'The Economy of Machinery and Manufactures' published in 1832, discussed these common issues, paying particular attention to such problems as calculating costs and establishing incentive systems.

General ideas conducive to the formulation of theory were brought forward and put into practice in the 1850s by some excellent American railroad managers such as Daniel C. McCallum, Albert Fink and J. Edgar Thomson. American engineers and manufacturers such as Henry R. Towne, Henry Metcalfe and Frederick A. Halsey started a movement in the late nineteenth century, now called 'systems management', which initiated the development of modern management technology and laid the foundation for Taylor's approach scientific management (Williams [1979] 1989). At the end of nineteenth century, with the transition from *laissez-faire* capitalism to monopolistic capitalism, the scale of industrial activity increased continuously and the capitalist economy entered a new development period, with competition between enterprises intensifying. Therefore, in order to win over both the indigenous and international markets, proprietors needed special technologies and experts to perform management 'scientifically', to improve production efficiency and reduce costs, which gradually brought about the separation of management and ownership. Frederick W. Taylor, the 'Father of Scientific Management', was the first person to consider management consulting to be a social occupation. Although Taylor continued to be an outstanding inventor (he possessed more than 40 patents), in 1890 he also established an independent consulting business under the rubric of 'an engineer of efficiency'. He thus became the founder of the management consulting industry. Taylor contributed much to the idea that management is a special discipline of science and technology. In 1911, Taylor published *Principles of Scientific Management,* which marked a milestone in the history of management theory and practice.

The development of business education played an important role in the systematic development and diffusion of management technology. In 1881, the University of Pennsylvania attempted to initiate business education in America. In 1898, the University of Chicago and the University of California both created colleges of business. In 1908, Harvard University established a graduate school of business, heralding the formal entry into universities of management as a discipline.

The twentieth century was an age in which management giants came forward in large numbers with plenty of innovation in managerial practice and management theory.

Henry Ford incorporated Taylor's scientific management principles into his factories and designed a production line for mass production in 1910, thereby influencing mass production in industry for the entire twentieth century. Not inconsequential for his time, Ford turned out one million cars by 1915.

However, neither Taylor nor Ford attached much importance to the organizational problem. In the 1920s, A. P. Sloan came up with the concept of 'division management' in his book *Organization Research*. This concept greatly influenced the organizational innovations of large enterprises. American companies such as Dupont and GM accordingly adopted this organizational mechanism to fulfil the demands of business diversification and cross-regional management of their products. The mechanism of division management was extensively applied throughout the world by the 1960s.

Harvard University applied the well-known 'Hawthorne effect' to one factory in Chicago belonging to Westinghouse Electronic in the middle of the 1920s. This application provided important theories on respecting interpersonal relations and human emotions and attaching importance to informal organizational functions.

Marketing development also emerged during the 1920s. Since the second Industrial Revolution, enterprises generally emphasized production and finance. The distribution of commodities was left to wholesalers. After World War I, American enterprises turned increasingly to their domestic markets and gradually began acquiring market shares and controlling the markets that they served. With respect to competitive strategy, after learning the lessons of focusing on pure price competition, enterprises began to pursue organized and systematic approaches centring on non-price competition. For instance, they began to use packages, advertisements and trademarks as weapons for controlling the market.

The early twentieth century signalled the beginning of marketing as a field. In 1910, R. S. Butler first introduced the term 'marketing'. In 1912, A. W. Shaw explained his marketing theory for the first time in *Some Problems in Market Distribution*. Following the end of World War I and the reduction of overseas demand for American products in the 1920s, the US government became actively engaged in the promotion of industrial efficiency, the application of standardized products and the mass production system. However, enterprises subsequently had to face serious business challenges, such as how to develop markets and how to promote the products they were now mass-producing. Consequently, the period of 1920s were years of high-pressure marketing.

This high-pressure marketing period suddenly downshifted in 1929 when the world economic depression occurred. In order to sustain market share, enterprises performed promotional activities, such as market research focusing on the attitudes of consumers. They also emphasized customer service, thereby promoting the notion that 'consumers are God'. The 1930s was the beginning of the downshifted, low-pressure marketing period (Robert Bartels, Robert Keith and Robert L. King are the source of this review of the history the marketing development divided into the above phases) (Tanouchi & Murata 1985). After

World War II, competition between enterprises focused on technology innovations and new products. The priority for competitors was comprehensive market management under an overall strategic framework in order to gain a competitive advantage, thereby entering the era of marketing.

Now, marketing has become an independent branch of management science. Marketing technology has been fully developed as a field for generating new market demands and consumption. It is aimed at the development of new markets, new clients and new market channels based on research about the behaviour and psychology of customers. For example, client technology offers means attracting customers, creating markets, developing products and services together with customers. At the same time, it may also provide ways of keeping customers and satisfying them. Sometimes customers are even incorporated into the enterprise as partners, trade agents or product agents.

Modern management reached another milestone in the 1950s. Around that time, American troops introduced many 'business and management technologies of American style' into Japan (Kishimoto 2000). In addition, quality management theories initiated and developed by American land forces during World War II evolved further in Japan as Japanese managers applied these theories, together with W. E. Deming's theories, in 1954. Later, these theories became the basis for the distinctive characteristics of Japan's enterprise management. During the 1950s, the Management Science Association (1953) was established, the journal *Management Science* began publication in America, Abraham Maslow published his *On Stages of Demands* (1954) and Douglas McGregor proclaimed his 'X and Y' theories, which provided the theoretical foundation for respecting people and paying attention to interpersonal relationships in enterprise management.

In the 1950s America began applying 'strategy', originally a military term, to enterprise management. During the 1960s, a number of relatively mature publications on the theory and practice of managerial strategy appeared one after the other. These included: *Strategy and Structure* by Chandler Alfred in 1962; *The Creative Manager* by Peter Drucker in 1964, *Corporate Strategy* by Igor Ansoff in 1965 and *Top Management Planning* by Steiner in 1969.

In the 1970s, the quality of enterprise leaders and how they functioned as managers became important issues in management theory. In particular, great emphasis was placed on the essential issue of helping enterprises improve management levels and succeed. Henry Mintzberg's path-finding book, *The Nature of Managerial Work* (published in 1973), is an example of a highly influential piece in this genre.

The 1980s was an era of total quality control. The 'economic miracle' that Japan achieved during the thirty years after World War II, and the many Japanese products that occupied the international market, shocked the developed countries, especially the United States. Before Japan's economic breakthrough, American enterprise had created the microelectronic industry and all the scientific and technological resources that had assisted in the development of this industry, such as the transistor, the semiconductor chip and large and small computers. But Japan's companies drove the large American enterprises, and companies that had previously controlled this industry, out of the market. America's Xerox

Company invented modern duplicators but Japanese counterparts conquered this market in 1979.

The history of the consumer electronics industry involves a continuous retreat of American companies from the market. For example, in 1955 95 per cent of all radios sold in America were American-made, but in 1975 there were no American-made radios sold in America. At its peak the total amount of sales of the American TV industry equalled six billion US dollars and this amount accounted for 22 per cent of the entire consumer electronic products market. However, in 1987 there was only one American TV manufacturer (Zenith), whose sales accounted for 15 per cent of the total sales in the TV market. Video recorders made in Japan enabled Japanese companies defeat the American companies that pioneered innovation in the field. The mass production of automobiles was invented in America but European and Japanese manufacturers eventually mesmerized American consumers with their high-grade and low-price automobiles, frustrating their American counterparts. This experience proved that even the best product concepts, technologies (hard technology) and patents might not be translated into commercial success if companies are not managed well.

In 1980, a programme called *It Worked in Japan, Why it Did Not Work Here with Us?* was broadcast on America's NBC television network. This stimulated a new wave of research and inquiry into the nature of Japanese management theories and managerial practices. In the meanwhile, *TQC* by Deming was publicized around the world.

Rapid development of the technology of 'competitive advantage' occurred during the 1970s and 1980s. Typical of that period were publications such as Michael Porter's *Competitive Strategy: Techniques for Analyzing Industries and Competitors* (1980); *Competitive Advantage: Creating and Sustaining Superior Performance* (1985) and *The Competitive Advantage of Nations* (1990). *The Competitive Advantage of Nations* was reprinted eleven times and translated into twelve languages. In addition, business philosophy and enterprise cultural theory fully evolved during this period. Many successful enterprises committed themselves to forming common values inside their companies and strengthened their competitiveness by encouraging the enthusiasm and creativity of their employees.

In the 1990s, the established management mode faced serious challenges with the new application of information technology and economic globalization. Various new management concepts and technologies were put forward in an attempt to adjust to the new management context of the twenty-first century, for instance, virtual technology, reengineering technology, etc. The typical works, such as *Reengineering the Corporation* (1990) by James Champy and Michael Hammer and *Fifth Generation Management: Integrating Enterprises through Human Networking* (1994) by Charles M. Savage, were produced.

Tremendous progress in science and technology was made in the twentieth century. A large number of achievements and inventions from the past several hundred years were applied to industry in that century. The reason why they were not applied earlier may be attributed to the emergence in the twentieth century of 'management technology'. The Japanese scholar Yoshituki Kishimoto encapsulated the innovation of representative management technology during the last hundred years of innovation by calling the twentieth century the 'Management Century' (Kishimoto 2000).

Facing the challenges of sustainable development of the Earth and human beings in the twenty-first century, it is necessary to redefine and even transform thoroughly human survival and its development mode – national development mode – the concept of economy, the meanings of value and even the positioning of enterprises. It will call for the emergence of new management theory and management techniques.

7. Stock Technology and Securities Technology

Stock system is a form of economic organization in which various people pool their funds by purchasing shares in a business as a whole, becoming entitled to dividends or bonuses according to the proportion of shares they hold. The stock system only emerges when the economy has developed to an appropriate level. Its historical development has been long and gradual.

The stock system and its organizational form came from Europe during the Middle Ages. At the beginning of the twelfth century there were many enterprises owned by a single investor or partnership enterprises in Italy.

During the sixteenth and seventeenth centuries the growth of large-scale industrialized production created the need for a great deal of investment capital in order to purchase the large quantity of machines and equipment required, but the funds available from family, friends and relatives were far from sufficient. Under these circumstances, stock technology was invented to extend the range of sources in society from where investment funds could be raised and the stock enterprise mechanism was established to systematically manage the funds of many shareholders.

Britain established the first joint-stock company, the Muscovy Company, in 1553. It was a chartered overseas trade company and its shareholders were all aristocrats, landlords, dukes, high ministers, merchants and churches. The first genuine British stock company, the Levant Company, which was also called the Turkey Company, was an overseas trading company established in 1581. 242 people bought shares in this company and it raised its capital by floating shares in the form of open stock (Yu: 62–248). Another stock company, the Dutch East India Company (at that time it sold special local products purchased in the Orient to the countries of Europe), was established in Holland in 1602 (*100 Questions...* 1988). Later, stock companies began developing rapidly in the banking, transportation, mineral, public services and manufacturing industries. At the end of the eighteenth century, stock companies replaced almost all the partnership enterprises and enterprises owned by single investors.

In the late nineteenth century, traditional industries in Britain slowed down, while those in Germany and America accelerated. At that time, the events that promoted the industrial change in the German and American economies stemmed from the shift from competition between the small internally-financed enterprises towards cooperation in establishing large-scale stock companies. These eventually developed into the gigantic monopolies that were the hallmark of the early years of the twentieth century. What was so fascinating about this age was that most chemistry and electronic industries were supported by scientists, while many strong industrialists were involved in the metal and machine industry. Kelvin, Edison, Siemens Brothers and Brunel (father and son) all shifted away from being primarily scientists towards being industrialists. By using the stock market, large-scale enterprises raised a large amount of funds required for their development. Sufficient funds promoted the extensive application of science and technology, including internal-combustion engine technology and transportation technology that had stimulated the Transportation Revolution. Another feature of this period was the extensive application of science and technology to war: submarines, torpedoes, high-quality explosives, cannons and the mechanization of war further promoted the development of technology (Bernal [1965] 1970: 338–9).

At the beginning of the twentieth century, stock companies contained one-fourth to one-third of the national wealth of Britain, America, Germany and France, thus occupying the prevailing position in the economies of these countries (G. Zhang 1988: 13–19). By the 1980s, stock companies played an even more important role in the market. For example, there were 2,710,000 stock companies in America, representing 16% of all the enterprises, but their assets made up 85% of the total; sales made up 89% and 80% of net income came from such companies. By the 1990s, 18.5% of all American companies are stock companies, accounting for 90% of the total sales (T. Zhao 1997: 5–6). In Japan, 53.5% of ordinary corporations are stock companies but almost all of those enterprises with capital funds greater than one hundred million Yen were stock companies in 1984 (Fukumitu 1986).

The invention and skilful design of securities technology provided an excellent resource for the development and application of stock technology. In fact, stocks are just one type of security. Securities technology is the most successful 'technology' in raising, using, appreciating and managing property. Securities are, legally, 'pieces of paper recording rights and obligations of property', and can be divided into 'evidence securities' that are simply used to prove a certain fact, and 'portfolio securities' that are used to explain the conditions for the transfer of rights and obligations. From the point of view of rights and obligations, portfolio securities can be further divided into 'commodity securities', 'monetary securities' and 'capital securities' (ibid.). Fund-raisers issue capital securities in order to raise funds from providers based upon equal conditioning. On the one hand, small sums of money that are held by individuals in society can be gathered together through the issue of securities to ensure the larger amount of funds the raiser needs for a certain purpose. On the other hand, from the perspective of providers, they can expect to receive a relevant reward after purchasing the securities with money that is temporarily left unused for various reasons but is gathered by the raiser for something significant. Capital securities can be divided, according to the nature of their income, into bonds and stocks. Stocks are a means of raising capital to fund the development of enterprises; hence, they are a type of revenue security and they also provide various rights for the shareholders.

With the prosperity of stock companies, stock merchants in Amsterdam established a stock exchange in 1613. In 1657, a relatively well-organized and stable stock exchange appeared in Britain (Zhiyong He 1997: 2–5). The securities market was extensively developed in industrial countries and regions by the end of the nineteenth century. The establishing years of the top ten securities markets in the world by the end of the twentieth century are shown in Table 4. At present the securities market can be divided into several markets, such as bonds, stocks, issues, circulation, exchange and over-the-counter markets.

The securities market became increasingly important to the capital market in the twentieth century. In order to minimize risks in the securities market, listing standards became much stricter, which made it almost impossible for small enterprises, especially small scientific companies with relatively great risks, to list their stocks. For the purpose of establishing capital markets to support SMEs (small and medium-sized enterprises), high-tech companies and other relatively risky enterprises many countries around the world began to establish a 'second market' for high-risk securities. In 1971 a brand-new stock market – Nasdaq – was established in the United States. The Nasdaq was an innovation in both financing techniques and operational mechanisms for venture capital. These innovations created the conditions for small enterprises to list their stocks. Since 1992 the funds requirements for listing enterprises has been reduced and the condition that the listed enterprises must make profits for three consecutive years has also been repealed. These innovations in the securities markets not only lowered the requirement of listing but also initiated the competitive system of the market maker.

Table 4: Global top securities exchanges and their establishment year.

Securities Exchanges	Establishment Year
German Stock Exchange	Beginning of the eighteenth century
Paris Stock Exchange	1724
New York Stock Exchange	1792
London Stock Exchange	1801
Australian Stock Exchange	1837
Toronto Stock Exchange	1878
Tokyo Stock Exchange	1878
St Paul Stock Exchange	1890
Hong Kong Stock Exchange	1891
Nasdaq (America)	1971
Japanese OTC	1991
EASDAQ (12 European countries)	1996
KOSDAQ	1996

*Source: *Science and Technology Daily*, 4 June 2000

It is evident that stock market innovations provided SMEs, especially high-tech enterprises, with a quite favourable environment for raising capital. Each country with a formalized securities market has customized its design to take into account the special circumstances of the domestic economy. By the end of 1999, 4829 companies had been listed on Nasdaq, among which there were almost 2000 science and technology companies and Internet enterprises. At the same period, this kind of market has developed greatly in twelve European countries (1996), the United Kingdom (1995), Germany (1997) and also in Japan (1991). Modelled on Nasdaq, securities markets in the service of small-size scientific enterprises were established in Asia's newly rising countries and regions such as Singapore (1987), Malaysia (1997), South Korea (1996), Thailand (1999) and Taiwan (1994) (Tao 2000). Hong Kong established a second market in November 1999, but the listing regulations were changed in 2001 to embody much tougher standards. The innovative points are as follows: the listing difficulties for SMEs have been lessened through attaching importance to the listed enterprises' future; reducing listing standards; advancing the market maker system in the stock market; reducing business costs; and improving efficiency and transparency by the application of e-commerce. Thus, 260,000 terminals have been established in 55 countries and regions, and Nasdaq has become an incubator for start-up enterprises.

With the development of the securities market, various crimes centred on the stock exchange have emerged, and the securities market has become a stage for illegalities of listed companies, which has required a new wave of innovations for the securities market institutions. This has prompted the Chinese Securities Control Committee to set up an investigation bureau for securities crimes.

The above analysis demonstrates that the technology of stocks and the securities market play indispensable roles in the production, commercialization and industrialization of hard technology. Meanwhile, just as is the case with other technologies, securities technologies also need constant innovation.

8. Merger and Purchasing Technology

The enterprise merger is another type of organizational technology. One enterprise buys the property of another (acquisition) or combines with other enterprises under single ownership (merger) to seek common development. The stimulus of this merger may be to strengthen market control ability, extend technological scale or economic scale, reduce business costs or disperse risks. All of these purposes centre on the larger goal of intensifying competitive power. The enterprise merger is a method of external growth for enterprises. Mergers in this chapter include annexation, acquisitions, consolidations, amalgamations and takeovers.

In order to understand further the function of merger technology and to explore the relationship between soft technology and institutional innovation through the example of merger technology history, let us analyze the four merger waves in American industry (Muramatsu 1988).

The first merger wave in the United States occurred during 1898 and 1902. After the civil war, owing to the establishment of the railroad system and the introduction of new production technologies, the general trend amongst American enterprises was to shift their focus from serving local and regional markets towards serving the national market. Those enterprises with the production capability to meet the demands of the entire nation, especially enterprises in the petroleum, tobacco and steel industries, avoided competition, improved efficiency and enjoyed monopoly profits through horizontal mergers. For instance, the American Standard Oil company, whose petroleum production made up 90 per cent of the production for the entire country in 1878, became an exceptionally large monopoly group whose petroleum production made up two-fifths of the entire world in 1880 through mergers with other enterprises in the same industry. However, there were some negative consequences on market power brought about by the giant monopolies such as Standard Oil, and in an effort to regulate the behaviours of these monopolies occurring in horizontal mergers, the US government promulgated new laws to curb what was seen as the extreme power of the corporations. The new law, which was known as the Antitrust Act, passed in 1914.

The second merger wave took place between 1925 and 1929. By then the horizontal merger had become greatly restricted owing to various anti-monopoly laws. However, managers and owners increasingly discovered the value of 'vertical mergers' (mergers with upstream or downstream enterprises, mergers between production and marketing enterprises) as a vehicle to enlarge the scale of their enterprises and vertical mergers became widespread. In 1929, before the world economic depression, stock investments and the second merger movement became popular in American business. The primary purpose of this merger wave was to expand markets and to reduce costs through the creation of monopoly markets and through vertical mergers. Once again, the US government felt the need to curb what was perceived as the excessive power of this new kind of giant enterprise through the imposition of new regulations. As a consequence, in 1933, Congress passed a set of laws – the Stock Act – forcing companies to become far more transparent to public scrutiny about their financial affairs. These regulations were reinforced further in 1934 by the passing of the Stock Exchange Act.

From 1966 to 1968, the third merger wave reached its peak against a background of economic prosperity and continuing enthusiasm for investment in the stock market. In order to get around the strictures of the Antitrust Act, companies found new forms of merger pathways through which they could grow. The new approach involved creating business conglomerates aiming to expand the market or diversify business through the merger of large corporations in the different sectors which are less interrelated with others, now called 'conglomerate mergers'. At that time there were a number of loopholes in corporate law, particularly in the area of corporate taxation related to mergers, which were exploited by companies as they pursued stock exchanges through non-taxation mergers. Under this kind of acquisition strategy, many companies also engaged in highly dubious financial tricks, involving large and complicated merger patterns, often designed to hide tax liabilities and other financial obligations. And with improper use of acquisition technologies, there existed even mergers with the mode that 'a small fish can nibble a whale'. Some companies actually carried out 125 mergers within eight years. This wave of corporate behaviour led to the US Congress passing the Williams Bill in 1968 and to new taxation reforms in 1969.

The merger fever that appeared from the middle of the 1970s to the 1980s could be considered the fourth merger tide. Though the number of the companies merging was small compared to the previous wave, most of them were very large in scale, exerting a great influence upon the industries involved. Some commentators called the mergers of this period 'mega-mergers'. During the climax of company merger in 1985, there were actually 3000 mergers occurring in the world.

Since the 1990s, the distinguishing feature of enterprise mergers was the rapid increase in transnational mergers. The techniques of merger and acquisition had existed in the developed countries of the West since the beginning of the nineteenth century, but as vehicles for extending domestic enterprises, the transnational merger did not fully emerge until World War II, along with the increase of direct transnational investment. The transnational merger and acquisition process began to rise in 1993 after being at low ebb during the period of 1991 to 1992. Up to 1995, the trade volume of the world transnational acquisition and merger reached 229.3 billion US dollars, which was twice as much as that of 1988; it was 334 billion US dollars in 1997, which increased 45 per cent from 1996 and accounted for 58 per cent of the world's direct foreign investment (J. Zhao 1998); in 1998, it reached 604.6 billion US dollars. It has been reported that the worldwide volume of trade for which companies involved in international acquisitions and mergers were responsible reached a total of 3300 billion US dollars by 1999 (among which the transnational purchase and annexation amount totalled 862.7 billion US dollars); and that figure exceeded the total amount for the six years from 1990 to 1995. The telecommunications industry, at 561 billion US dollars, accounted for the highest proportion of the total, and the banking industry was second at 297 billion US dollars. With 45 exchanges totalling 13 billion US dollars, Microsoft was responsible for the largest number of trades. Intel came in second with 35 exchanges totalling 5 billion US dollars.

From the above we can see that the transnational merger technique is not only a force that has promoted the development of the world economy and increased direct foreign investment, but it is also an important symbol of economic globalization.

Entering the twenty-first century, global mergers and acquisitions have continued apace and most companies wanting to maximize corporate value seek to do so through mergers and acquisitions, making the scope and scale of mergers and acquisitions larger and larger. Hewlett-Packard merged with Compaq in 2001 to form a new company with assets of 56.4 billion US dollars, and the annual turnover of the new HP reached 87.4 billion US dollars; in 2002, the world's largest pharmaceutical company, Pfizer Inc. merged with Pharmacia Corp. of the US by stock-for-stock exchange with a value of about 59.5 billion US dollars. In December 2004, China's Lenovo merged with IBM's PC business with a combined value of 1.25 billion US dollars, which caused a stir in the industry as a case of transnational merger and acquisition with 'the immediate mode'; Procter & Gamble Company announced an acquisition of Gillette Co. for 57 billion US dollars in January 2005, forming the world's largest consumer products company; in April 2007, a consortium led by British bank Barclays signed an agreement with Dutch bank's management group to purchase the Dutch bank for about 90.8 billion US dollars; but thereafter a consortium led by the Royal Bank of Scotland announced that they were willing to purchase Dutch bank for 97 billion US dollars.

Within a hundred years, the merger technology has developed from that of the horizontal merger and then the vertical merger, through a series of new forms including downstream combinations, upstream combinations and the conglomerate mergers, and finally to the transnational merger. Mergers also can take place between alliance partners and between competitors.

The merger is now no longer a simple external organizational technology of the enterprises: transnational mergers have led some multinational companies into the global business with unknown nationality, thereby impelling strongly the economic and technological globalization, changing the world economic structure and even affecting the political pattern. The history of the enterprise merger indicates that the wise use of the merger technology can be a favourable means of improving competitiveness. The development of the merger technology has also promoted institutional innovations relevant to enterprise competitiveness, such as the Antitrust Act, the Securities Exchange Act and various taxation laws. One important lesson that may be garnered from the history of merger is that institutional innovation must keep pace with the application and development of soft technology.

9. *Venture Capital Technology*

'Venture capital' refers to investments projects which may have high risks by normal business standards but which are likely to yield high profits or high business growth rates sufficiently large to offset the high risk. These investments are sometimes also known as 'venture investments' or simply as 'ventures'.

The earliest investment company known to have been created specifically for risk investments, the ICFC (Industrial and Commercial Finance Corporation), was founded in Britain in 1945. The American Research and Development Corporation (ARD), founded in the USA in 1946, was the first formalized risk investment company engaged in public exchange and managed by a financier.

In 1946, J. H. Whitey founded the first private risk investment company with a starting capital of 10 million US dollars, inventing the new term 'venture capital'. In addition, Whitey was the first one to make a risk investment by way of stock sharing.

In 1952, Canada's first venture capital company was established. In 1958, the US Congress promulgated the Small-size Enterprises Investment Act. Many investment companies devoted to small enterprises emerged in the wake of the creation of this Act and became the main source of venture capital for American firms at that time.

The Draper & Johnson Investment Co., founded in 1962, pioneered the administrative mode of the venture capital fund that is still widely used today.

In 1973, the National Venture Capital Association (NVCA) was founded, indicating that the venture capital industry had become a new American industry in its own right (M. Liu 1998: 1–51). In 1983, 40 European venture capital foundations set up the European Venture Capital Association (EVCA).

The level of development of a country's venture capital industry is a reflection of the level and prospect of its high-tech and related industries. For example, although the British venture capital industry started very early, it developed at a slow pace. It did not improve until the 1980s after the British government adopted a series of policies and measures – such as tax privileges, a loan guarantee programme and an enterprise expansion programme – that encouraged the use of venture capital. According to the BVCA (British Venture Capital Association) statistics, British venture capital totalled 0.8 billion pounds in 1979 and nearly 2.5 billion pounds in 1995, which was 0.4 per cent of the British GNP (Ren 2000: 20–58).

America, with the highest rate of development of high technology and the strongest technological competitive power in the world, is also home to the premier venture capital industry. The first peak in the history of American venture capital appeared in the 1950s and 1960s. According to the *American Venture Capital* magazine, more than 13,000 high-tech companies supported by venture capital were established from 1970 to 1979. The 1980s trends in the venture capital industry were stimulated by the tax reform system and by the rapid development of high-tech companies; this increased the amount of venture capital invested from 1.4 billion US dollars in 1975 to 11.5 billion US dollars in 1985, and 40 billion US dollars in 1995. By 1998 there were nearly 2000 venture capital funds in America, totalling more than 60 billion US dollars in investments. Venture capital still supports about 10,000 high-tech projects every year (Jiangling Li 1999: 16–21).

Since the beginning of the 1990s, the venture capital industry has progressed impressively throughout the world. According to an OECD estimate, the total amount of venture capital sum invested worldwide in 1996 reached 100 billion US dollars. Countries such as France, Germany, Australia, India and Israel all initiated formal programmes and policies aimed at promoting the development of the venture capital industry. In order to encourage venture capital the French government instituted a number of policies such as loosening taxation policies and corporate institutions, and founders were allowed to keep their preferential shares for up to fifteen years after the new company was established (prior to that the limit was only seven years). The French government also introduced a number of improvements to the securities system that required greater transparency in arrangements for preference shares, together with a requirement that 5 per cent of the stocks invested by life insurance companies must take the form of venture capital. In addition, they delayed taxation on the part of capital appreciation of innovative enterprises that was reinvested. In Germany, venture capitalists investing in high-tech companies have received tax exemptions on shareholder gains since April 1997; in addition, two high-tech enterprise stock exchange markets, Frankfurt Neuermarket and EASDAQ, were established. The larger, established German technology-oriented enterprises such as Siemens, Deutsch Telecom and Daimler-Benz also became involved in the high-tech venture capital industry (ibid.).

Venture capital technology is undergoing continuous innovation in both theory and practice. For example, a series of theories and typical cases about raising venture capital arranging investment systems, establishing operating mechanisms and administrating venture capital companies and support systems have taken shape. Most countries have now worked out laws, regulations, financing systems (including the stock market), policies and corporate institutions relevant to venture capital. However, the concept of venture capital has changed. It now covers all stock rights and financing activities aimed at private stock-holding companies. Meanwhile, many new trends have appeared during the development of venture capital technology. These include the globalization of venture capital, the increasing annexations and acquisitions among the venture capital companies, the increasing involvement of securities in entry and exit mechanisms of venture capital, the increased public regulation of venture capital management and an increase in the diversity of sources for venture capital.

10. *Logistics Technology*

In the traditional sense, the logistics technology refers to the technology that makes raw materials and products add greater value in the process of circulation between enterprises → suppliers → customers by improving service levels and reducing the cost.

From the perspective of the most original meaning of logistics (in the sense of physical movement), at its core the history of logistics technology is as long as human history itself. Logistics technology developed along with the birth and development of human civilization. As far back as the appearance of the circulation of commodities, the movement of labour tools from one place to another by the human body is a nascent form of what we call 'physical distribution'. With the arrival of commodity production, however, consumption activities and production activities gradually separated from each other. This created the need for linking production and consumption, and the need to develop physical distribution techniques that involved managing the interspaces between commercial exchange activities and the activities of transporting goods.

The development and elaboration of early physical distribution technology should be credited to the applications of logistics technology in the field of military. At that time, logistics technology functioned as the technology that supplied the necessary material, conditions and services to maintain a normal and efficient working environment for an organization. Originally, the term 'logistics' was an abbreviation for the term 'logistic service in the rear' that was used by the army. Whether the ancient wars, World War II or the modern Gulf War (1991), it's completely inconceivable to carry out military action without logistical support.

The early application of logistic theory and technology was evident in Sun Tzu's *Art of War*, written in 520 BC. Logistical support has been raised to the level of determining the survival of the army and victory or defeat in war.

Contents of military logistic theories could also be seen in *The Epic of Homer* and the West's earliest military work, Xenophon's *The Anabasis*. Fixed logistic organizations were included among the troops of some European and American armies during the eighteenth century (Meng 1997: 1–15).

In the late eighteenth century Britain established an international distribution and circulation network that linked the consumption market and the supply of raw materials into an organic whole. These developments made an indelible contribution to Britain becoming the centre of the first Industrial Revolution. (See the section of 'Soft Technology and Thrice Industrial Revolutions' in this chapter).

A US marine officer, Lieutenant Colonel Cyrus Thorpe, published his *Pure Logistics – The Science of War Preparation* in 1917, which is so far recognized as the first monograph on the theory of military logistics in the world.

In 1918, during World War I, the Instant Delivery Co. Ltd. was founded in Britain, the purpose of which was to deliver goods to retailers, wholesalers and consumers in a timely fashion throughout the country.

However, most people knew almost nothing about logistic activities before World War II, and few enterprises or organizations established specialized logistics branch. Towards the end of World War II American military departments, which at that stage were grappling with the challenges of guaranteeing ammunitions supplies for the war, developed formal plans using operations research and computer technology to organize expenses, purchases, transportation routes, weapons and storage. The resulting system for providing and maintaining military supplies came to be called 'logistics'. The demands of war furthered the development of military logistics technology. Logistics technology played an important role in both the European and Pacific battlefields. The demands of war promoted the development of military logistics technology, making the determination, procurement, warehousing, transport and inventory management of the demand for military supplies be treated as a system, and be able to plan using logistic model.

As a result of the intensified competition of the world economy after the war, the West's large enterprises began applying military logistic theories and approaches to their production and management departments, thus reducing their logistics costs. However, enterprise logistics of that time emphasized mainly cost control of transportation and storage functions. With the development of logistics technology, it is no longer limited to logistical services in the field of enterprise or the military, but applied to all areas related to the circulation of materials, and then it has gradually been renamed as 'physical distribution technology'.

Yet it was not until the publication in 1912 of A. W. Shaw's book *Some Problems in Market Distribution* that the term 'physical distribution' appeared formally for the first time. Shaw stated, 'Goods and materials will have additional value due to their transfer through time or space'. The term 'physical distribution' mentioned here to refer to what Shaw described as the transfer through time and space of goods and materials.

In 1935 the American Marketing Association established the earliest definition of physical distribution as the service activities that take place as goods and materials circulate between the place of production and the place of consumption.

In 1956, in their book *The Role of Air Freight in Physical Distribution*, Professors Howard T. Lewis, James W. Culliton and Jack D. Steele of Harvard University introduced the total costs analysis into logistics management.

In 1963, the National Council of Physical Distribution of the United States was established, which further promoted the development of this field.

The period of 1970 to 1980 is generally considered to be the period in which enterprise logistics was formally institutionalized. Owing to the energy crisis and to hyperinflation in the American debt market, corporate revenues decreased greatly and economic competition was intensified, making it more difficult for companies to manage their businesses. These circumstances placed even more demands for results related to enterprise logistics functions. Thus, the standardization of logistics service and management for various organizations was promoted (Wang & Luo: 1–7).

With the change of times and technological advancement, logistics technology has become one of the basic functions of government, enterprise, social communities and all organizations, while physical distribution itself also has different contents and requirements. Therefore, the 'logistics sector' replaced 'physical distribution by logistics'. In 1986, the National Council of Physical Distribution Management of the United States (NCPDM) was renamed 'The Council of Logistics Management' (CLM).

In 1964 Japanese organizations began using the formal concept of physical distribution. In the *Physical Distribution Handbook* compiled by the Japanese Comprehensive Institute, physical distribution referred to 'the physical movement of materials from providers to demanders and the economic activity of creating the time value and place value. Physical distribution includes such activities as packing, loading and unloading, storage, stock management, circulation processing, transportation and distribution according to a plan' (Cui 1988: 1–3; Z. Wang 1995: 26–59).

Toyota Motor Corporation's real-time logistics management strategy is to integrate Toyota Just-in-time (JIT) production system and sales network to establish a 'flexible marketing system', which divides the products into small quantities to sell-out faster, further reducing the cost in the field of circulation (C. Liu 2003). Meanwhile, Toyota provides training (in fields such as product knowledge instruction, sales training, business management, financial guidance, shop design, advertising guidance, etc.) for all dealers, and gives appropriate guidance about promotion policies and operational problems according to market information feedback in order to improve sales efficiency. Toyota also assists them in sales and after-sales service from the personnel and technical aspects. Toyota's real-time logistics system can speed up the velocity of circulation of goods and reduce inventory levels, making the replenishment time more precise so as to actualize the objective of reducing costs and improving the service level, thereby manifesting the refinement of enterprise management.

Emergence of e-commerce has developed broader business opportunities for logistics.

Japan's retail trade and transportation industry are very creative, driving the Logistics Revolution in the whole business circles through e-commerce.

With regard to online shopping, for example, customers became most unsatisfied with online shopping – despite the fact that online orders may have been executed very quickly – when the delivery of products and services was very slow, or when the purchasers were absent when merchandise was delivered. Japan's 24-hour stores are located in the residential areas of major cities, and are able to receive orders and even complete banking services online. There are more than 40,000 convenience stores (supermarket chain stores) in Japan, which have formed the logistics network for the Internet end-consumer. For example, the retail chain store Lawson has more than 7400 stores and has become gradually equipped to facilitate customers ordering goods online, after which an e-commerce enterprise provides the goods to the distribution centre of the nearest retail store, which will then ship out the goods in the same day. Although e-commerce enterprises need to pay consignment fees to the stores, it nevertheless saves a lot on transport costs.

Take the transportation industry as the example. Yamato Transport Co., Ltd. in Japan not only provide 32,000 delivery vehicles for e-commerce enterprises, but their mail service sector itself also operates a successful Internet platform where users can open up their own online business for free on this platform. Yamato Transport is able to enact transactions for the business of all e-commerce enterprises – warehousing, packaging, delivery, payment of transactions and even dealing with customer complaints. Sagawa Express Co., Ltd. (the competitor of Yamato Transport), Japan's second-largest parcel delivery service company, invested 45 billion Japanese Yen in a new information system during the last three years, making all of their delivery trucks equipped with a small computer terminal so that customers could track the location of their mail ('Logistics Revolution' 2000).

Today, logistics has developed into a discipline of science and technology. Application of the concept of logistics begins not only at the point where the products leave the factory, nor considers only the physical distribution process from the producer to the consumer, but it also includes a series of activities that occur in the physical circulation process, such as the purchase of raw materials, processing and manufacturing, sales of products, post-servicing and recycling. Owing to the great profits produced by reasonable logistics, the logistics field has been praised as 'the third profit source' of natural resources after raw materials, fuel and human resources. Logistics is called as the largest industry in the twenty-first century because of the increasing difficulties of extracting profits from the other two sources.

Age-old logistics technology is fresh with new vitality due to the rapid development of e-commerce, supply chain technology and third-party logistics as an intermediate market.

So-called 'first-party logistics' (1PL) here refers to logistics processes, such as trucking and warehousing, handled in-house by manufacturing enterprises; 'second-party logistics'

(2PL) refers to offering the capacity to carry out transportation and warehousing by using an external team and warehouse; 'third-party logistics' (3PL or TPL) refers to coordinating and offering one-stop logistics solution for its clients by the intermediary service providers. A 3PL service company could act as a service centre which provides all the value-added logistics activities according to the needs of clients, such as warehousing, distribution facilities, maintenance services, electronic tracking, etc. Therefore, logistics technology includes client service, demand forecasting, order processing, distribution, inventory control, transportation, warehouse management, layout and site selection of factory and warehouse, transport handling, purchasing, packaging, intelligence service and other hard and soft technology. With the development of globalization of supply chains and Internet technology the concept of 'fourth-party logistics' (4PL), which concerns the role of the supply chain integrator, has emerged.

The level of the logistics industry has become a fundamental determinant for the business development of every country or region. China's logistics costs accounted for an average of 19.57 per cent of GDP for the period 1996–2005, while the equivalent figure for the US during the corresponding period was 9.45 per cent (Juan 2007). The proportion of total logistics spending accounted for by 3PL is also on the rise.

11. Supply Chain Technology

During the early stage of its development in the 1980s supply chain technology was confined to a logistics process within the enterprise, mainly involving the functional coordination between those business units like materials procurement, inventory, production and distribution, in order to optimize business processes, reduce logistics costs and improve operational efficiency. With the rapid development of information technology, manufacturing technology and transportation technology, supply chain expanded to cover the entire 'movement' process of the product. It thus came to focus on the evolution of the core enterprise into a whole chain within the functional network within which suppliers, manufacturers, transporters, distributors, retailers and even end-users are combined together through control over the flow of information, logistics and capital.

Supply chain management has also been upgraded from logistics management to the management of the entire supply chain, as well as to integrated supply chain management. It has also risen from being concerned primarily with management of interactions between partners to the level of corporate strategic management. In seeking to integrate the whole supply chain, enterprises should not only strengthen the integration of internal resources, but also pay attention to integrating the outside resource advantages so as to improve the level of resource utilization, reduce costs of production and operation, improve the ability of market reaction and enhance the competitiveness of the supply chain. Therefore, a successful supply chain design must solve the problem of how to make all the stakeholders a community of interest. Supply chain member agencies need to form a 'virtual enterprise' which can optimize comprehensive performance through information sharing, as well as through financial and material coordination and cooperation.

Supply chain technology is also in the process of development. Examples include, jointly developing and formulating benefit-sharing plans through sharing detailed information with suppliers and customers, as well as early involvement of suppliers; reducing the number of suppliers; streamlining customers to build partnerships with selected customers; cooperation with competitors to expand new ideas and to enlarge the space and approach to innovation, etc. At present, success stories in supply chain technology can be found everywhere, and in some of the best cases stock in the organization has been reduced by 50 per cent; in some cases the proportion of deliveries that are punctual has been increased by 40 per cent; in some, the inventory turnover has been doubled; and in others, the incidence of items being 'out-of-stock' has been decreased by 9 times (and so on). Procter & Gamble, Wal-Mart, Chrysler, Dell, IBM, Siemens, 3M and Xerox are examples of some of the most successful enterprises in supply chain management.

Supply chain technology is actually one kind of business model.

12. *Incubator Technology*

Incubator technology emerged through the process of providing systematic support services for the early stages of the development of SMEs. The original meaning of business incubation derives from the analogy of a bird brooding – or incubating – its eggs. Incubator technology can provide new companies with the appropriate growth environment needed to improve their chances of survival.

The vigorous support of incubator technology symbolizes the commitment of a country towards enhancing its prosperity and the development of its economy. This is because promoting the growth of small and medium-sized enterprises is an important means of maintaining the vitality and innovative ability of any county and of adjusting its economic structure.

An American, Joseph Mancuso, first conceived the concept of the enterprise incubator. He also founded the first incubator, the Batavia Industrial Center (BIC), in Batavia, New York in 1956. By 1995 a total of 750 incubators, exhibiting different ownership structures (government, academic institutes, private and joint state-private ownership), different management goals and different configurations of services were operating in the United States. The incubator has been popularized in 42 states in America, and the content of services include office space, office services, commercial planning, assistance with raising money, training and education, employment and administrative support. The National Business Incubator Association (NBIA) was founded in the United States and the members include the owners and administrators of incubators, incubator founders, economic development experts, elected officials, real estate developers, venture capitalists and investors and enterprise consultants, among others.

Besides helping draft business plans and providing advice on commercial strategies, most incubators also help businesses with their finances and assist in selecting outside experts because obtaining appropriate talent and funding are fundamentally important when starting up a company.

Through abundant research and practice America's Laura Kilcrease of Triton Ventures divided incubators into five groups: open-type commercialized incubators; company-style incubators; private-type incubators; risk investment incubators; and virtual incubators. Tatsuno Sheridan established a venture company in the Silicon Valley, developed the virtual incubator technology and operated eight virtual incubators synchronously (Sheridan). He incubated entrepreneurs through all-round services such as the seed capital, business plans, legal services, management, training, technology transfer and other services. Some others have classified incubators according to other criteria, such as their business domains and the degree of support provided to incubator tenants.

Today, incubator technology has spread from incubating enterprises to incubating entrepreneurs, and from incubator professionalism to incubator globalization. About one thousand incubators have been established all over the world. In 1987 the first Chinese incubator was established at the Wu Han East Lake Founder Centre. By the end of 2000, 110 incubators existed in China with 1785 enterprises incubated and more than 5000 enterprises in the process of becoming incubated; by the end of 2001, 465 incubators with 3887 enterprises had been incubated, 15,449 enterprises were in the process of becoming incubated and 32 enterprises had already entered the market.

13. Tactics Technology

Tactics has a long history with a strong and fresh vitality from the ancient times to the present. The term 'tactics' is most commonly used in military area such as the tactics thinking in China's 'Art of War' and 'Romance of Three Kingdoms', which is widely commended by the common people. In China, the classic tactics technique, 'Besiege the state of Wei to rescue the state of Zhao', was created by a famous strategist Sun Bin in the Warring States period (fourth century BC). When Sun Bin received orders to lead corps of the state of Qi to help the state of Zhao fight against the state of Wei, which was stronger than themselves, he avoided the enemy's main forces and struck the weak point, as well as waited at ease for an exhausted enemy so that they not only rescued Zhao, but also weakened the strength of Wei.

In the long course of human history, people have accumulated rich historical experience for resolving or dealing with conflicts of complex interests. And stemming from the interests of their own different interest groups, people summed up numerous common experiences, rules and wisdoms to predominate and control 'interests game system' (Luo 2007), which is commonly referred to as 'tactics'. The essence of tactics is the wisdom to settle the complex conflicts of interests and power struggles, and its basic function is to convert the competitive pattern.

Now, tactics and tactics technology have already gone beyond the military area and have been widely applied to strategic, campaign and tactical dimensions in various fields like politics, diplomacy, economy, management, sports, etc. For instance, the book titled *100 Techniques of Management Tactics*, published in 1999, incorporated 100 successful stories of tactics technologies in the world (S. Zhou et. al. 1999). The book included advice on how to apply those tactics in business operations such as 'to think deeply and plan carefully'; 'to kill two birds with one stone'; 'to be pre-emptive'; 'to take action after the enemy had struck'; 'to win by novelty'; 'to know one's own strength as well as that of one's opponent'; 'to compromise out of consideration for the general interest', with which the people were familiar. Whether the application of tactics technology is appropriate or not, is not only related to the outcome of a war but also the key to success of an undertaking or incident: a coup could win a war, an idea can save a business, a good policy can be a cause of success, some scheming can win instead of lose and come out safely from danger.

Chinese scholar Zhihua Luo has carried through an in-depth discussion in his 'Introduction of Tactics Technology' (Luo 2007). The context of his research is to address military scientific research and teaching but it makes sense to understand the essence of tactics technology:

- Tactics technology is a common triumphing technology of 'interests game'. It refers to the operable knowledge system such as general approach, techniques, rules, procedures of controlling interests system, which are summarized in the process of resolving complex conflicts of interests and power. Tactics technology takes human psychology, thinking and behaviours in the process of interests conflict as the operational object, and serves for obtaining final victory by dominating the competitive situation and its trend.
- The research on tactics has opened up a new research field of thinking. In this area, the process of tactics thinking will be available to be measured and characterized with a series of indicators so that its content, direction and correctness will have a clear criterion and the expressions of logical relationship. It will certainly be able to accelerate the application of modern scientific knowledge to reorganize and normalize the tactics thinking of the classical Art of War, and create a good basic condition for learning and training of tactics, as well as build a foundation for the relay-like development of tactics technology.
- From the practical point of view, the study of tactics technology can reorganize, summarize, standardize and sublimate the experience, methods and procedures of solving the complex conflicts of interests. And when we encounter similar situations, we can control and straighten out the interests relationship with clearer ideas through putting it into practice, and also find better ways to create more favourable conditions, while minimizing the cost or the price of the competition. In extreme cases, a momentary slip caused by the level of generals' tactics technique will lead to bloody, serious consequences.
- It is well known that piloting a battle plane is a 'professional technical issue'. If we treat 'controlling the complex interests game system' as a 'professional technical issue' rather than the one that will be able to deal with experience, general knowledge and passion, then

developing and upgrading tactics ability will be transformed into an issue of 'professional learning, training and advanced studies'. Therefore, the research and cultivation of tactics technology in strengthening tactical training and improving the level of tactics technology has vital significance for the decision-makers in military circles, the business community and the political and diplomatic community.

- The emergence of concept of tactics technology will establish the correct procedures and standards for tactical learning and training. As a painter or musician must master the drawing techniques or musical techniques at first, and then form different styles of music or painting through the creative use of these techniques. Fostering tacticians also needs to train first tactics technology. Through the creative application of tactics technology, they would be able to grow more rapidly as artists of tactics who could put into practice flexibly based on the actual situation. Certainly, tactics technology and the art of tactics are tactical knowledge systems at different levels and different angles, but they are mutually reinforcing.
- Thousands of years of human history proved that the basic rule for victory or defeat of competition is 'the weakest goes to the wall', and resulted in the basic approach of victory 'using force to subdue the weak/the defeat of the weak by the strong'. But in the process of interests struggle, the circumstance in which the enemy is weak while we are strong would not naturally occur. It is generally necessary to actualize 'the transformation of strength'. In fact, the approach of 'making the strong or weak relationship counter-transform' is the process of 'strengthening one's own side and weakening the opponent'. In this sense, as a technique of winning, tactics can also be understood as technology of 'the transformation of strength'. Certainly, it's necessary to consider standards for how to identify the strong and the weak.

As the world has entered the new century, the ideal of sustainable development, harmonious society and a harmonious world poses new challenges to tactics technology. New world order and new development modes of the future would ask that both sides of contradiction follow a new code of conduct, new ways to coexist and new development modes. How to integrate these ideas and values into the design and implementation of tactics technology for achieving win-win solution is new task of the research in tactics. In 2001, China has put the 'military tactics' into the system of military science, and the Association for Military Coordinating of Chinese People's Liberation Army has also set up a Tactics Research Center.

14. Business Model and Management Pattern

The so-called business model refers to comprehensive solutions aimed at maximizing created added-value in business activities, and it is also the outcome of the comprehensive integration of various commercial technologies. As the social, economic, and cultural environment undergoes great change, business transactions need new rules, new forms and new channels. If it can be confirmed that such innovation yields superior benefits to the old methods, then it will be followed by enterprises or organizations and be gradually standardized and improved so that a new business model is built up which becomes the key to determining the sustainable survival and development of enterprises. Any successful company such as Samsung, Disney and Wal-Mart, has its own unique business models or management models, which are often at the core of the enterprise's business secrets and intellectual capital.

The well known types of management methods for supermarket and restaurants, including the concept of self-service fast food promoted by McDonalds, are all examples of business models in service industry. The present trend is that due to the added value created by business model innovation gradually surpassing the value of products and (hard) technological innovation, business model innovation has expanded beyond the service sectors to the agriculture industry and manufacturing industry, as well as all soft and hard industries. Namely, the driving force in business model innovation is the transformation of the value-added centre. Creative industries, the current hot topic, also are examples of business model innovation.

IBM provides a successful story of business model innovation within a multinational corporation (S. Tu 2008). IBM believes that the global integrated enterprises (GIE) will replace the multinational companies to compete in the flat world of the twenty-first century, and has articulated three stages of development: from the nineteenth century to the start of World War I was the period of the international company in which international business was characterized by import and export trade, and independent enterprises carried on foreign market development; 1914 to 2000 was the period of transnational corporations in which regional centres replaced the headquarters' functions and 'copied' massive branches with 'complete functions' across the world in order to expand business; and, finally, after 2000, GIE emerged. GIE spans national boundaries, the company's logistics, marketing and manufacturing, and services are decided by the advantage of different areas. Its principle is to do the most appropriate thing in the most suitable place. In line with this principle, IBM's global purchase centres have been reduced from 300 to three, 31 networks were reduced to one, and 155 data centres were reduced to ten centres, etc. IBM therefore improves efficiency, reduces costs and forms a virtual company by 'coordinating all the worldwide corporate activities like moves in a chess game', so as to actualize the restructuring of enterprise from a hardware-oriented business into a services and solutions-led enterprise.

The success of IBM's GIE model, in addition to the pioneering, keen insight and excellent strategic implementation ability of its leadership, owes to the following three points from the perspective of soft environment and soft technology.

Firstly, the progress of globalization: as globalization deepens (see the section of 'The Characteristics of Soft Technology are Suited for Coping with the Challenges in the Twenty-first Century' in Chapter 1) the space for business model innovation grows and its created-value gradually exceeds that created by product innovation and technological innovation. Thus, people carry on business model innovation consciously and accelerate its institutionalization. In the United States it is even possible for a large number of new business models to obtain intellectual property protection. Secondly, the contribution of Internet technology: information technology and Internet technology provide effective tools and convenient channels for moving, sharing and allocating resources on a global scale. Now that we are in an era when almost anything can be digitized and interconnected, humans will maximize the use of information resources with unprecedented ability, and will have insight about building smart systems to adapt to the requirements of the whole society and users, emphasizing such priorities as high efficiency, low energy consumption, low-risk, low-cost, a green-business orientation, etc. Thirdly, the mechanism and institutionalization of soft technology (solution): in essence, the integrated innovation of soft-tech and hard-tech dominates this business model revolution. However, the most critical factor is the conscious innovation of soft technology and institutionalization in a timely manner; namely, making the procedures for finding solutions turn into the institutions of enterprises or organizations, and for these to be syncretised into corporate culture.

Huawei Technologies Co., Ltd. is a private company founded in 1988 and headquartered in Shenzhen, China. According to the World Intellectual Property Organization (WIPO), Huawei was ranked as the largest patent applicant in the world, with 1737 international patent applications published in 2008, taking first place from second-ranked Panasonic. Philips Electronics ranked third place. The number of LTE (Long Term Evolution) patents accounted for more than 10 per cent of the world. At present, Huawei's products and solutions have been applied to more than 100 countries worldwide and now serve 36 of the world's Top 50 operators. The reason why this company has been able to grow so quickly is that Huawei profits from its business model transformation. In other words, the company has changed from being a traditional technology company engaged in the research, development, production and distribution of communication products, into a telecom solutions supplier, which assembles traditional products led by client requirements to form a one-to-one 'solution' for providing customers with a fast, high-quality service.

In general, 'business model' refers to the pattern of external transactions between enterprises, while 'management pattern' refers to the method of operation within the enterprise. While both of them are closely intertwined, management pattern also belongs to the category of business models in a broad sense. The successful business model must take a good management pattern as a foundation and promote innovation in management pattern, and management pattern innovation is also required consequentially to promote business model innovation.

The following is an example of business model innovation in traditional industry. Early in 2000, a remarkable piece of news in the Japanese business world appeared. '7-11', the convenience store, overcame the unfavourable influence of continuous economic stagnation and the dramatic depreciation of the Japanese currency and managed to obtain a retail value and increase in profits of 4 per cent and 15 per cent respectively, compared with the previous year. It became the biggest retail sales outlet in Japan, taking the place of

DAIEI, the supermarket magnate, in 2001. Apart from strict management, the company benefits from flexibly using various kinds of online sales, connecting with more than 8500 stores throughout Japan, tracking and satisfying customers' demands and forecasting the new trends of each day. It has also helped its suppliers improve efficiency and has helped commercial networks and manufacturers to control stock. This is called 'the revolution of supermarkets'. We can see that, whether it is a traditional or an IT industry, an enterprise must change its old mode of management in order to meet the changes of economic and social environments.

Hereinbefore we mentioned cases of business models focusing on the industry and service sectors. In the era of industrial economy, the business models of agriculture were not given much importance.

The above are the cases of business model involving the industry and service industry. However, in the era of the industrial economy, the business model of agriculture has gotten no attention. Business model innovation in agriculture will be essential to increase the added value of agriculture, give full play of the multiple functions of agriculture and avoid the agricultural crisis, as well as the establishment of the future human civilization. Now, there are many successful cases of innovation in agricultural business models, future agriculture and the new life style worldwide (E-Square Inc. [2007] 2008). In the section of 'The Softening of Primary Industries and the Agriculture Service Industry' in Chapter 5, we will discuss several successful examples of agriculture business model.

Indeed, an excellent business mode for the twenty-first century should be a green business mode (see the section of 'Changes in Corporate Culture and Values – Shaping "Good" Enterprise of the Twenty-first Century' in Chapter 4). In addition, the new transaction model based on e-commerce has brought good opportunities for enterprises, while, at the same time, the characteristics of the Internet, such as its emphasis on openness, sharing and inter-linkages, have created big pressure for changing management modes of enterprises. For example, movie producers have to join efforts with high technology, and manufacturing industry is moving towards 'IT-oriented manufacture'. The complexity of the network is a great challenge to the management of traditional enterprises.

It is noteworthy that business model innovation similar to that in the successful examples of GIE described above (in addition to its significance in the economic world) will have an impact on the future global hard environment and soft environment, and even play an invaluable (positive and negative) role in the establishment of future civilization. On the one hand, as the gradual expansion of GIE's financial, human resources and organizational capacity, it generally forms a giant enterprise controlling a large number of various resources, such as interests, power and finance, technology, managers, even politicians, intellectuals, beyond national borders and national sovereignty, to invest, set up factories, transfer technologies and carry on trade in the world. GIE becomes the protagonist on the global stage and even controls the globalized market. In some areas its capacity may exceed some countries, even sponsors' national power and influence. On the other hand, along with the progress of globalization and information technology, it is not necessary to use international relations as all kinds of information can spread via Instant Message around the world only as the internal information of GIE, and this kind of information dissemination is organized. These effects will be potential factors for global governance and the establishment of an international order.

B. A Retrospective of Social Technology Development

1. Research about Relevant Social Problems in Industrial Countries

After World War II, the economies of western capitalist countries rapidly progressed. However, together with material abundance, various social problems became more prevalent and even overwhelming. Take the United States as an example. Since the end of the 1960s, American society has been plagued by problems such as inflation, unemployment and the social effects of pollution. A variety of experiments in social policy have emerged in an attempt to address these and other social problems.

On the other hand, for a long time American science and technology has been conducted in the traditional government-dominated mode. For instance, a number of projects have been implemented by the US government as national plans, with the simultaneous goals of pursuing general American social ideals and building up the national reputation. Examples include the TVA (Tennessee Valley Authority) Plan in the 1930s (President Roosevelt authorized the Tennessee Valley Development Plan in 1933), the Manhattan Project in the 1940s, the Defensive Weapon Development Plan in the 1950s and the Apollo Plan in the 1960s. The achievements of these plans did have significant influences on industry, provided many employment opportunities and promoted the progress of America's basic science.

However, many people observed that the efficiency with which the great technological achievements of the public research institute were transferred to civilian applications was low. In particular, the 'systems technology' developed by the Apollo Plan, which lasted from May 1961 to July 1969, and which cost the American taxpayers 24 billion US dollars, did not achieve the expected results vis-à-vis transfer to civilian enterprises. This caused many people to seriously rethink their previous opinions regarding the purported 'trickle-down' effects of the enormously expensive military and space technology programmes; rather, it was felt that in the new economic and social environment, science and technology should be targeted more directly towards the multiple needs of consumers, with less burden placed on tax payers. The challenge of how national projects that are directly applicable to the needs and demands of consumers may be developed was added to the policy agenda of the United States. In addition, enterprises that were perceived by the public as too being ruthless in the pursuit of profit, to the neglect of social needs, became the targets for increasing public criticism.

At around the same period a new international debate ensued, encapsulated by the claims published by the Club of Rome in its wave-making *The Limits To Growth: A Report for the*

Club of Rome's Project on the Predicament of Mankind (Meadows et. al. [1972] 1997). This best-selling book observed that the economies of the developed countries were encountering a new class of problems, while the developing countries were developing rapidly in the resource-consuming mode previously followed by the now developed countries. Various conflicts of interest and conflicts of values now faced the world centred on issues such as the speeding-up of industrialization, population growth, the mode of economic development and the progress of science and technology development goals on one hand, and the constraints of natural resources and the environment on the other.

Against this backdrop of the 1950s and 1960s, the United States began to invest in basic research aimed at solving social problems. The so-called 'intellectual technologies' needed for solving social problems such as forecasting, evaluation and planning techniques developed rapidly. In some academic circles this new field, focused on the analysis of policy issues associated with social problems and social demands, was called 'policy science' (The Institute for Future Technology 1973), which stressed the importance of interdisciplinary and interdepartmental research and focused attention on comprehensive inquiries at the interfaces of science and technology, the social economic system, the cultural system and the environment system, placing special attention on the interactions of these domains.

This new approach to inquiry required considerable restructuring of the research system. For example, the so-called interdisciplinary and interdepartmental research are forms of cooperative research conducted across the boundaries of traditional academic divisions such as physics, biology, social science and new types of science, but they left the actual boundaries of traditional disciplines essentially intact. However, in order to solve the social problems at the centre of the new research, we now know that the coordination of research needs to go beyond the now established interdisciplinary modes; it must enable the conduct of research from completely new fields and allow established disciplines to move beyond the constraints of their orthodox viewpoints and paradigms. The kind of research needed to address major contemporary social issues tends to undermine the old university research system based mainly on the division of departments and disciplines, and to stimulate the gradual establishment of interdisciplinary and inter-field education and research systems.

Some representative organizations that have moved in this direction are the Research and Development Corp. (RAND 1948), the Battelle Memorial Institute (BMI 1955), the Stanford Research Institute (the Behavior Science Advanced Research Center was set up in 1952 by Stanford University) and the Technical Military Programming Organization (TEMPO). These organizations have completed numerous effective research and development studies regarding national defence planning technology, management technology, systems development, communication systems that supported the Apollo Plan, management technology in the diplomacy field, the redevelopment of cities and many other topics. Consequently, these organizations have strongly furthered the reformation of research in the education system and attached more importance to generalist areas of interdisciplinary and inter-field studies.

The policy sciences that developed in the 1960s and 1970s are divided into two categories: first, R&D for the so-called intellectual technologies – technology forecasting, evaluation and planning – aimed at improving the efficiency of science and technology; and, second, research on the influence of progress in science and technology (hard technology) on human life, human emotions and inquiries into methods of solving social problems related to science and technology. A report of the US Senate entitled 'Applying Systems Analysis and Computer Technology to Social Science and Social Problems' (1969) was an example. It listed some major projects on forecasting, evaluation and planning that reflected the understanding of social technology prevalent at that time in America:

1. The Public Opinion Inquiry technology and online Delphi project located at the RAND Institute and MIT. MIT was also the location for the development of the 'Technology for Groups Dialogue and Social Choice' theory.
2. The Parallel and Sequential R&D Strategy project based on the publication by Professor William Abanasie of Stanford University on the 'Parallel and Sequential R&D Model' in 1969.
3. The New System of Social Development Plan managed by Stanford University's Professor Bruce B. Lusignan, who was also responsible for the implementation of Stanford's Interdisciplinary Engineering Course.
4. The Problem Solving course, which was the responsibility of Moshe F. Rubinstein of the University of California Los Angeles.
5. The Survey Research Center at the University of California at Berkeley.
6. The ICPR – Inter-University Consortium for Political Research – the development of which was led by the University of Michigan.
7. Others.

During this period the application of social technology to solving domestic and international political, economic and social problems was emphasized worldwide and various modern 'think tanks' were born. An example of this phenomenon was the birth of the Club of Rome in 1968 and its research plan regarding the human crisis. The Club of Rome, which is still active today, gathers together many famous economists, politicians, environmentalists, natural scientists and social activists to form a comprehensive research institute that addresses important international issues and global development strategies. In addition, the International Institute for Applied Systems Analysis (IIASA) was founded in December 1972 in Austria, with funds from twelve countries. Its research covers a range of issues associated with the entire human race – resources on the Earth, human resources, human society and economic technology methodology. The London Strategy Institute – another example – was founded in 1958 in the United Kingdom, changed its name in 1971 to the International Institute for Strategy Studies (IISS) and became a branch of the International Relations and Security Network (ISN). The main business of this network is to formulate explicit judgments on various international political, economic and social problems. Roland

Berger & Partner GmbH International Management Consultants, with its headquarters in Munich, is a global advanced management consulting service company that was founded in 1967. Further examples from the Orient are the Nomura Comprehensive Institute, which was founded in 1965, and the Mitsubishi Research Institute (MRI), founded in 1970; these are both famous intellectual centres and research institutes in Japan.

2. Innovation in Social Sciences and Social Technologies is Extremely Urgent

Social sciences have developed quite rapidly in the past 200 years. Behavioural science developed in 1949 after a scientific conference at the University of Chicago. Behavioural science has had such an impact on the social sciences that the boundary between the different disciplines has blurred and now the different theories, views and approaches flow freely between them, fusing this scientific complex together. Despite the great progress that has been made, the achievements of the social sciences and social technologies are overshadowed worldwide by the significant achievements of the natural sciences and technology. The lag between these two general domains of science and technology has become a bottleneck for sustainable socio-economic development.

In 1955, Jyuji Misumi, a Japanese scholar, pointed out, 'The misfortune of modern society lies in the comparison of the outrageous development of natural technologies and the backward development of social technologies. Furthermore, many roots of modern misfortune lie in the failure to strike a balance between the two technologies' (Misumi 1955).

In 1966 the American scholars Olaf Helmer, Bernice Brown and Theodore Gordon stated that when expatiating the mission of social technologies the following happens:

> It has been said that many of the difficulties that beset our world today can be explained by the fact that progress in the social science domain has lagged far behind the physical sciences [...] we who are in the social science field are faced with an abundance of challenges: how to keep peace, how to alleviate the hardship of social changes, how to provide food and comfort for the poor, how to improve the social institutions and the values of the wealthy, how to cope with revolutionary innovation, etc. However, unlike the physical sciences, where failures normally mean mere delays, the social sciences cannot afford to fail in their major aspirations; to do so could have a direct and catastrophic impact on society. (Helmer et. al. 1966)

In China, although failure is evident everywhere, it was caused by inefficient decision-making, forecasting and macro management, and we acknowledge these well-known causes. Social sciences are often misplaced with politics and disregarded by many people as a discipline of science, let alone as the basis for developing social technology. This makes it difficult to systematically study social sciences and technology from the perspective of 'solving' realistic problems of economic and social development. Since 2007, in China 221 major national labs and 141 national engineering and technology research centres under the ministry of Science and Technology, and 124 national engineering research centres under the National Development and Reform Committee have been established (State Science and Technology Commission: 246–268), but none of these focus on the important application technology or application engineering related with social sciences. However, none of the interdisciplinary research centre specializes in social technology or social engineering related to major strategic decisions.

Today, while hard technology has been making rapid progress, innovations of social technology lag far behind. The reasons are analyzed below:

- The knowledge of social sciences is one of the vital sources of soft technology, but social sciences still have not been fully developed as scientific disciplines and the existing classification of social science disciplines is obsolete. Although interdisciplinary research activities have become more popular in recent years, these responsibilities still belong to the inner academic circles of traditional social science departments and they tend to be conducted with indifference to the need for application-oriented research.
- Natural scientists and engineers who were educated in the old education system generally lack knowledge of social sciences, and many economists and sociologists are not aware of the developments of modern technologies. This means that the 'different trades are separated by mountains', not just by the small hills of academic boundaries.
- The largest mental barrier is the prejudice against 'technology'. In most situations within the social sciences people focus on studying its various structures and territories as disciplines of science, but they fail to devote attention to summarizing and integrating those methodologies, means and rules of soft technology, let alone regard them as 'technology'. Thus, 'the conscious application of social sciences' to practice is continually held back. The traditional and widely accepted understanding of technology tends to mislead people about the potential relevance of social sciences and blinds them from seeing the technology that is, and can be, derived from non-natural scientific knowledge.

3. Distinguishing Social Science from Social Technology

Along with the further development of science and technology, especially the development of information technology, science and technology become increasingly difficult to be distinguished. On the one hand, the cycle from the presentation of a scientific theory and technological breakthroughs to the production applications becomes shorter and shorter; new disciplines are emerging unceasingly which are interpenetrating, and the new frontier disciplines are emerged in interconnected nodes. Moreover, it comes into being through the mutual integration of comprehensive subjects such as sciences of information, materials, environment, life, etc. On the other hand, the meaning of the concept of science has been expanded; not just social science, thinking science, cognitive science, etc. are subsumed into the concept of science, but the standard of differentiating science also has new developments. For example, the requirements on a rigorous repeated verification for scientific theory may need to be relaxed; general principles would gain an important place in science as law and theory; value orientation would play a key role in science as logic and experience (W. Jin 1997).

It is generally considered that science of the twentieth century has evolved from 'small science' to 'big science' (Dong 1996), which has four important features compared to the traditional understanding of science. Firstly, science with the traditional understanding reveals only the knowledge that can be duplicated by any scientific explorers, but a new type of science takes the behaviours that can not be reproduced as an important target of scientific exploration. Secondly, the former takes social application of science as the social issue out of science, while the latter incorporates it into the process of scientific exploration. Thirdly, the former neglects or despises value factors, making the exploration for the direction of value freedom simplified, while the latter must take value factors into account. Fourthly, the scientific knowledge system with the traditional understanding does not relate to or involve in the system itself, but the knowledge system of new type of science relates to (or is involved in) the knowledge of the system itself. If this new type of science is regarded as a basic form of science then science with the traditional understanding is the limiting form that is severely restricted. The growth of such new types of science can be regarded as an omen of the transformation of scientific paradigms in general. This new feature of science means the trend of convergence between science and humanities. The traditional scientific rationality is confined to instrumental rationality and experimental rationality, while the new type of science takes the value rationality as an important factor of scientific rationality, which has greatly changed the structure of scientific rationality.

In the context of above, the conceptual confusion of science and technology will not be conducive to their development. A Japanese scholar has lamented that Japan's biggest mistake was the invention of an abbreviation which merges 'science and technology' into a word. Science and technology have essential differences in their essence, characteristics, roles, development ways, laws, etc. There are all-round adverse consequences to confusing science with technology from its concept to applications. For example, for the issues in scientific ethics, on the one hand, one-sided pursuit of material interests and value of science resulted in the weakening of its spiritual values, making scientific concepts, scientific thinking and scientific attitude lost; on the other hand, to regulate science using the policy, measures and ethical principles for technology and engineering led to the loss of the scientific spirit of free exploration, setting up restricted areas for science. There exists the distinction between scientific ethics and the moral responsibility of scientists, and scientists should assume the moral responsibility that everyone in the society should take. They know more about the evil consequences of abuse of scientific and technological achievements than most people; therefore, scientists have greater responsibility to reveal the dangers to the world (W. Jin 2000). As for the code of ethics and criteria needed to develop technology, we will discuss below the direction of technological advances and soft technology.

Chinese scholar Guangbi Dong posited the differences between science and technology succinctly as follows: 'From the perspective of knowledge, science is the theoretical knowledge, while technology is the operable knowledge; from the perspective of methods, the scientific means is to discover, while the technological method is to invent; from the perspective of activities, the purpose of science is cognition, while that of technology is practice' (1998: 59).

Take the social sciences and social technology as examples – no matter internationally or in China, there is the lack of research on the essential issues of social sciences.

For example, with the scientific and technological breakthroughs in the twentieth century – especially the rapid development of contemporary robotics, genetic technology, biotechnology, etc. – humanities and social sciences have been facing unprecedented challenges. So, the development and application of science and technology led to an argument about social, ethical and legal issues, which pressed for a solution of humanities and social sciences to solve these issues or at least to provide possible ways forward. However, the research of humanities and social sciences on these issues has seriously lagged behind, not only in the careful study of the social consequences of modern science and technology using multi-disciplinary and cross-disciplinary approaches, but also by limiting its ability to put forward regulatory strategy for development and application of hard science and technology. The research level of the humanities and social sciences needs to be improved, taking it to the high level where research results aim at complex social relations, various value choices and non-linear social trends, as well as where the incisive issues concerning values, beliefs, ethics, environment, security and others, may be produced so as to provide adequate knowledge for the solution of above issues, including the means of early warning, prevention and control.

In another example, frequent financial crises in modern society, especially the worldwide economic crisis which began in 2008, not only brought forward serious challenges to contemporary economics, political science, sociology and so on, but also strongly urged the humanities and social sciences circles to provide high quality responses on the new world order, human development mode and other fundamental issues. A lack of theoretical guidance holds back our ability to find the right solution.

Nevertheless, in the fields of social sciences, social technology is confused with social sciences or is treated generally as 'policy or strategic issues', affecting not only the development of social sciences but also the development of social technology. It is imminent to separate those knowledge systems belonging to technological category from the social sciences, so as to be 'perfectly justifiable' to systematically develop, innovate and, furthermore, institutionalize them.

Certainly, technology rooted in the knowledge of social sciences is not only social technology. A large number of commercial technologies, such as market exchange technology, currency technology, accounting technology, stocks technology, business contract technique, financial derivatives and so on, which are 'technological' inventions having far-reaching effects on human society, also belong to operational technology of social scientific knowledge. It is taken as commercial technology in this book depending on the application areas.

4. Social Technology and Its Value

1. The awareness of Social Technology

In the section of 'A Fresh Understanding of Technology' in Chapter 1, I pointed out that social technology should not simply be summarized as the application technology of social science. Proper consensus on the meaning of social technology has still not occurred and, in any case, after half a century of development its meaning continues to evolve.

In *An Introduction to Social Technology – Group Discussion*, written by Jyuji Misumi in Japan in 1955, the technologies that control human relationships and spiritual phenomena in a society are described in general as social technology. Jyuji Misumi also conducted research on the two themes of social science and social engineering based on the concept of group dynamics put forward by K. Lewin in the 1940s. He labelled the application of group dynamics and group engineering as 'social technology' in the sense of social engineering, and discussed the operating technology of various meetings, including the formal meeting, the informal meeting, the small-scale meeting and large-scale public meetings. It also discussed the functions of group and individual in course of discussions, types of problem-solving functions and the process management function of group function, etc.

In 1966 the American scholars Olaf Helmer, Bernice Brown and Theodore Gordon co-authored the book *Social Technology*. The authors believed that social technology is a form of social science methodology, which mostly includes the methods of operational research, the Delphi method and Experts System, etc. They described how to apply these techniques to long-term forecasting in the research report *Long-Range Forecasting Study* sponsored by the RAND.

Erich Jantsch, the OECD Adviser, cited Olaf Helmer's concept of social technology in 1967, supporting the idea that social technology is the generalization of technologies that are essential to society and which are based mainly on social inventions. 'Social inventions' refers to inventions that exert a strong impact on the social system and on social technology transfer; 'social engineering' refers to human activities that are effective and directive in the transfer of social technology (Jantsch [1967] 1968: 15).

Guangbi Dong, a Chinese scholar, also discussed social technology in a 1989 article entitled 'On Social Technology'. He stated that social technological knowledge is solidified in social organizations in a way similar to how natural technological knowledge is materialized in tools, mechanics and equipment. Political organizations (governments, congresses, courts, etc.), economic organizations (factories, farms, stores, banks, insurance companies, etc.) and cultural organizations (schools, hospitals, etc.) are all transporters of social technology. The operating process of various social organizations is social technological knowledge, namely, operable social knowledge.

When the Japanese Monbusho (the Japanese government's Ministry of Education) approved the research project entitled 'Social Technology' in 2000, an explanation of social technology was added: 'Social technologies are technologies that comprehensively apply natural science and social science to the construction of a society and which make people feel at ease and substantial' ('The aim of social technologies...' 2001). Saikawa Hiroyuki, chairman of the Japanese Theory Convention, believes that the object of social science is the nature or behaviour of individuals or groups. Therefore, the significance of social technology R&D lies in 'coordinating science and technology and society by applying humanities and social science knowledge'. Monbusho started the research project entitled 'Social Technology' in April 2001, and the budget for the year 2001 was 1.5 billion Japanese Yen. The first group of projects may focus on the design of the accident prevention system of atomic energy equipment.

We can see that although the same term is used, the understanding and explanations of social technology differ greatly.

I believe that social technology can be summarized as the process technology by which various social resources are developed and applied, that is, the process, method, procedure and system of creating or embodying the value (wealth) of social resources. Social technology is mainly applied in improving social efficiency and solving social problems.

2. Types of Social Technology

Social technology involves two interrelated and interactive dimensions.

The first dimension is the process and mode by which technology develops social resources. This generally refers to the social activities of various organizations, such as meeting technology, discussion technology, coordinating technology, alliance and cooperation technology, public relations technology, human relation technology, organization technology, service exchange technology, communication technology, complementary currency technology, operational mode and mechanisms of social groups and communities, etc.

The other dimension is the methodology of solving social issues and handling social affairs, including all types of systems technology, planning technology, evaluation technology, forecasting technology, decision-making technology, strategy choosing technology, city technology, social simulation technology, monitoring technology for social crises, institutional innovation techniques, etc. The institutional innovation techniques in social technology refers to the design technology, analysis technology, evaluation technology and operation technology on the system, mechanism, law, regulation, policy and standard focusing on social issues. Today, many soft science methodologies mostly belong to social technology from the perspective of soft technology and they mostly provide services for decision-makers at different levels. The social technologies of Helmer, Brown and Gordon start with the methodology of solving social problems.

From the perspective of social engineering, which is at the implementation stage of social technology, there are two ways to solve problems. The first is by stressing each social problem and to develop or search for a corresponding technology (hard technology) appropriate to solving that problem; the other is by conducting an overall design and planning process according to space and region limits while ignoring individual problems and looking for solutions for the effective operation of an entire system. The former belongs to the hard-tech application in social development, which has been our usual way for hundreds of years

– and, strictly speaking, it is not social technology. The latter is the methodology of social engineering. It involves first of all searching for solutions from the level of the totality or the 'whole' and then at the level of the part. The answers to the questions about what kind of hard technology may or not be appropriate, and whether or not it is even appropriate at all to use high technology depends on how the technology might fit within the larger solution. This approach captures the charm of social technological innovation. Ecological planning technology and engineering is an example. At the stage of social engineering, what merits our special attention is that we can not implement social engineering under the framework and ideas of hard technological engineering. And it needs to make clear that social technology is not simply an operable system rooted in the knowledge of social sciences. Stemming from its basic purposes, the development and application of social technology must stick to the principle of balancing economic, social resources and environmental development in a sustainable manner.

Social technology, as an important component of soft technology, has significant differences with hard technology. According to environmental changes, it needs to continually improve and even involve revolutionary transformation. Moreover, in view of the dualism of social resources, we should pay attention to restrict the negative side of social resources, and to expand the positive side will be one of the main tasks of developing social technology.

Today, social technology is facing the arduous challenge of moving from being a noble idea towards becoming a workable method. For example, until now formal decision-making technologies tended to be comparatively 'hard', with many emphasizing the analysis of objective factors only; and often the more quantitative format was considered to be the more scientific approach, even though it produces erroneous results because it ignores essential subjective factors. This issue was elaborated upon further by Theodore Gordon in a letter to me:

The decision-making process should be improved throughout the whole world in order to solve problems at present and in the future. I wrote about this in the 1997 State of the Future report. Today, decision-making technology is taught from the standpoint of economics using techniques like cost benefit analysis and mini-max theorems. However, in the future, I think, good decision-making will go well beyond economics. A course of decision-making in the future might include, for example, the role of human intuition, risk taking propensity, values (consideration of what is the right thing to do and models for seeking the right decision) and the consideration of how the mind works in the process of making decisions, particularly how the mind can distort data and information and reach for decisions that are illogical and inconsistent. The field of cognitive science will also have a role by showing how the mind creates models of outcomes; small and fast scenarios that are accepted or rejected when arriving at a decision. (Letter dated 29 October 2000)

As to the role of psychology and the significance of subjective factors in the decision-making process, Kimindo Kusaka has noticed that the high-speed growth of the Japanese economy since 1963 has much to do with people's confidence in the national economy at that time. He remarks that 'although high speed economic growth is an economic term, the key is whether it is believed or not. Many Japanese managers at that time invested in equipment with their personal property as guarantee. They did this without any hesitation for the belief in the return of investment' (Kusaka 1985).

3. The Value of Social Technology

One characteristic of social technology is that it empowers public participation in its design and implementation, and relies on people's creativity to make society more efficient. Such progress of social technology must be the goal that we need to pursue. However, in view of the dualism of the value of social technology, it needs to identify the direction for innovation, promotion and application of social technology.

The economic value of social technology is the value created by social activities that can be measured by currency. Social activities themselves have always been accompanied by economic activities in the traditional sense of being funded through the redistribution of public resources and investments, the expansion of charitable donations or self-financing. These economic activities can only be maintained through the support of public investments and donations from traditional economic sectors, the main sources of which are business transactions.

The practices of recent decades have proved that to solve social problems, as well as promote sustainable development, the economy, society and environment can generate a lot of business opportunities (Grayson et. al. 2008). In fact, the contemporary social enterprise creates its economic value by utilizing the above opportunities. And, moreover, during the last seven decades social activities have gradually formed another type of business and become increasingly enlarged: various types of transactions within social activities, such as career services, social services and networking exchange services, can achieve their aims by using another kind of exchange tool – complementary currency (see the section of 'Currency Technology' in this chapter) – and by forming an economic domain completely different from the traditional economy (which uses financial capital to develop material resources). As society develops, these economic domains take up a larger proportion of the national economy.

The social value of social technology is expressed through various approaches: handling a greater volume and greater complexity of social affairs; enriching people's lives; acquiring more knowledge, skill and opportunity, so as to satisfy people's social aspirations; solving problems that cannot be solved by the government or market, including moral issues; and providing a favourable environment for creativity, originality, responsibility and initiative.

According to the author's definition of social technology, the dualism of social resources must be considered when we discuss the value of social technology. The development of social resources may bring high quality economic growth, employment opportunities and multifarious social benefits but may also have a negative effect or even a devastating effect in some cases.

Portes describes four aspects of the negative effect of social capital (Portes 2000):

1. 'The same strong ties that bring benefits to members of a group commonly enable it to bar others from access [...] social capital generated by bounded solidarity and trust are the core of the group's economic advance'.
2. 'The group or community closure may, under certain circumstances, prevent the success of business initiatives by their members', namely, it restrains the creativity. 'Many successful entrepreneurs are constantly assaulted by job and loan-seeking kinsmen. These claims were buttressed by strong norms enjoining mutual assistance within the extended family and among community members in general'.
3. 'Community or group participation necessarily creates demands for conformity'. This feature is also a source for generating the capability of social control. 'The level of social control in such settings is strong and also quite restrictive of personal freedoms (the privacy and autonomy of individuals were reduced accordingly), which is the reason why the young and the more independent-minded have always left'.
4. There are situations in which group solidarity is cemented by a common experience of adversity and opposition to mainstream society. In these instances, individual success stories undermine group cohesion because the latter is precisely grounded on the alleged impossibility of such occurrences. The result is downward levelling norms that operate to keep members of a downtrodden group in place and force the more ambitious to escape from it.

I would like to add another negative effect:

5. Depending on the standards of internal resources followed by organizers and handlers of different networks, the social value of social resources may tend towards negative. For example, it is frequently effective to use internal standards for instigating behaviours not conducive to social stability, including the idea to solve social problems through violence, even to seek interests of a minority or minority groups through illegal means, etc. Once the core layer of a social organization is manipulated by those who violate the public interest and social morality, or when the inherent characteristics of forming social networks are based on anti-social goals, it the will play a disruptive role on the community. The extreme examples include sectarianism, racism, criminal syndicates, organized crime, Mafia and heresy organizations.

The attitude to the negative effect of social capital has impact on the basic system of public associations and social organizations of a country. For instance, in view of a policy of 'social stability first', a strict restrictive policy for developing a social community or NGOs has been taken in China.

5. Social Resources

According to the definition in this book, social technology is 'the process of technology to carry on the development and application of various social resources'. Namely, social resources cannot create value automatically, only the application of social technology can make social resources create or embody value. Then, what are social resources?

As to social resources, many western scholars conduct research from the perspective of social capital: Pierre Bourdieu's, *Le capital social* (1980), James Coleman's *Social Capital in the Creation of Human Capital* (1988), Robert Putnam's *Bowling Alone* (1995), etc. The book *Knowledge and Social Capital* (2000), compiled by Eric L. Lesser, reviews and summarizes the research achievements of typical contemporary social capital. Lesser indicates that, according to the Merriam-Webster Dictionary, the term 'capital' refers to 'accumulated wealth, especially as used to produce more wealth', hence the notion of social capital is maintained.

However, there is a confusing array of definitions of social capital in the literature of political science, sociology and economics. For instance, after analyzing the definitions of eighteen scholars, Paul S. Adler states, 'social capital is a resource for individual and collective actors created by the configuration and content of the network of their more or less durable relations' (Adler & Kwon 2000).

Lesser, on the other hand, believes the following:

Social capital is the benefits or wealth that exists because of an individual's social relationships. Within those social relationships, there are three primary dimensions that influence the development of those mutual benefits: the structure of the relationships, the interpersonal dynamics that exist within the structure and the common context and language held by individuals in the structure. (Lesser 2000)

Regarding the resource of social capital, there is considerable confusion between the sociological literature, the political science literature, the development economics literature and the organizational research literature. For instance, experts in organizational science believe that social capital is composed of three dimensions: 'under the structural dimension, they list network ties, network configuration and appropriable organization; the relational dimension includes trust, norms, obligations and identification; the cognitive dimension includes shared codes, language and narratives. Adler concludes by emphasizing that the resources of social capital comprise networks, norms, beliefs, formal systems and institutions (ibid.: 97).

According to the definition provided in this book and also for the convenience of researching social industries, we will examine social technology from the perspective of social resources, which is divided roughly into two domains: internal and external.

Since the essence of social resources is the various personal and collective relations, so the various networks among individuals and groups, as well as various social organizations or social entities, can be treated as 'external social resources'. These resources include schools, social groups, communities, associations, chambers of commerce, academies, hospitals, political organizations (governments, congresses and courts, etc.), NGOs (non-governmental organizations), enterprises, families and public resources, e.g. public facilities.

'Internal social resources' are generated by internal factors that, in turn, make it possible for social networks and organizations to be maintained. For example, internal factors such as trust between people within the organization could be generated or enhanced by various organizational activities, and they are also the basis for the formation of all kinds of social organizational activities and networks. It is the same with common beliefs and values. Therefore, the following are treated as internal social resources: institutions (including social standards and laws); beliefs; values; morality; customs; all kinds of desires of people (the desire to work, the fulfilment of their ideals or dreams, the enjoyment of life, a high quality of life, peace, security, trust and an environment that is stable and friendly, etc.); and human emotions.

We will further elaborate the role of social technology through taking the research institute, virtual technology and public relations technology as the examples.

6. The Mechanism of Research Institutes

From the perspective of the development and utilization of social resources, the mechanism by which research institutes conduct their work is one kind of social technology to organize the resources of R&D, and is also one of the most successful examples of organizational technology in history. Just imagine how Einstein invented so many technologies.

Before the nineteenth century, scientific research was the personal behaviour of scientists based on the interest of the nobles; therefore, R&D always depended on independent personal conception and invention. The great scientific achievements in this era as well as many inventions and technological innovations were basically personal achievements of genius scientists and great inventors (e.g., James Watt, Thomas Edison). In 1863 Alfred Krupp, the German steel magnate, set up the first chemistry lab in the world (von Braun [1997] 1999). In 1876 Edison set up his lab and decreased the habit of scientists engaging in research all alone; he organized a group of specialized talents, allocated tasks himself and devoted the group to one invention collaboratively, which helped him to obtain the patent rights for 1093 inventions (*CEB* 1985a: 256). This ushered in the new age for scientific research.

At the end of the nineteenth century and the beginning of the twentieth century, big companies in Europe established a number of industrial R&D laboratories (e.g. the Siemens Company in Germany) fashioned after the practice of Edison. Later on, the Dupont Company, AT&T, Westinghouse, Kodak and Standard Oil set up R&D institutes. In fact, the research institutes that appeared in industrial companies came about as a result of the growth of the above companies' personal workshops and testing places after inventors like Siemens and

Edison became industrialists. Industrial research in the strict sense began after the 1920s. Between 1920 and 1960 research results from industries in industrial developed countries increased nearly a hundred times: 80 per cent of the industrial scientific technologies were controlled by research departments of the above monopolized enterprises and most researchers belonged in the research institutes associated with the industrial and military sectors (Bernal [1965] 1970).

As science and technology was increasingly used for military purposes, governments of various countries invested more in science and technology and interfered more directly in the conduct of science and technology research, and this was one of the reasons why national research institutes were established. German physical engineering institutes were the first national research institutes to be established, commencing in 1887. After they were established, many governments worldwide set up a number of state-sponsored research institutes based on variants of the German model.

Research institute mechanism helped build the interaction of science and technology and social development. They have played a key role in the industrialization of the great results of scientific and technological achievements over the past several hundred years, thereby speeding up the process of economic development in those countries involved in the game.

At the end of the twentieth century, in order to adapt to in the environment of globalization and trends of open innovation, virtual technology was introduced in the organizational innovation of research institute. As to the virtual institute, we will probe into the details in the next section.

7. *Virtual Technology and Organizational Innovation*

Virtual technology is now one of the most extensively and successfully used types of soft technology. The term 'virtual' is derived from the notions of the virtual machine and the virtual address that are widely used in the field of computing. For example, virtual corporations, virtual institutes, virtual banks, virtual offices, virtual stores, virtual manufacturing systems, virtual purchasing, virtual universities and virtual scientific parks, etc. are just some of the possibilities. They presage a bright future for the development of virtual organizations.

1. New Environment and Organizational Innovation

Since the world has entered the 'Knowledge Age', information technology is gradually turning the world into an information highway network. In addition, the development of modern transportation technology has greatly accelerated the international flow of information, capital and commodities, and has sped up the rate of economic globalization. At the same time, the cycle of technological development and commercialization has become shorter and shorter. Consumers have more choices in the market so that world market competition is consequently becoming fierce. The pressures of external market competition increasingly influences the internal affairs of organizations in fields such as technological innovation, the concept and meaning of working and the distribution of profits after the application of new technology, including new management technology. In the meantime, companies' technology innovation, the efficiency of technology transfer and the cycle of technology development depend upon how well these organizational capabilities are coupled with appropriate organizational structures, flexibility, knowledge management and personnel management (including internal possessor of technological patents, senior managers, entrepreneurship, etc.). The ability of an organization directly affects its competitiveness. Organization itself has become an important factor in competitiveness. In such environment, traditional organizational systems have been proven to be increasingly incompatible with competitiveness.

Now, organizational reform is the top item in the world's new agenda as it faces the reality of the fierce competition that characterizes the new society, the economic environment and the new international context. We need to reconsider the definition of 'organization' and to discuss the following question: what kind of organizations are best able to adapt to changing market conditions and are most suitable for survival and development in a competitive environment?

Therefore, on the one hand, the management academe supports the organizational innovation in theory through the centralized research on organization, the regularity of human psychology and behaviours in the organization. For instance, as early as 1938 Chester I. Barnard had conducted the in-depth research on the essence of the organization, organizational characteristics, organizational management, formal organizations and informal organizations, etc (Barnard [1938] 2007). Now, 'Organization Behaviour' has become a complete theoretical system, thereby emerging a large number of excellent organizational behaviourists. On the other hand, the business circle has been continuing to develop the means and methods (organizational technology) by which organizational resources may be restructured and resource allocation may be optimized through practices, thereby improving organizational efficiency.

Organizational modes, such as the straight-line organization, functional organization, straight-line plus staff organization, project organization, matrix organization and fishing net organizations, are well-known types of internal organizational technologies. Enterprise merger techniques are examples of external organizational technology. No matter from which country they come, enterprises in today's world are forced to experiment with a wide variety of organizational innovations with a variety of labels including: down-sizing; restructuring; re-engineering; changes in scope, scale and level, to absorb external resources; virtual organization; new forms of inter-organizational cooperation; and various types of mergers and corporate annexations. All of the above phenomena reveal the contemporary importance of organizational innovation.

Virtual organization, which is widely promoted now, is a new mode of organizations which meet the character of a new era.

Let us now take the virtual corporation and virtual institute as example, and further probe into the significance, trends and issues faced by virtual technology.

2. Virtual Corporation

Virtual corporations have attracted increasing attention as a comparatively mature example of the application of virtual technology. In 1991, Dr. K. Preiss, Steven Goldman and Roger Nagel produced a report entitled, '21st Century Manufacturing Enterprise Study: An Industry-led View', in which the role of the virtual technology in manufacturing enterprise was proposed.[1] The 1992 monograph by William Davidow and Michael Malone, entitled *The Virtual Corporation: Structuring and Revitalizing the Corporation for the 21st Century*, further enriched our understanding of virtual corporations (Davidow & Malone 1992).

In March 1994, Michael Malone and Bill Davidow wrote an article entitled 'Welcome to the Age of Virtual Corporations':

> The Virtual Corporation, a mere theory on a scrap of paper three years ago, has now become a common phrase in daily business life. [...] Similar corporations temporarily join together into meta-enterprises. Manufacturers, suppliers, distributors, and even customers are linking together in enduring relationships built on mutual trust. [...] The Virtual Revolution is the defining business transformation of our generation. (Malone & Davidow 1994)

The Taiwan *Economic Daily* has described virtual corporations as being able to fulfil functions such as production, sales, design and finance, etc. but without having internal organizational capabilities to carry out these functions. That is to say, the virtual enterprise retains only its necessary key functions and relinquishes or 'virtualizes' the remaining functions, owing to its limited resources or insufficient competitiveness in those functions. The virtual company must use every means at its disposal to borrow strength from the outside to increase its competitive edge. Whatever form it may take, the guiding principles of the virtual corporation are that it breaks past the conventional limitations of enterprises and extends its scope by using a strategy called 'integration of external resources'. In general, the virtual corporation is a type of dynamic union of enterprises for responding quickly to market changes.

In fact, Englishman David J. Skyrme pointed out the necessity of 'virtual working' in modern societies earlier in 1988, and he published an article entitled 'Virtual Teaming and Virtual Organizations – 25 Principles of Proven Practice' in 1997. However, the virtual corporation has a longer history in a sense. Over the past several decades, the Japanese type of virtual production and virtual organization has been widely applied in some Japanese enterprise groups, the manufacturing industry, and with cooperation between large enterprises and SMEs. Perhaps this is a success experience in Japanese enterprise management that has not been adequately summarized. The operation of some big engineering projects in China also has the characteristic of virtual organization, but they are only organized and managed by the administrative way and have not adequately reflected the advantages that contemporary virtual organization expected.

In short, the development and application of virtual technology has paved the way for optimizing combinations of resources by corporations. For instance, virtual cooperation breaks the boundaries of traditional enterprises and makes use of the external resources, including overseas resources. By virtualizing the functions or resources that it does not possess or in which it has few advantages a corporation is able to generate additional competitive advantage for itself. It is a good way for enterprises to avoid becoming 'large and complete', or 'small and complete', and to share resources and complement the superiority qualities that each possesses. It may be wise for small and medium-sized enterprises with weak R&D capabilities to take advantage of virtual research institutes, virtual manufacturing systems and global manufacturing networks.

3. Virtual Institute

International competition today is centred on economic competition, and economic competition is dominated by competition in technology. Virtually all governments in the world realize that only by strengthening technological innovation, possessing their own intellectual property and grasping high-tech resources can they take the initiative in international economic competition. Technologically advanced countries therefore adopt various policies to protect their intellectual property rights and to achieve market monopolies through technology monopolies. R&D institutes – as the main agencies of producing, accumulating and spreading knowledge and technology – hold the primary responsibility for providing technological sources for technological innovation. As a result, the capability and efficiency of research institutes affect directly the comprehensive competitiveness for a nation. In these circumstances, the significance of studying future-oriented research institutes, including its best structure, management and functions, as well as a healthy research atmosphere and a new culture for research, go far beyond the organizational innovation itself.

The Virtual Institute and Virtual Research Center will completely challenge the definitions, boundaries and forms of research organizations and bring about a revolution to research organizations. Research institutes are a type of collective in which research activities are organized according to certain research goals and systems. They should provide an environment in which researchers can make best use of their individual creativity and collective wisdom.

In a virtual research institute, various research resources and functions are organized into a new 'flexible institute', focusing on certain research goals and contents involving close cooperation across time and space. The virtual institute can involve experts from different fields, different communities and different departments, and employ researchers from all over the world in addition to entrepreneurs, government officials and even users from different fields. The institute can also operate regionally and internationally. Its research scope and scale can be adjusted according to demand.

The virtual institute transcends the boundaries of traditional research institutes. It can infiltrate and extend the functions of different institutions. It can integrate a large quantity of internal and external resources. Theoretically, there are no limits to the scope, resources and options of a virtual institute. The virtual institute is thus able to achieve organizational flexibility at a low cost. Functions, other than key research functions such as supply support, can be relegated to outside organizations for greatest efficiency.

Though the work mode of the virtual institute is somewhat similar to that of conventional project groups, the virtual institute itself is more compact and retains the same functions of an institute. It is not a temporary collective group that will be disbanded upon completion of the project; it is a union where researchers trust each other and have a common understanding of the goals; they share common knowledge and working conditions. All this is directed at achieving high levels of success in long-term research by making optimal use of limited resources.

In short, the purposes of the virtual institute are as follows:

- To organize research resources in such a way that the institute can deal appropriately with change at a minimum cost and with maximum speed.
- To attract researchers by offering them mutual exchange, flexibility, the integration of wisdom, new concepts of institutional culture, open policies and glamorous research goals.
- To promote the principle of combining team spirit with individual creativity so as to increase competitive edge and create new conditions for innovation.

Background and Significance of the Virtual Research Institute Summary

- **The age of collaborative innovation.** In the international and information era of the twenty-first century the costs of creativity and of adopting new technology are rising. Uncertainty and high risk are important factors in the research processes of science and technology. Promoting pre-competition technological cooperation has become a necessary means by which companies may reduce the risks and costs of their R&D. As Debra Amidon has commented, the twenty-first century is the age of fifth R&D generation and the core strategy of this generation is to take knowledge as an asset to carry out collaborative innovation through interdisciplinary learning and knowledge flow (Amidon 1997).
- **An environment that integrates intellectual endeavour through various fields of cooperative and interdisciplinary research.** Alongside its increasing complexity, the increasing integration of fields is a defining characteristic of contemporary technological development. All parties involved in any important invention or creation must rely on the integration of different technologies and upon interdisciplinary cooperation. Technological integration or knowledge integration involves not only technology, however, but also the integration of minds. Virtual institutes gather people from different fields and technological backgrounds and organize teams according to specific goals and systems to create superior environments for interdisciplinary research and for the integration of knowledge and intellect.
- **The age of breaking away from the boundaries of the old research system.** In a knowledge society, the differences between basic and applied research, and between science and

technology, tend to blur. The distance from research, development and application to the market has become considerably shorter than was the case in the past, and the different stages of technology development have become harder to distinguish from each other.

The traditional division of institutes by academic disciplines and the rigid separation of research, development and production from commercialization are now obsolete. Virtual institutes combine research, production and application into one system. They also enable mutual exchange and mutual extension of activities between their members; they break the boundaries between systems, encourage the development of new ideas and make it easier to overcome the obstacles to knowledge transfer between traditional organizations. By these means they speed up the application and dissemination of new technology.

- **A 'flexible organization' that facilitates the flexibility.** One of the problems of traditional organizations is their inability to make quick decisions in the face of ever-changing markets and opportunities. An ideal organization should be able to readily respond to surrounding changes. However, because modern technology develops so quickly, it is impossible for new enterprises to be created for each change in demand or products. It is not feasible to continuously establish new institutes to adapt to ever-changing research requirements created by the emergence of new topics and concepts. The core of an organization is people. Unreasonable organizational adjustments and personnel exchange usually results in huge, invisible costs. It is also very difficult to quickly transform an existing organization into a larger or more comprehensive one. This problem is compounded by the fact that the bigger an organization becomes, the weaker its ability to deal with change becomes.

Because a virtual institute is not a hierarchy, its reorganization does not affect the social status or welfare benefits of its researchers. The establishment and dissolution of a virtual institute does not follow the traditional procedures for determining management so the 'human' obstacles become fewer than is the case in traditional organizations. A virtual institute can accordingly make swift changes at minimal cost to changes in external demands, and it can increase or reduce staff according to the task at hand, guaranteeing a free flow in its activities and achieving the best distribution of its resources.

- **A collective body that encourages people to learn from each other and share common knowledge.** In modern times the amount of extant knowledge increases rapidly, as does the need for people and organizations to keep up to date with new knowledge. An increasing number of jobs require knowledge as their foundation. The worker who relies only on what he or she learned in school will be left behind. For employee training and the updating of knowledge, we must rely on on-the-job training and actual work experience. An employee's ability to apply new knowledge cannot be determined just by looking at his or her age, education, qualifications, title, rank or position. Virtual institutes encourage people to share their common culture and to be open to the entire society and the world.

Virtual institutes provide a basic infrastructure for ongoing study and research. They may become an organizational base for retraining, a place where people share knowledge and a medium for life-long learning. Work no longer means only the means by which people make a living and a place where people make that living. Rather, work is an important vehicle by which people may seek greater self-realization and pursue happiness. Updating knowledge is expensive in rigid organizations and, as a result, stagnation through 'inbreeding' easily happens.

- **A network-style parallel organization and the new concept of 'leadership'.** One of the greatest dangers facing an organization with the passing of time and with the growth of its achievements is that the expansion of authoritarian leadership proceeds to such an extent that the organization gradually becomes a bureaucracy. In such cases leaders resist democratic initiatives, preferring instead to make autocratic decisions. The core of modern organizational reform is to simplify management procedures; that is, reduce hierarchy, shorten information channels and quicken the pace of decision-making. A virtual institute is a network-style parallel organization. As the most basic research unit, it no longer contains any internal hierarchy, enabling it to prevent deterioration and bureaucracy. The leaders of research institutes should therefore change the emphasis of their leadership style from that of a 'director' to that of a 'facilitator, coach, mentor, advisor and indeed a peer in the exchange of knowledge and experience' (Skyrme 1997).
- **As a combination of a relaxing environment and team spirit, leadership must not only give rein to individual creativity but also insist on team spirit.** As society develops and people's ideas change, the connotation of this principle evolves. This is the eternal subject for organizational innovation. For research work, it is not sufficient only to have strict discipline. As in composing music, a free environment is necessary for enthusiastic thinking, as well as a concentration of energy to complete the task. Researchers in virtual institutes avoid the binding of traditional organizations. They are not protected or pressured by others. They will discover their own ability in a completely new environment and find a place suitable for themselves. However, the process of technological innovation becomes more and more complex; any major invention or creation cannot be removed from the team. A virtual institute is not a paradise for individualism; it survives on team spirit. The purpose of this type of institute is to seek a balance between the close alliance and the relaxant environment to fulfil the task that no individual can fulfil alone. As David Skyrme put it, 'It is an outcome that organizations today try to take a balance of the highly innovative and the tightly coordinated' (ibid.). As for the form of team activities, besides a face-to-face exchange, many depend on advanced communications technology to achieve the goals of research without regard to time or place.

In brief, a virtual research institute meets the needs of organizational innovation and a trend of social development.

Since China commenced carrying out it policy of openness and reform, there have been many brave attempts to seek new ways to organize and manage research projects. We can divide these attempts into two basic types.

First, there are interdisciplinary research organizations and academic associations for which changes in administration must be led by government or must involve intervention by government. When their task is fulfilled, the group is dismissed, or if the result is a success and the scale becomes large, these institutes are often contaminated by the virus of big companies, big units or big institutes. For instance, once the institute has been approved by the pertinent government administrative organizations, the leader of the institute is appointed by the administration. Soon, hierarchical management departments and administrate structures emerge. The new institute then gradually adopts the mode of traditional institutes.

Second, there are a variety of civil or private institutes in the broader society and there are a number of independent research bodies and associations backed by certain universities or institutes. Many of them have the features of a virtual institute. As the institutes pass through their initial stages of growth they often become hindered by traditional concepts, and their further development process becomes difficult. However, the virtual institute is becoming a new force in research and is winning more and more attention.

There are still many issues facing China for the further development of virtual institutes:

- It should be recognized as an innovative organizational model by whole society.
- The nature of research institutions is no longer an authority.
- The leadership position of the research institutes of grass roots is no longer a commander.
- Knowledge itself has won the recognition and respect from society that it deserves, and it receives rewards appropriate to the value.
- The level of the social welfare and the socialization of logistics system should be comparatively high.
- The 'informatization' of society has enabled convenient circumstances (in both time and expenses) under which people may work efficiently in different locations.
- Standardization has been established to a certain extent for people holding concurrent jobs.
- The evaluating system of professional titles has been reformed.
- Objective and equitable standards have been implemented for evaluating research achievement (especially the results of soft sciences and soft technology).
- Virtual institutes should maintain long-term and healthy relations with traditional research institutes.

It is obvious that there is still a long way to go for the development of virtual research institutes in China.

In short, the formation and development of virtual organization is an unstoppable trend of social progress and an organizational revolution raised by technological revolutions. However, it will only be able to realize their true potential when the society has matured in terms of its 'informatization' and network level of economy and society; in terms of socialization levels of social security and welfare functions; as well as the conceptual change of the meaning about work and life; and, moreover, most people understand that it is imperative to change the organization – the implementation system in which they live and work – in order for virtual organizations to be accepted as a popular and official form of organization.

8. *Public Relations Technology*

PR technology belongs to relationship technology. Relationship technology regulates, channels and deals with all kinds of relationships and provides the norms that guide the relations between technology and technology, equipment and people, and people and people. It incorporates the norms of international relations, interpersonal relations, diplomatic relations, family relations, conjugal relations, etc. All these technologies are directly related to morality, social ethos, cultural background, habits and knowledge levels, and are therefore regionally specific.

Among relationship technology, public relations have already become a kind of discipline theory, thereby promoting PR technology to bloom into a specialized technology and industry. PR technology can be summarized as 'the means and approaches of changing human behaviours through influencing and affecting their psychologies'. It is applicable in many fields, ranging from business operations of enterprise to the political life of a country, from commercial activities to social activities, and even to dealing with international political relations. Therefore, PR technology has evolved from commercial technology to social technology and intellectual technology, including sales promotion technology (sales promotion of products, services, people and policies) and lobbying techniques in politics, etc., and the business scope and scale of PR industry is also growing.

Encyclopaedia Britannica defined 'public relations' (PR) as the 'communication, or the extension of good will, of an organization that seeks to draw attention from a public that is already, or probably, interested in the institution' (*CEB* 1985b: 422).

Modern PR originated in America at the end of the nineteenth century and the beginning of the twentieth century. American attorney Dorman Eaton gave a speech entitled 'Obligations of Public Relations and the Law Occupation' at Yale University in 1882 in which he applied the concept of PR in the real sense for the first time. In addition, the term PR first appeared in 1897 with the publication of America's *The Yearbook of Railway Documents*.

In 1889, George Westinghouse, the owner of American George Westinghouse Electric Appliance Company, employed a reporter from Pittsburgh by the name of E. H. Heinrichs to help keep people from using direct current. This company was the first to use PR.

In 1903, Ivy L. Lee and George Michaels established the first PR consultation office in the world in New York, and one year later, with George Parker as a new partner, they opened a PR company in the same city.

In 1908, the official position of PR Manager was established at AT&T. T. N. Vail, an expert in news public relations, was the first holder of the position. AT&T used PR advertisements to promote the image of the company and the need for a full PR department was later established. Later, many other businesses in America followed AT&T's footsteps and established their own PR departments.

Crystallizing Public Opinion, the first PR book, written by Edward L. Bernays, appeared in 1923. Dr Rex F. Harlow, a famous American PR scholar, opened the first PR course at Stanford University in 1937. The first PR degree, bachelors or masters, was available through Boston University in 1947. In 1948, the PRSA (Public Relations Society of America) was founded, which established the regulations of association and rules for publicists.

PR spread to Britain in 1920 and by 1948 the Public Relations Association of Britain was established in London; since then it has grown considerably. London's Public Relations Association has more than 2000 members in over fifty countries and regions and is the biggest professional PR organization in the world. PR spread to Canada in 1940 and by 1947 the first PR association was established there.

After World War II, PR made new progress in theory and practice with the recovery and improvement of the world economic order. Members of the war information office in the United States turned to business after the war and became active PR experts. Thus, PR experts began entering management positions in industry and the PR field became referred to as one of the four pillars of modern enterprise together with funds, equipment and talents. PR experts participated in building relationships between America and other countries. Furthermore, PR now plays a role in politics and domestic presidential elections.

In 1955, the International Public Relations Association (IPRA) was established in London. More than sixty countries have become active members (Wang & Zhang 1989: 10–11). The participating members, signalling that public relations technology had begun to institutionalize internationally, adopted the international rules of conduct of PR in 1961, and then adopted the Athens Regulations on PR in 1965.

PR activities began to appear in the coastal regions of South China in 1982 (W. Gao 1998: 27–29).

According to the statistics provided by the American *Fortune* magazine in 1980, 436 out of the 500 largest enterprises had established special PR organizations. America's PR field contains over 2000 PR advisors, 20,000 PR managers and 140,000 employees in 1990s (Qin & Zhou 1995: 49–54). Over 300 universities in America have added PR courses to their curricula.

Carl Byior and Associates was established in 1930 in New York, and is still the largest multinational PR company in the world. In 1997, Byior provided a new definition of PR: 'PR technology is in fact perception management; through managing the public views of things, enterprises and individuals, PR changes the public's behaviors and decisions and is finally gaining acknowledgement' (Yifei Li 1999). Many large enterprises established high-level PR positions such as CPO (Chief Perception Officer) after establishing CIOs (Chief Information Officers) and CFOs (Chief Financial Officers).

The International Public Relations Association (IPRA) World Congress is the world's largest professional conference of public relations, which has been held in different countries every three years since 1958.

C. Soft Technology and Thrice Industrial Revolutions

For hundreds of years the Industrial Revolution has made unparalleled contributions to the material civilization of human society. However, there exists wide misunderstanding about the driving force of the Industrial Revolution. A general understanding of the Industrial Revolution is that due to a major breakthrough in science and technology making significant changes in industrial structure, it enabled various aspects of economy and society to appear brand-new. Thus, the descriptions of the thrice Industrial Revolution are as follows: the first Industrial Revolution is characterized by cotton textile technology; the second one is symbolized by electronic technology; and the characteristic of the third one is information technology. These descriptions paid great attention to the breakthroughs of science and technology but neglected the force that ultimately impelled the emergence of the Industrial Revolution – soft technology.

Take the first Industrial Revolution as an example. The Industrial Revolution in Britain first began in the cotton textile industry. At that time, India is the first cotton industry power in the world. The competition between British and Indian textiles led to a series of technological inventions and innovations such as the 'flying shuttle' invented by John Kay in 1733; the 'spinning jenny' invented by James Hargreaves in 1765; the 'water frame' invented by Richard Arkwright in 1769; the 'spinning mule' invented by Samuel Crompton in 1779; the 'power loom' invented by Pastor Edmund Cartwright in 1785, and so on. The contradictions of spinning and weaving in the United Kingdom were mitigated and, as a consequence, the textile industry rose rapidly.

The extending fibre industry and its consumption caused a great shortage of power and transportation supply. The invention of the steam engine by James Watt in 1769 satisfied these demands. The application of the steam engine furthered the development of energy and transportation technologies; the invention of steam trains and steamships, facilitated the expansion of the Industrial Revolution into the fields of machinery and steel.

First, the reason why so many technological inventions and innovations took place in Britain lies with the Britannic soft environment of the time – the innovative culture (see the section of 'Making soft-tech institutionalization and "mechanism-ization" keep pace with technological innovation as well as socio-economic development' in Chapter 4), government policy on technology transfer and industry protection, including loose immigrant policy and religion policy (see the section of 'Patent Technology' in this chapter). In addition, the Royal Academy, established in Britain in 1662, promoted scientific and technological research for the entire nation. Many excellent scientists came to the fore under the leadership of Newton. Thus, Britain became the centre of the Science and Technology Revolution in the seventeenth century. Moreover, positive conditions were cultivated for Britain's Industrial Revolution.

Second, the elementary circulation network, which integrated the consumption market and the supply of raw materials, was first accomplished in Britain in the late eighteenth century. This helped Britain efficiently import raw materials from other countries and export textile products to the entire world. Simultaneously, Britain established the most advanced logistics system worldwide to support international trade through advantages in marine transportation, international remittance and finance (see the section of 'Soft Technology and Institutional Innovation' in Chapter 4).

Evidently, not only textile and steam engine technologies prompted the first Industrial Revolution. The global economic order that ensued under the auspices of the British Empire was made feasible because of the advantages enjoyed by Britain in its soft environment (culture, institutions and policy, etc.) which favoured the introduction of advanced technology and talents from abroad, and favoured technological invention and innovation. It also favoured the creation and exploitation of the advantages of soft technology through vehicles such as the patent system, the international circulation system, financial institutions and logistics systems, etc. – thus enabling Britain to become the centre of first Industrial Revolution.

Consider the contribution made by soft technology and soft environment to the second Industrial Revolution. It is commonly believed that revolutions in energy technology and transportation technology brought about the second Industrial Revolution which focused on electricity, organic chemistry and the internal combustion engine. It is thought-provoking that the second Industrial Revolution was no longer centred in Britain. Commercial success resulting from a monopoly of direct trade with India and other countries gradually caused Britain to ignore science, technology and the patent system, as well as the protection of scientific and technological talents. The inventor of synthetic dye did not receive attention in Britain but was invited to establish a chemistry laboratory in Germany, which laid the foundation for the German chemistry industry that strengthened over time. In addition, during this period Britain did not pay attention to taking their advanced technologies as the direct investments in other countries. This was unfortunate for Britain because there were many excellent intellectuals who were enthusiastic about the financial industry and overseas securities investments. In this period, business was everything. However, in the meantime, Germany and America absorbed advanced technologies introduced from Britain and accelerated their speed of development. America and Germany surpassed Britain in the occupancy ratio of industry in 1881 and 1906.

The example of America is instructive. With rich natural resources and a small population Americans paid more attention to soft technology in the hope that it would improve efficiency and gains to scale. Taylor's production mode and Ford's batch production technology are examples of the approaches that ensued. Moreover, because America had been essentially a set of British colonies for a long time, American patent law, which was promulgated in 1790, imitated Britain's law. In 1836, America's patent laws underwent many changes. The United States established a formal patent bureau that examined independent inventions and foreign issued patents. The patent auditing system first began in America. Further alterations to the patent system were made in 1870, greatly increasing the value of patents and making the patent system an astonishing facilitator of development in American science and technology. At the same time, many research institutes specializing in science and technology R&D were established, increasing the quantity of technological inventions and the efficiency of technology transfer.

To summarize, the secret of the second Industrial Revolution, the centre of which shifted to America, lay mainly in advanced soft technologies such as modern management technology, the institutionalization of R&D, a flexible immigration policy, the reformation of the patent system, the new application of stock technology (see the section of 'Stock Technology and Securities Technology' in this chapter) and rich natural resources. Without such innovations in mechanisms, law, institutions and policy supporting economic reform, or without management technology, organizational technology and production technology raising the efficiency of technology transfer, the market throughout the American continent would not have appeared.

Let us now make some observations about the third Industrial Revolution. As it entered the 1990s, the United States experienced as long as 110 months of steady and forceful economic growth characterized by low inflation, low unemployment rates, low financial deficits and prosperous science and technology stocks, leading to what some economists have called the 'new economy' (it was just an economic phenomenon). On the other hand, people involved in science and technology have been paying attention to the contribution of the IT Revolution and high-tech industries to world economic development and globalization. Certainly, the development and wide application of high technology taking the lead by IT has played a key role in the persistent economic growth of America. Therefore, the new economy has been called the network economy, the digital economy or the knowledge economy, while the industrial part of the economy, focused on product manufacturing and marketing, is referred to as the 'old economy' in spite of the multitude of high technologies that it employs. Events of recent years have demonstrated that the 'new economy' cannot escape the basic laws of economic life. The American economist and Nobel Prize winner Milton Friedman has heralded this theme by stating that the 'economy is still "old" and that it is running according to proven economic rules' ('Interview with Milton Friedman' 2001). However, the present development trend in the economy has raised far-reaching challenges to traditional economic viewpoints and traditional management modes.

No matter what the above phenomenon should be called, it is clear that we are experiencing a change similar to that of the Industrial Revolutions at the end of eighteenth and the beginning of the nineteenth centuries. We are witnessing a great change at the end of twentieth century in human society from an industrial economy era to a service economy era. Some experts have already described this revolution as the 'third Industrial Revolution' caused by IT and the Internet, arguing that it is deeper and wider than the first two. What it brings to the world is not only the expansion of high-tech industries typified by information technologies, networks and telecommunications but also increases in the productive efficiency of all industries, stemming from the wide infiltration of high technologies and changes in the human living and working mode. More significantly, it facilitates a variety of institutional reforms involved in the function of governments, as well as the function and boundaries of nations.

As Alvin Toffler said, capital, currency, as well as the mode and essence of the labour force have changed. This series of transformations led eventually to substantial economic growth.

What, then, is the driving force of the third Industrial Revolution? What propels the breakthroughs in the development of information technology and biology technology? What helps to create new companies, new industries and new business models so that enterprises can organize their production in more effective ways? What drives the emergence of new products and new technologies and their worldwide application to push forward economic growth?

Just like the misunderstanding many people have had about the main driving forces of the first and second Industrial Revolutions, few people have noticed the driving force behind the third Industrial Revolution. That force is the soft technology wave of the late twentieth century.

Let us take the 'new economy' in the US as an example. On the surface we can see that the integration of micro-electronics, computers, telecommunication and Internet technology produces many new companies and new industries, helps to increase the productive efficiency of traditional industries and, in turn, quickens the social life tempo with dazzling new products that enrich people's lives.

However, if we look below the surface we can see soft-tech innovations, such as global management, transnational companies and transnational corporate annexations, venture capital technology, virtual organizing technology, etc.; together with complementary institutional and policy innovations, such as removing government controls, opening markets, solidifying financial markets and implementing new currency policies, etc. On one hand, they have promoted innovation in hard technology represented by information technology and biological technology (for example, Internet technology has been invented in the 1960s). On the other hand, they have also created conditions for innovation in traditional industries which enable more effective operations and continuous innovations in labour utilization, products and capital markets; and they have helped create global capital markets, global trade and global flows of technology and talent. With its economic strength and its influence in the world, the US enjoys the benefits of globalization, realized through optimal resource allocation globally, including human resources. Furthermore, soft-tech innovation and the rapid expansion of soft industries have quickened the adjustment and optimization of economic structure, assisting the transition from an industrial economy to a service economy.

Interestingly, the hard technologies that are serving as the driving forces for the third Industrial Revolution are themselves technologies derived from natural scientific knowledge and are gradually 'softening' (see Table 5).

Japan serves here as a sharp contrast with America but can match America in the field of hard technology. The Japanese economy was declining in a number of ways during the 1990s and faltered in its efforts to get on the right track with regard to the new economy. The key to Japan's decline was not a lack of high technology or a lack of intellectual property but rather mistakes it made in macro soft-tech operations; in particular, Japanese institutional innovation failed to keep pace with the times and with changes in the international environment. For example, the Japanese venture capital system is rather conservative and Japanese universities do not have a system encouraging innovation like those in contemporary America.

This section ends with an attempt to summarize concisely the history of our commercial technology. The first period in the commercial technology development lasted until the end of the eighteenth century. During this period such commercial technologies as accounting technology, banking technology, stock technology, logistics technology, circulation technology and patent technology had been developed, and it was these technologies that facilitated the birth of the first Industrial Revolution.

The second period in the development of commercial technology occurred at the end of the nineteenth century and the beginning of the twentieth. This was the period of the institutionalization phase of commercial technology. It was also the period in which the reformation of patent technology, R&D institutes, scientific management methods, mass production technology, securities market popularization, monopoly concerns, horizontal annexation techniques and the Anti-trust regulations developed.

	Time	Soft tech	Hard tech
The 1st Industrial revolution	The end of 18th century	Patent tech Circulation network of integration of consumer-market and raw materials World logistic system	Cotton textile tech (hard) Steam tech (hard)
The 2nd Industrial Revolution	The end of 19th century	Scientific management tech Research institute Patent tech innovation Stock market tech Monopoly enterprises system	Electronic tech (hard) Telecom tech (hard) Transportation tech (hard) Organic chemistry (hard)
The 3rd Industrial revolution	The end of 20th century	multinational management tech. Transnational annex tech. Venture capital tech Virtual tech Incubator tech New business model	IT (soft/hard) Network tech (soft/hard) Biology tech (soft/hard) Robot tech (hard/soft)

Table 5: Soft Technology and the Three Industrial Revolutions.

Figure 9: Waves of commercial technology.

The third period in commercial technology development took place in the 1950s and 1960s. During this period venture capital technology, modern management accounting, scientific management technologies, social technology and mega-merger techniques, among others, were developed. In addition, a variety of remarkable institutional innovations came into effect in a large number of countries during this period.

As the world entered the information era – in the 1980s and 90s – the scope of markets surpassed traditional limitations of time and space, current transaction modes were transformed thoroughly and thereby moved towards an age of overall innovation in soft technology. The fourth wave of commercial technology incarnated by the emergence of transnational management, total quality management, innovations in the stock market, transnational mergers technology, virtual organizational technology, e-business, incubator technology and modern logistics technology, etc., helped promote the third Industrial Revolution, as well as facilitate the age of the intellectual service economy (see Figure 9). The interaction between above soft technologies and information technology has not only helped sustainable economic development but also promoted the comprehensive informatization of society.

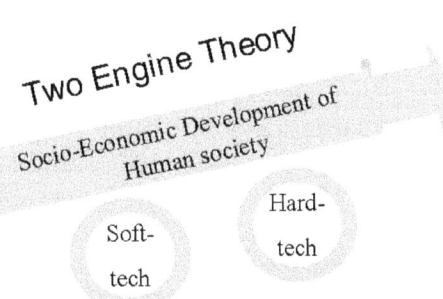

Figure 10: Two Engine Theory.

It is no exaggeration to say that there are two engines for driving human society: one is hard technology, another is soft technology.

It was unexpected that a global financial crisis would occur during late 2008. Despite the continuing rapid development of various hard technologies supporting the third Industrial Revolution, the financial crisis, caused by the errors of soft technology operations (innovation of financial tools) and the subsequent global economic recession, were not avoided. In fact, the culture of excessive consumption, loss of business ethics, mad pursuit of money and wealth, etc., were the underlying causes of the financial crisis as well as the rootstock for the crisis of today's capitalist financial system, the world management system and the world development mode. Thus, it is necessary to recur to soft-tech innovation from a global scale to the national level and its sound institutionalization in order to be redeemed from the present predicament.

The challenges we face (see the section of 'The Characteristics of Soft Technology are Suited for Coping with the Challenges in the Twenty-first Century' in Chapter 1) such as the widening gap between rich and poor, social polarization, unceasing regional conflicts, depletion of natural resources, degradation of natural and ecological environment, climate change and other human survival crisis – in particular, taking the financial crisis as an opportunity – are spurring an emergence of a huge historical change: human society is moving gradually from a civilized society with the pursuit of materialism toward a society with 'new civilization', that is, a society with coordinated development between material civilization, spiritual civilization and eco-environmental civilization. And I believe that this new civilization must be based on new Industrial Revolution: the building of a fourth Industrial Revolution.

In fact, today some transformational trends in values (Jin & Bai 2009), industrial structure, profit patterns, business modes, enterprise positioning, global technological change and thinking modes, etc., are urging the world to make essential changes to its economic growth mode, industrial structure, lifestyle, labour-style, working mode, the concept of rural and urban, and so on, and it indicates that human beings are entering a new era. The financial crisis has given us more specific information: the Industrial Revolution that will be take sustainable development as its main theme is inevitable.

As was the case with the first three Industrial Revolutions, the driving force of the next Industrial Revolution will also be hard technology (including new-type information technology, intelligent robots, nanotechnology, life technology, Internet technology, etc.) and soft technology (which supports future Industrial Revolution through the fifth wave of commercial technology, including innovation technique in a broad sense, innovation methods in global governance structure, globalized operating management technologies, new business model, green business model, 'Internet of Things' technology, 'softening' technology of manufacturing and 'softening' technology of agriculture, etc.).

However, the difference between it and the first three Industrial Revolutions (inevitable) are that the new economy or green economy will become the mainstream of economic activities (otherwise it will be just a continuation and deepening of the third Industrial Revolution), and the survival and development of industries will require the coordinated and sustainable development of the economy, society, environment, ecology and resources as prerequisites so as to make great strides towards the ideal of 'new civilization', which human beings have been longing for all the time.

Even though the fourth wave of Industrial Revolution will probably not take place until around the 2030s, most of the world – regardless of its level of economic development or social system and ideology – has already felt that the train towards the new economy has already left the station.

Now, human beings need a more profound conceptual revolution than at any time in the past.

D. Soft Science, Soft Series of Science & Technology (SSST) and Soft Technology

1. Research Course of Soft Science

With the continuous high-speed development of Japan's economy since World War II Japan is now facing many of the social tensions and problems that are now common in western countries, so that the research on social problems have been on the agenda. As to the study in these issues, Americans are using terms such as 'intelligence engineering' and 'policy science' in their research; however, Japan has instead invented the concept of 'soft science' and has conducted centralized research on social problems within the framework of soft science.

The Soft Science Seminar set up by the Planning Bureau of the Science and Technology Agency (STA) in 1970 is the starting point for the development of soft science in Japan (NISTEP 1988). The seminar's report points out that the 'research object of soft science is not limited to natural phenomena and technology and it includes activities pertaining to the human race, social affairs and knowledge. Therefore, soft science advocates the application of natural science methods, such as systems theory and information processing, in the solution of the above mentioned wide-spread comprehensive problems'. After the 1971 Soft

Science Seminar (of the STA), the Institute of Future Technology was entrusted to conduct specialized research (1971–1973) on 'the science and technology policy and the research and development system with Japanese characteristics'.

Researchers from the Institute of Future Technology travelled to the United State in 1971 to investigate research on social problems. As a result, the Institute published a series of research reports entitled, *Basic Design of Japanese Type Science and Technology Development System*, pointing out that soft science is the new trend in science and technology development: if soft science is not properly developed 'in the near future, the gap between Japan and the United States in soft science will result in major social problems' (The Institute for Future Technology 1973a). The Institute's reports also stressed the urgency of developing soft science in Japan. Firstly, it argued that Japan needed a different and more comprehensive scientific method if it wanted to solve the complex social problems associated with the environment, energy, regions, cities and transportation; secondly, it showed that developed industrial countries had already conducted R&D and application in this field; and thirdly, it concluded that 60 per cent of enterprises in Japan had already applied methods that could be considered soft science in these fields, but that the gap between Japan and advanced countries such as the United States was still large.

After the report from the Institute of Future Technology was examined and discussed at the Fifth Meeting of the Council for Science and Technology, high-level experts from all fields were required to conduct centralized research on the concept of soft science; its necessity, its characteristics and the fields requiring research and basic knowledge from the vantage point of soft science.

From 1971 to 1987, the STA treated the 'promotion of soft science' as an important R&D field and listed it in its *White Paper on Science and Technology* each year. To further promote soft science and training in the civil think tanks, STA listed it as a specific topic in the R&D plan of the Economic Planning Agency and Ministry of International Trade and Industry (Science & Technology Agency of Japan 1971–1987).

The Comprehensive Research and Development Organization was set up in Japan in 1974 to enable research in soft science to be combined with analysis of various policy problems. The basic understanding of soft science that prevailed at that time was as follows: soft science is a science based on information science, behaviour science, systems engineering and social engineering; the means for solving problems of soft science are mainly forecasting, planning, management and evaluation, etc. The primary characteristics of soft science were considered to be that: 1) its research objects were not only natural phenomena and science and technology but also issues that contained human and social factors; 2) that understanding the above issues from a systematic viewpoint and putting the emphasis of research on 'soft' intellectual technologies could help solve real problems; 3) that soft science organically combines a wide range of fields of knowledge and systematically synthesizes theories and methods that can contribute to the solving of different problems; and 4) that the basis and background of the discipline is information science, systems engineering, management science, behaviour science and social science.

The classification of soft science provided in the report, *Basic Design of Japanese Type Science and Technology Development System*, reflects the understanding of soft science that eventually dominated Japanese academic circles. The ten categories proposed in the report were: 'general' soft science; information soft science; energy soft science; material soft science; system soft science; environment soft science; behaviour soft science; policy soft science; life soft science; and others (IFT 1973b). Explanations of the above ten categories are below:

- *General* refers to general questions pertaining to soft science.
- *Information soft science* is divided into information patterns, information media, information processing, information theory (including semiotics theory), signal theory, forecasting theory, signal detection theory, automatic machine theory, learning theory and decision-making theory, etc.
- *Energy soft sciences* conduct research on energy in four areas: energy of the Earth; biology; technology; and society.
- *Material soft sciences* research materials within a social economic system, e.g. social economic problems, such as population problems, employment forecasting, family and consumption forecasting, urban planning, the development of the Earth and industrialization; technology economic problems, such as industrial forecasting, technological forecasting, products planning and market forecasting, as well as operational research.
- *System soft science* includes the research, design and application of systems.
- *Environment soft science* divides the environment into the physical, technological, economic and social environments.
- *Behaviour soft science* examines and deals with human beings as energy systems, which have the following basic characteristics: (human body) structure characteristics; (human) function characteristics; and (human and environment relations) environment characteristics.
- *Policy soft science* is based on social engineering, management engineering and futures engineering. It also applies to various methods that are beneficial to the understanding of phenomena, planning and control.
- *Life soft science* mainly addresses the problems of biological science, ecological science, medicine and pharmacology, etc.
- *Others* refers to other soft science problems not included in the above nine categories.

This report also listed major soft science R&D problems, such as the social effects of pollution countermeasures, urban problems, disaster prevention, transportation and communication, transaction, medical treatment, precautionary public security and consumer protection.

In 1977, the dominant understanding of soft science in Japan evolved. Soft science came to be viewed as comprehensive science and technology whose principal aim was to develop and apply theories, methods and tools that can explain and solve various complex problems we are facing, and further help decision-making processes to become scientific

(NISTEP 1988). Under this view, soft science would research and synthesize dynamic analytical frameworks, methodologies and skills in the fields of information science, systems engineering and management science, as well as the new theoretical models or knowledge in behaviour science and social science. Compared with the 1971 perspective, this definition removed the limitation of 'solving policy issues' and changed the framework from 'applying natural science methodology in human beings and the social system' to 'synthesize various achievements and knowledge in natural science, humanities and social science'.

The above reviews reflect that the research of Japan's academia in soft science went into considerable depth, but the top ten categories of soft science showed that the understanding of soft science at that time was still limited in applications and methodology and, moreover, it mixed technology (means) up with science.

The concept of soft science was officially introduced to China from Japan in the early 1980s. In July 1986, the State Science and Technology Committee hosted a forum on national soft science research. After that forum the development of soft science and the need to conduct systematic research on the meaning, objects, relevant disciplines and methodologies of soft science were promoted. For example, the theoretical method of soft science was divided into general theory (systems theory, information theory, cybernetics, dissipative structure theory, synergetics, mutations, indistinct mathematics, 'scienology' [the science of sciences], behaviour science, etc.), the methodology of systems science, planning and optimization methods, forecasting and evaluation methods, management and decision-making methods, simulation method, econometric methods and others. Particularly, a series of Oriental systems methodologies were developed, such as the Comprehensive Integrated System Methodology put forward by Xuesen Qian, the Processional Triangle Cycle of System Engineering proposed by Huanchen Wang and the System Methodology of 'Wuli-Shili-Renli' developed by Jifa Gu and Zhichang Zhu (Xu & Gu 2000). At the same time, there were corresponding soft science institutes established at both the local and national levels, ranging from government agencies to specialized institutes. The National Soft Science Research Planning Projects Fund was set up to encourage relevant research by the Ministry of Science and Technology.

Soft Science in China, the Chinese soft-science book series, reflects China's focus:

There are two main threads in soft science: one is the quantitative analysis and research containing Operational Research, systems engineering and techno-economics at the centre; the other is the study of the development strategy and the policy and management methods that are centered on scienology and futurology, etc. The characteristics of this science are that it follows systems thought and makes the decision-making method more scientific and democratic by integrating social science with natural science, quantitative analysis with qualitative analysis, working experience and technique with scientific method and means, and researchers with decision-makers. (S. Gan et. al 1989)

In other words, the goal of soft science research is to make decision-making more scientific and democratic; the main content is centring on methodology of strategies, decision-making, policy-making and management. Obviously, this kind of conclusion is not comprehensive enough.

However, China's soft science research and application have made great contributions to national economic and social development, taken its own place among national scientific and technological research, and The Association for Soft Science of China is active in related fields as a first-level academic community. Even after thirty years, soft science research has failed to distinguish the concept of soft science and soft technology; that is, the focus of its research and applications still emphasizes particularly on the methodology of social technology and the field of decision-making, and has therefore not broken out of the initial framework of American intelligence engineering and policy science, and, moreover, it has been unable to support the systematic study of soft technology.

2. Soft Series of Science & Technology and Soft Science

In the late 1980s the general rubric under which research on soft science in Japan was conducted was transferred to the Soft Series of Science and Technology (SSST). The main idea behind this shift was reflected in two reports both entitled *A Survey on the Present Situation and Future Trend of R&D in the Series of Soft Science and Technology* reported by NISTEP (National Institute of Science and Technology Policy, under the auspices of STA). In 1988 and 1989, the reports elaborated on the notion of SSST, its application techniques, the scientific fields it supported and its research system. It pointed out that the 'soft series of science and technology' is the corresponding notion of the 'hard series of science and technology', which was developed as natural science and engineering. The research on SSST enlarged the notion of science and technology, which in turn enabled the systematic study in this cutting-edge field and new methodological groups with practical significance.

The major themes discussed in the reports are described below:

1. Science and technology includes natural science and technology, humanities and social science and technology.
2. The basic disciplines of the soft series of science and technology are systems theory, information processing, cognitive science, behaviour science, organization science, management science and policy science.
3. The relevant disciplines include all fields of humanities and social sciences and natural sciences.
4. Soft industries that can be developed include the talent industry, information industry, education industry, the think-tank and consultation industries, etc.
5. The 'hardness' of technology means that it has the physical world as its operational object, which includes natural systems and man-made physical systems (like machinery); while the 'softness' means that its operational object is the 'representation world' presented by the psychological activities, thinking and actions of human beings, which includes

the man-made abstract system (objects abstracted by the human psychological activity process, such as information, knowledge, system, mode and notions) and the system of human activities (behaviours realized by internal processes rooted in human self-consciousness, such as services) (IFT 1973s).

Two particular facts signalled the prominence and respect that the SSST had garnered within Japanese society by the late 1980s. In 1992, the report, *A Survey on the Present Situation and Future Trend of R&D in the SSST*, which qua report entitled *On R&D Basic Project of SSST*, was submitted to the Council for Science and Technology under the name of the Prime Minister (Miyazawa Kiichi, who was then Chairman of the Council concurrently). In addition, *the White Paper on Science and Technology*, published annually by the STA, listed SSST – alongside the fields of information science, materials science, life science, space science, ocean science, geosciences and cutting edge basic technologies – as one of the eight major R&D fields every year during the decade from 1988 to 1998 (Science & Technology Agency of Japan 1988–1998).

Further insight as to the meaning and essence of the soft series of science and technology in Japan during that period may be found from four angles.

1. Projects

Reviews of the major research projects of SSST listed in *White Paper on Science and Technology* in 1998, as follows:

- *STA*: investigation of science, technology and human and social issues; study on R&D management (NISTEP); study on the thinking function (Institute of Physical and Research); and study on human features (Japan Atomic Energy Research Institute).
- *Environment Agency*: comprehensive research on the management and maintenance of the natural environment (trial research funds of national organizations for pollution prevention, etc.); research on Earth environment issues from the human and social perspective (comprehensive Earth environment research promotion expense).
- *Ministry of Finance*: research on 'sensory measurement' techniques for evaluation of wine quality; research on the improvement of wine production techniques (Research Institute for Brewing).
- *Monbusho*: software composition principles which are essential for development organizations (relevant universities).
- *Ministry of Agriculture, Forestry and Fisheries*: comprehensive research on the development of basic technology for suitable light labour agricultural practice (corresponding research on crops); evaluation technique and adaptive technology development for the health of agricultural, forest and aquatic products (general individual research); the quantitative evaluation of the rural living environment; and development of conditioning technology (National Research Institute of Agricultural Engineering).

- *Ministry of International Trade and Industry*: human media (industrial science and technology R&D); environmental system technology (institution) suited to human behaviour; measurement and evaluation technology for the care and behavioural characteristics of the aged (Research Institute of Life Engineering and Industrial Technology).
- *Ministry of Posts and Telecommunication*: research on the 'friendly communication society' (Communications Research Laboratory, communication and broadcasting organization).
- *Ministry of Transportation*: research on physiological aspects of human beings driving automobiles; the technology of preventing man-made mistakes; logistics structure analysis; and Logistics Information System necessary for their new system (Ship Technology Research Institute).

2. Status and Achievement

Reviews and forecasts of fields and issues considered to be at the forefront of research (which were listed in the reports *A Survey on Present Situation and Future Trend of R&D of Series of Soft Science and Technology* in 1989). These reports divided the development of SSST into three stages. The first stage of the development of SSST took place before 1960, the second was from the 1960s to the first half of the 1980s, and the third began with the second half of 1980s.

- The representative fields at the forefront of research in the first stage focused on large-scale systems development without experience (like the Apollo Plan); the second stage focused on research about resource and environment issues; and the third stage focused research on creative behaviour itself and the development of knowledge resources.
- The primary research objects for the first stage were complex technology and man-made systems that the aim of research is clear from the beginnings. The primary research object of the second stage was the complex ecological environment system which may adjust the target. The primary objects of research during the third stage were objectives of creation and complex self-organizing systems.
- The development goal for the first stage was the choice of tactics, e.g. the optimization of management skills aimed at improving efficiency (minimize resource input). The second stage was the choice of system, e.g. system adjustment through planning, forecasting and evaluation technology based on systems models. The development goal of the third stage was the choice of strategy, e.g. choosing the structure of factors according to strategic goals.
- Computing was seen as a basic field of technology. During the first phase the emphasis was on basic practice in computing. During the second phase the emphasis was placed on the shift from batch processing to online processing, and from centralized processing to decentralized processing. During the third stage the emphasis was placed on compound batch processing and knowledge information processing, with terminal users as the locus of attention.

- The notable achievements in the field of SSST during the first stage were found in operational research, industrial engineering, brain-storming methods, the Delphi method, relevance tree methods, matrix methods, network methods, scenario analysis, creative techniques, linear programming, dynamic programming, game theory and factor analysis methods, etc. The second stage's primary achievements took place in systems analysis, systems methods, decision support systems, value analysis, graph theory and systems dynamics. The representative achievements of the third stage were found in the comprehensive soft series of science and technology itself, such as social engineering, policy science, strategy information system, the ABC science (artificial, brain and cognitive science), knowledge and creative engineering.

3. Research Institutes

The same report, *A Survey on Present Situation and Future Trend of R&D of Series of Soft Science and Technology*, also listed those National Institutes that conducted researched on SSST and their major projects in 1989:

- *Research Institute of the Population Ministry of Health and Welfare*: demographic research.
- *Institute of Statistical Mathematics of Monbusho*: investigations and research in the field of forecast and control.
- *National Institute for Educational Research of Monbusho*: research on educational practices and basic theory.
- *Construction Research Institute of the Construction Ministry*: residence environment, city planning, construction and infrastructure structure.
- *National Research Institute for Agricultural Economics of the Ministry of Agriculture, Forestry and Fisheries*: agricultural economic problems.
- *Economics Research Institute of Economic Planning Agency*: investigation and research on the structure of the economy.
- *Research and Training Institute of the Ministry of Justice*: research on criminal policy.
- *National Institute for Environment Studies of the Environment Agency*: comprehensive research on the prevention of the social effects of pollution.
- *Institute of Financial and Monetary in the Ministry of Finance*: domestic and foreign finance and economy.
- *Institute of Posts and Telecommunications Policy in the Ministry of Posts and Telecommunications*: utilization, storing, insurance and the information research of telecommunication matters.
- *NISTEP of STA*: research on science and technology policy.

4. Application Fields

The application fields of SSST during this phase were divided into the fields of policy, management and administration, society, family and individual life and the general field of knowledge activities such as R&D.

In 1990, Japan's SSST investigation committee newly defined the soft series of science and technology as a new field of science and technology with the aim of clarifying human knowledge activities, such as cognition, thought, reasoning, judgment, innovation and the corresponding motivation mechanisms (the part of science), handling and operating the means that support or partially replace the above activities, and the information and experiences produced by them (the part of technology).

SSST can be divided into hardware, human-ware and other new fields according to the characteristics of their research objects. SSST also can be divided into basic fields (cognitive science, psychology, thinking psychology, behaviour science, economics, politics, system theory, information science, mathematical science, linguistics, organization science, etc.) and application fields (value engineering, social engineering, software engineering, policy engineering, systems engineering, management science and engineering, urban engineering, etc.) according to their conjunction with the previously accepted fields of science and technology. Namely SSST is divided into two parts: basic fields (science part) and application fields (technology part). But it does not advocate defining them respectively.

Approximately in the same period, Chinese scientist Xuesen Qian advocated dividing modern science and technology system into eleven major branches (Qian 2001) – natural sciences, social sciences, mathematical sciences, systems science, thinking science, somatic science, military science, behavioural sciences, geographical science, architectural science and literary theory – and he believed that 'this is an active system which is evolving in the activities to unceasingly understand and change the objective world'. With this social development and scientific progress, it is not only the structure of this system which has been developing but also its content has been enriching and, moreover, new scientific branch will be constantly emerging.

There is also another point of view which advocated incorporating science into three aspects: the knowledge system reflecting the laws of nature (natural sciences); the knowledge system reflecting the social laws (the social sciences); and the knowledge system reflecting the laws of human thinking (thinking science). However, systems science and cognitive science, etc. can no longer be simply incorporated into the social sciences or thinking science.

In short, according to different standards, science can be classified with different ways. If science is divided into two big camps according to the nature and characteristics of the object of study, it can be divided into soft science and hard science. Soft science is correlative with hard science. Despite the fact that in modern society the trend is towards mutual penetration and integration between soft science and hard science, the essential differences between them ought not to be ignored. Hard science refers to the knowledge system reflecting the laws of nature which studies the physical form, structure, nature and laws of motion of nature; while soft science refers to the knowledge system which studies social laws, human behaviours, the laws of human psychology and thinking. Integrating the above two kinds of representative Chinese-Japanese points of view, we may say that soft science may include social sciences, cognitive science, psychology, thinking science, behavioural sciences, systems science, information science, mathematics science, linguistics and organization science, etc. Moreover, just as is the case with hard science, it is a system which is ever growing and changing.

In discussing soft science in the study of soft technology we may say that, on the one hand, it can develop and enrich soft science (from practice to theory) from the perspective of soft technology; while, on the other hand, it can continuously identify and develop an operable knowledge system – soft technology in soft science knowledge system – and expand the space and approach for soft-tech invention, creation and innovation.

3. *From Soft Series of Science & Technology to Soft Technology*

National level support in Japan for research on SSST has effectively stopped since the 1990s. For instance, in the Science and Technology Basic Plan that was submitted by the Science and Technology Policy Bureau to the Science and Technology Convention on 26 December 2000, SSST was not included as a field. The field of basic social infrastructure, which mainly included disaster prevention technology, crisis management technology and technologies concerning basic human livelihood, etc., replaced the SSST field. This is clearly a setback to the cause of soft-tech research.

I encountered Professor Hayashi Yujiro, the former director of the Institute of Economic Research at the Economic Planning Agency of Japan, by chance in 2000 in Tokyo, when he was the first director of the Institute for Future Technology, the main institute of studying soft science and SSST at that period. When he learned that I was studying soft technology he sighed with regret that, 'Stopping the research of soft sciences is Japan's failure'.

No matter whether it is called 'soft science' or 'SSST', the reason why it eventually disappeared in Japan is worthy of reflecting and summarizing.

First of all, soft technology was not recognized as another paradigm of technology. SSST laid particular stress on studying social technology and the softening part of hard technology, especially on applications in the field of decision-making, and the understanding of soft technology in the commercial and cultural field was insufficient; SSST focused on the methodology of solving social problems by applying natural science and technology as well as the knowledge of systems science, and the awareness of its broad function as part of technological innovation, institutional innovation and industrial innovation was even more deficient. For instance, the notable achievements in the field of SSST, no matter during the first stage, the second stage or the third stage, took place in the aspects of methodology and engineering.

Secondly, soft science was confused with the soft technology. Soft technology was not taken as an operable knowledge system in soft science. Even in the age when the soft science research was most popular in Japan, soft technology had not been raised and it was even deliberately declared that SSST is not soft technology.

Thirdly, SSST was studied within the framework of the relationship between the natural sciences and hard technologies, thus the understanding of its 'technical part' was limited to value engineering, social engineering, software engineering, policy engineering, systems engineering, management science and engineering, urban engineering, and so on. In the field of soft technology, the approach to creating value or solving problems is completely different than the approaches of traditional science and technology. If the approach to solving problems follows the thinking mode of hard technology, it will inevitably reach an impasse. The case in point is that Japan's research in soft science and SSST over 20 years finally had to be stopped, and was ultimately reduced – in their *White Paper on Science and Technology* – down to only 'disaster prevention technology, crisis management techniques and technologies related to national life, etc.' as the social technologies.

To sum up, Japan's SSST was similar in its understanding of technology's 'soft' features as the understanding portrayed in this book; but there are significant differences between the two with regard to the basic concepts, connotations, features, research directions, research aims, composition of research fields, and meaning of research, etc.

Nonetheless, the research and practice of SSST in Japan bears significance for the R&D history of soft technology. Firstly, the understanding of technology in SSST, broadly defined, and the discussion within SSST of the soft nature of technology, were pioneering contributions to the field. Secondly, no matter what labels were used, it was particularly meaningful for research to have been conducted on a type of technology, including SSST, which is different from traditional technology and for this to have been listed in the Japanese *White Paper on Science and Technology* for a number of years.

Note

1. In 1991, Dr. K. Preiss was selected by the Iacocca Institute of Lehigh University to analyse the US's role in the changing structure of world-wide industry. He was one of the facilitators and a co-editor with Steven Goldman and Roger Nagel of the resulting report – '21st Century Manufacturing Enterprise Study: An Industry-led View'.

Chapter 3

Soft Technology and Technological Competitiveness

Almost every developing country wants to catch up with and surpass the developed countries as quickly as possible. The reality, however, is that along with the economic globalization and 'informatization' the gap between developing and developed countries has become larger and the gap between the poor and the rich has grown. In the hope of narrowing the gap as soon as possible developing countries constantly adjust their strategies and policies, invest more in science and technology and seek actively to develop their high technology capabilities.

During the last fifty years, China has allocated a large amount of human resources, materials and funds to developing high technology and high technology industries. The purpose of this activity has been to pursue international economic competitiveness which focuses on high technology. Since the 1980s the Chinese government has started a series of national programmes aimed at the development of high technology and industrialization of high technology. In addition, more than fifty state level High-tech Development Zones and Experimental Zones have been established. While there have been many great accomplishments through these initiatives, the overall position of China still lags behind advanced global standards in most technological fields. In some fields the gap continues to widen. Indeed, the Chinese government has placed a great deal of importance on the advancement of technology and Chinese scientists have put great effort into pursuing the vision of global technological competitiveness. What, then, is the problem?

What is the key to narrowing the gaps and to enabling developing countries to achieve the leapfrog development? What is the essential element in the gap between developed and developing countries?

A. Knowledge and Technology are Merely Potential Sources of Competitiveness

It has been said that high technology is the core of competition for comprehensive national power and that knowledge is the kernel of international competitive capacity. However, many cases which make one reconsider illuminate that knowledge or 'high' technology (hard technology) may not always lead to success.

1. The United States is the most advanced country in terms of technology. However, in the 1980s, it lagged behind Japan in many industrial fields (see the section of 'Patent Technology' and 'Management Technology' in Chapter 2). In middle of the 1980s

the Japanese semi-conductor industry held 50 per cent of the world market. The US government, on one hand, criticized the support policies of the Japanese government and forced Japan to buy US products; on the other hand, it strengthened its own support policies and implemented a series of strategic measures including preferential trade, amendments to the Patent Act, etc. By the 1990s, the semi-conductor and other important industries of the United States were restored to leading positions.
2. During the three-quarters of a century following its coming in to power, the former Soviet Union managed to leap from being one of the weaker countries, with low literacy levels to become an industrialized power with advanced science and technology, a well-developed education system and a complete high-tech system that could compete with that of the US. However, the Soviet socialist model was neither able to keep pace with the times – by being innovative in its theory, ideology and political and economic system – nor was it able to reform gradually. Its rigid economic planning system, in particular, exhausted the vitality of the Soviet society. Its subsequent attempt to blindly import western 'democratic values' eventually led to the collapse of Soviet Union. The Soviet type of soft environment proved insufficiently flexible to allow its high-technology to evolve fast enough to sustain its competitiveness and capabilities in activities such as human space travel, producing valuable products for the market, or making an adequate contribution to economic development.
3. Japan built the so-called 'Japanese miracle' out of the ruins of World War II through thirty years of hard work, and managed to move ahead of many western countries that possessed more advanced technologies and stronger economic foundations. Besides the favourable international environment, the first reason for Japan's great productivity improvements was the passive, but effective, institutional reform of Japanese industry and agriculture that was carried out under the supervision of the American occupation army.

The second reason was that the right development road was chosen. The mode suited Japan's national situation and incorporated a realistic and down-to-earth strategy for the development of technology and the economy. Japan introduced a great amount of suitable technology from the US, and also hired consultants from technologically advanced countries to help with technology transfer, thereby paving the way for Japan to eventually shift from imitative innovation to independent or endogenous innovation.

The third reason lay with the fact that the management and administration technology employed in Japan was imbued with Japanese characteristics. In particular, Japan excelled at applying integration technology and organization technology, transforming foreign advanced ideas into product technology of Japanese enterprises, and even integrating the doctrines of Confucius and Mencius and Sun Tzu's *Art Of War* of China with modern management techniques. Many new American technologies and inventions were first commercialized in Japan, making Japan a country with one of the largest numbers of new products and patents each year. We can say that the key to Japan's success was resolute institutional reform, an appropriate development mode and the fact that its management technology and managerial art was imbued with Japanese characteristics. However, institutional innovation, development mode, management technology and organization technology are not 'technology' as traditionally understood. It is unfair to claim, as some people have done, that the Japanese are simply experts at imitating western technologies. In reality, Japan's advantage and potential stems from its prowess at developing and applying soft technology.

Today, Japan retains first-class technology. In order to prepare themselves for the coming of the twenty-first century, Japan has accumulated a lot of future technologies in the field of super-robot, brain-based computers, brain science, bioengineering, micro-machining, optics, as well as energy-saving, environmental protection, etc. As a result, it is certain that Japan will be on an unmovable position and high tide of technological innovation in the future. However, since the late 1980s it has been mired in the consequences of it's 'bubble economy', and has failed to arrest the economic decline during the subsequent ten years, the so-called 'lost ten years'. There are various reasons to cause Japan's continued economic downturn in the late twentieth century but, fundamentally, its root cause lies in Japanese own rigid economic system and innovation system which no longer meet the domestic and international environmental changes. The reason is not that Japan's hard technology lost its competitiveness.

1. Microsoft did not achieve its success by merely depending upon advanced technologies (Wu 1999). Microsoft has been a master of applying soft technology. Dr Lee Kai-fu, Microsoft's former global vice president, summarized the reasons for Microsoft's continued success as involving four key factors: 1) *Technology* – grasping the pulse of technological development with an innovative spirit, bolstering the company's efficiency through a reasonable R&D system, and focusing on the development of the software industry with tenacity; 2) *The art of leadership* – top decision-makers have outstanding ability and excellent insight, there is a clear division of work, and there is superb cooperation; 3) *Talent* – finding and hiring people with the right talents through multiple channels, establishing an effective talent screening mechanism for the continued cultivation of talent, as well as ensuring that a person is employed according to their ability; 4) *Corporate Culture* – the traits of being willing to challenge, to self-criticize, to be flexible and to treat others with equality, as well as maintaining a responsible attitude to customers, have been developed as a tradition and institutionalized at Microsoft (Lee 2006).
2. The Haier Company, which started in the middle of 1980s by introducing refrigerator technology from Germany, is one of the very few Chinese enterprises that have been successful in both domestic and foreign markets. As of the end of 2005, Haier Group had formed 96 series of products with more than 15,100 varieties, holding China's largest household appliance group; and they had a total of 2799 patent applications for invention published up to the end of 2009. The company has carried out successful strategic adjustments and innovations at different stages in its development. For example, at the

beginning it emphasized creating a prominent and effective brand; during the next phase it emphasized the diversification of its product range; and it is currently emphasizing a strategy focused on internationalization. As their CEO Ruimin Zhang has concluded, Haier manages to seize opportunities in a timely manner by having conceptual innovation as the forerunner, technological innovation as the means and organizational innovation as the guarantee (G. Wang 2002). To sum up, the success of Haier has depended more on continuous innovation in soft technology than on the advancement of refrigerator technology.

3. Science and technology research in China over several decades has been very fruitful. Particularly since the adoption of its reform and opening-up policies, the central government has attached great importance to the development of science and technology. The number of people receiving science and technology awards is increasing, and the quality of their work is improving. However, these advances have not reversed the backward situation of the county in either industrial technology or high technology fields. China has not yet managed to occupy a leading international position in any high technology or industrial technology field. The technologies driven by the effort of the 'two bombs and one satellite' project of the 1950s and 1960s, such as electronics, materials, chemistry and machinery, failed to lead to the sustainable development of appropriate industries. According to an evaluation by IMD, Chinese international competitiveness in science and technology ranked 20th out of 46 major countries in 1997, 13th in 1998, 25th in 1999 and 28th in 2000 (IMD 2000). Compared with the previous overall evaluation framework of the world competitiveness, IMD has incorporated 'Science and Technology' into the main competitiveness factor of 'Infrastructure' in recent years, which includes 'basic infrastructure', 'technological infrastructure', 'scientific infrastructure', 'health and environment' and 'education'. The world competitiveness ranking of China by the factor of 'infrastructure' has always been in the middle and lower group, in which China ranked 28th in 2007, but it dropped to 31st place in 2008, and 32nd in 2009 (ibid.). Although the authenticity and comparability of IMD's index of scientific and technological competitiveness are questionable, and the criterion of evaluation is inconsistent, it nevertheless reveals that China's scientific and technological capability lags far behind that of developed countries. This runs counter to China's effort, inputs and expectations.

(We should emphasize that the 'technology' that has been the focus of the above discussion, and of China's policies, is 'hard technology' as defined in this book).

B. Where Does Technological Competitiveness Come From?

1. Strong R&D Capacity is the Source for Creating Competitiveness

R&D is the process of creating core technology as well as the means for 'producing' problem-solutions. Only by constantly researching and developing 'solutions' (hard and soft technologies) with independent intellectual property, competitiveness in the market and with corresponding products can a country provide rich technological sources for the development of its national economy. Otherwise, so-called 'technology transfer' and 'service' shall become rootless, just like water without a source or a tree without roots. Lucent Co., for example, maintains its leading position in the telecommunications industry through maintaining the research capability of its Bell Laboratory. It is reported that during 1996 one patent application per day came out of the Bell Laboratory, and that by 1999 the number had increased to four per day. Huawei Technologies Co., Ltd. was ranked as the largest international patent applicant in 2008 (see the section of 'Business Model and Management Pattern' in Chapter 2) and, in fact, more than 43 per cent of employees are engaged in research and development.

There are two essential conditions for creating core technology: human capital input and financial capital input. If a country does not have a comparatively strong industrial base and economic strength it cannot possibly afford the necessary R&D expense for the development of critical technology. The United States ranks first in scientific and technological competitiveness but, not incidentally, its R&D expenditure is also the highest in the world. For example, US spending on R&D in 1997 was 211.9 billion US dollars, which exceed the sum of the R&D input of Japan, the UK, France, Germany and Italy combined. This figure reached 264.2 billion US dollars in 2000 (Guan 2000). However, capital alone will not suffice. In order to obtain the human capital with high-quality, it is not enough to only depend on financial support and should be guaranteed by soft environments, including institutions, and cultures that are favourable to innovation.

What needs to be stressed is that R&D mentioned here must include hard technology and soft technology, as well as the integrated field of soft and hard technology as its objects. As elaborated in the second chapter, added value brought by an outstanding business model to enterprises is by no means analogous to the value added by a hard technological patent. For more details about soft-tech R&D, please see the section of 'Strengthening the R&D for Soft Technology' in Chapter 4.

With the increasing complexity of modern science and technology, the integration and merging of fields has become one of the main features of technological development. No important inventions or creations can be produced without interdisciplinary integration and coordination of expertise from many fields. However, few countries have placed interdisciplinary R&D on their research agenda. Hereinto there are problems, such as the old classification of disciplines and R&D fund system, interdisciplinary talent, as well as equipment devoted to projects. The main obstacles to the success of interdisciplinary

R&D, however, come from weaknesses in the communication of ideas between experts, from problems in the academic language of experts and through the intrinsic difficulties of managing interdisciplinary cooperation.

2. Soft Technology is the Tool for Creating Competitiveness

It is clear that the competitiveness of a country in hard technology is expressed in the number of competitive technologies that emanate from its enterprises and which are then transformed into the industrial technologies of that country. That is to say, the competitiveness of knowledge and technology is embodied in markets through commercialization (including military applications). All market applications of hard technology occur by means of soft technology. Therefore, soft technological competitiveness is the key; namely, the implementation capacity of innovation or executive ability in problem solving, which is manifested in the process of technological competitiveness as the efficiency of technology transfer. Here are some examples.

1. From the angle of approaches to obtaining hard technology

The approach to obtaining hard technology involves three processes: independently developing core technology and then transferring it into product technology in some companies and then diffusing it as industrial technology; the second approach involves introducing or purchasing core technologies from outside, transferring them into technology within companies and then translating them into industrial technology; the third approach involves developing or obtaining generic technology, transforming it into enterprise technology by recombining it with other hard technologies, through the process of soft technology innovation, and then translating it into industrial technology. All three of the above approaches are the process of soft-tech operation and innovation.

2. From the angle of the inherent process of technology transfer

In the case of manufacturing, hard technologies may be transformed into organizational technologies (e.g. enterprise technologies) through the process of being converted into products and then being commercialized through the development of customers and markets. They may then be further disseminated to form an enterprise group, build greater market shares and eventually form industrial technologies.

In the case of service industry, it is necessary to first of all understand the needs of service objects – soft targets – and then to design the appropriate system for meeting demand – solution. The following phases of activity are typically required for success: seeking hard and soft technologies from various technological fields which are growing 'explosively' and then adapting them for soft targets; forming the core technology of the enterprise through the successful operation experience of one-to-one service for specific clients; then, through various customer services, building the capacity to adjust the system and increase the scope of services to expand the market. The process of designing the system and 'assembling' the corresponding hard-tech and soft-tech for the soft targets appropriate to the client's needs is one of the most important aspects of developing new technology in the service sector. It is possible and appropriate to form the core technology of a service industry with unique characteristics according to the distinctive solutions required for specific clients (see the section of 'Business Model and Management Pattern' in Chapter 2).

3. From the perspective of the efficiency of technology transfer

Most of the developed countries experienced more than a hundred years of industrialization before reaching their current state of development. During the long history of the market economy, a great number of soft technologies which encourage innovation and which may be flexibly applied have been created, and, moreover, soft technology itself has innovated continuously. Gradually an appropriate and favourable macro environment has been formed by soft technologies innovation, creating a more complete environment than has emerged in developing countries. For example, those contemporary advanced soft technologies such as modern management techniques, venture capital technology, virtual technology, incubator technology and Nasdaq-type stocks, etc. were first developed by the US and UK.

Therefore, the speed of absorbing and applying advanced hard technologies is far higher in developed countries than in developing ones. The same may be said of the efficiency and the profitability of technology transfer. This means that once companies in developed countries have an appropriate high technology with good market prospects, they are able to transfer it rapidly into products and commodities, thus enabling them to form a technology-intensive/ knowledge-intensive industry and occupy a favourable position in the new process of global industrial structure adjustments and the international division of labour. Hence, among the three key factors of competitiveness, hard technologies are able to make a more direct contribution for those developed countries (see Figure 12); namely, when a community has a strong foundation in soft technology and a positive soft environment, then high-tech is able to act as the core of a country's competitiveness. This explains why international competition for leadership in high technology has become white-hot.

This is the main reason why developed countries absorb most of the global foreign investment (73%) worldwide. The British weekly news magazine *The Economist* estimated in 2001 that the United States, the United Kingdom, Germany, France, the Netherlands, Belgium and Canada would account for 59.2% of average annual foreign direct investments worldwide, while the United States would account for 26.6% alone during the first five years of the twenty-first century ('Advantage' 2001). In China, for example, although the western areas have attempted to attract foreign investment by proposing preferential policies, by the end of 2005 the number of projects absorbing foreign direct investment, the amount of contract foreign capital and utilized foreign capital in the western areas of China only accounted for 6.34%, 5.28% and 4.46% of the national totals respectively. The equivalent figures for June 2001 were 7.3%, 6% and 5.3% respectively. Obviously the foreign investors are heavily influenced in the investment decisions by their expected return on investment and by their expectations of the effect of technology transfer.

Hong Kong's *Sing Tao Daily* has published an article by ZhengPing Li in which an analysis is provided the reasons for the slow down of inflow of foreign capital in to China. Besides the obvious problems, such as the instability of investment environment, taxation and infrastructure, etc., he pointed out the backwardness in the services of government departments, the corruption of certain officials in mid-west area and, in particular, the lack of high-quality and high-level professional and modernized enterprises in China available to take on large foreign enterprises as strategic partners. Clearly the preferential policies in the 1980s and the early 1990s have already lost their charm.

Developing countries are obviously trailing in their R&D capability and technological prowess compared with developed countries. The main barriers to their becoming technologically competitive, however, are failures in technology transfer and low efficiency in absorbing advanced technologies (translating to enterprise technologies), which, in turn, result mainly from the incompleteness of their macro-environments and from their backwardness in developing soft technology. This is particularly true for China where the main tasks are still technology introduction, digestion and innovation. Therefore, technology transfer is the key for success in the coming two decades. However, soft-tech R&D and its application do not gain the positive attention that they deserve in China, even though its international position in soft technology lags farther behind that of its global competitors than its international position in hard technology. The biggest headache in the development of high-tech industries in China lies with the contradictions between knowledge and capital and between technology and market; these contradictions also call for the operation of soft technology. When Martin Kenney talked about the second economy in Silicon Valley, he pointed out the following, 'In many developing countries, the critical institutions of convertible currencies, the ability to sell a company and ease of firm formation and legal

Table 6: Foreign Direct Investment in eastern, central and western China by the end of 2005.

Regions	Unit: a hundred million US Dollars					
	Number of Projects		Contract Foreign Capital		Utilized Foreign Capital	
	Number	Ratio %	Amount	Ratio %	Amount	Ratio %
National Total	552942	100.00	12856.7299	100.00	6224.2531	100.00
Eastern areas	457944	82.82	11174.7601	86.92	5383.7139	86.5
Central areas	59947	10.84	1003.0724	7.8	562.9589	9.04
Western areas	35051	6.34	678.8974	5.28	277.5803	4.46

Source: Foreign Investment Statistics of Ministry of Commerce of China

transparency are not fully operating, which are the main obstacles to learn the model of Silicon Valley' (Kenney 2000).

Take the difficulties of Chinese state-owned enterprises as an example. Whether they exhibit a deficit or a profit, the hard-core issue is not high-tech but those issues related to institutions, property rights, management, incentives and benefit allocations. Although some issues have been presented as technology problems, in essence the problem lies within human factors. According to the investigation of the failure rate of large-scale equipment systems in China, more than 50 per cent of the faults may be classified as human or management factors rather than technical factors. According to an announcement by the Ministry of Information Industry of China about complaints from telecommunications users during the second quarter of 2001, the problem of service quality has ranked in the top of the complaints, exceeding even disputes over charges. Out of 92 official appeal cases on the record, 44 cases related to the quality of services.

Weikun Zhou, the former CEO of an IBM China branch, commented during a discussion about the ambitions of Chinese enterprises that, 'rather than certain areas of high technology, Chinese enterprises are wanting in modern management concepts, talent competition mechanisms, reasonable capital structure and a complete and sound financial system, including financing, investment, budget and planning. When those defects are remedied, a steady flow of talents will emerge and bring along technology, which can gather together the needed capital' (*Science & Technology Daily* 1999).

Fang Xuanjun, the General Manager of Chuangshiji Transgenosis Tech Co., Ltd., summarized the three bottlenecks of Chinese biotech industrialization as: insufficient intellectual property protection for biotechnology; the difficulty in finding financing for R&D, owing to the fields of genetic engineering in medicine and agriculture being described as 'a sword being sharpened in ten years' (owing to very low short-term economic benefits); and imperfect modernized enterprise institutions and governance structures of corporations.

Japan began investigating US science policy at the beginning of the 1970s. Japanese scholars of insight at that time realized that the gap between Japan and the US did not lie in hard technology. In fact, after World War II, Japan introduced lots of technologies from America almost free of charge, and many R&D achievements of America were first commercialized in Japan, thereby greatly narrowing the gap between the two countries in hard technology fields. The report of a Japanese investigation team in the 1970s said, 'There is almost no gap in technology between Japan and the United States. What we should learn from them, just like the situation with European countries, is how to close the gap of efficient management. If we do not improve the efficiency of R&D, the gap between Japan and US in soft science will result in major social issues for the near future'.

It is evident that no matter which way hard technology is developed – either through obtaining technology from external sources, or through the endogenous process of building competitiveness, or through the efficiency of technological transfer – all the pertinent processes are related to soft-tech operation. Hard technology may not produce real productivity improvements and competitiveness without successful soft technology operation and innovation no matter how advanced it may be. Accordingly, research by David Sawers shows that only one-fourth of the failures in technology commercialization projects stem from technological causes, and that the balance of failures are caused by business-related factors. From this point of view, we can say that soft technology represents implementation capacity and executive ability in problem solving.

3. Hard Environments and Soft Environments are the Basic Conditions for Competitiveness – Beyond Technology: Institutions, Culture and Values

There is no doubt about that technology is very important. Nevertheless, technology does not constitute the complete picture. Even though soft technology is the key to determining capability for technology transfer, it should nevertheless still be seen as a set of tools for problem-solving and as representing implementation capacity and executive ability. For creating competitiveness, there is one factor that has an influence on the direction and feasibility of implementing soft-tech design that is even more significant than the 'technology factor' itself. That is 'the rule of game' of soft technology. The main factors restricting or encouraging human behaviour are institutions, culture and values. These are the core of the soft environment that we have defined in this book.

With economic globalization, the macro-environment and the macro-level of management place greater restrictions upon technological competitiveness. A favourable macro-environment is like soil for the cultivation of innovative capability. It is the premise and essential condition for creating competitiveness. The macro-environment can be classified into hard environments and soft environments.

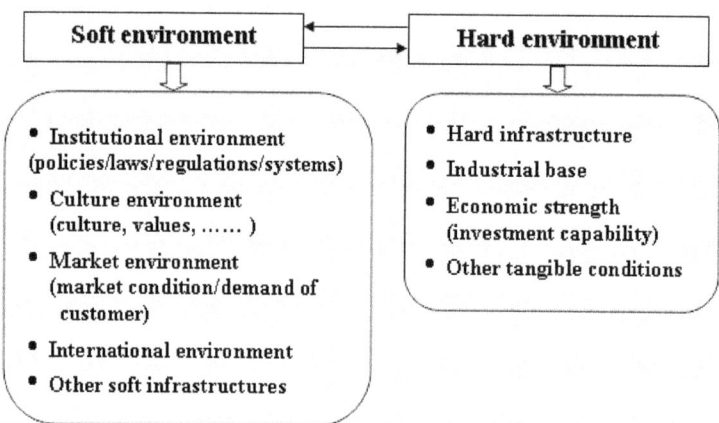

Figure 11: Soft environment and hard environment.

The hard environment includes tangible conditions such as infrastructure, the industrial base and economic strength (the capacity to provide capital or investments). Since innovation consumes more resources than research, a suitable hard environment is beyond any doubt one of essential conditions for creating competitiveness.

The soft environment includes invisible fields such as the institutional environment, the cultural environment, market conditions and customer demands (depending upon the educational level, living conditions and cultural background of consumer), the international environment and other soft infrastructures. Institutions here contain law, regulation, policy, standard, etc. The so-called soft infrastructure, which is correlative with hard infrastructure, can also be regarded as one of the important contents of establishing soft environment; it includes not only the structure of laws and regulations, rules and conventions, standards and the code of conduct in other business practices, but also cultural environment, the electronic and interpersonal network as well as natural infrastructure (ecological diversity) (E-Square Inc. [2007] 2008). They are used to control or encourage, guide innovations and high value-added economic and social activities.

Success is inseparable from a healthy institutional environment and appropriate soft mechanisms at either the national level or enterprise level. However, the soft environment tends to be neglected in developing countries, which also exhibit low levels of industrialization and a weak economic capacity. Especially in China, issues associated with the institutional environment are still the main factors which delay the sustainable development of the economy and society. Moreover, most people still worry that a single mention of 'institution' will raise political concerns, thereby further delaying institutional reforms.

Although construction and innovation of institutions is important, it is sometimes even more important for supporting reforms to take place in correlative or complementary institutions, in which some institutions may involve a deeper level of ideological emancipation and operable capability in soft technology.

The institutional environment will therefore now be discussed to illustrate the significance of the soft environment.

1. China's changes before and after the reform and opening up in 1978 provide a good example. The reform and opening up policy enabled China's productivity to increase at an unprecedented rate, and a high average economic growth rate of 9.88 per cent was achieved over the last three decades, while economic output grew eleven-fold. These results came not from the contribution of advanced technologies, but rather because of the transformation of economic system and policy reforms, such as the 'Open-the-Door-to-Outside' reform. In short, institutional innovation played a key role. At the macro level, the real solution required for the problem of how to maintain China's rapid development lies with soft technological operations at the national level.

The following are observations of the Chinese scholar Dexiong Zhan about the experience of maintaining sustained economic development in China during the past thirty years of reform and opening up (and modified by my own observation partially): it is important for a country to adhere to its 'own way' in following a strategy for development; stability is of overriding importance; leadership is the key requirement; economic development is of primary importance, but social development and the protection of natural resources and the environment ought not to be neglected; a proper democratic model suitable for the Chinese situation is needed; development needs to be pursued progressively ('fumbling the way to across the river') (Zhan 2008). Certainly, China's reform and development are still in progress.

A requirement for the next step in China's development is a second emancipation of the mind at multiple levels, involving ideas and concepts, as well as for systems to carry on in-depth and comprehensive reform and opening up. Four relationships which relate to structural adjustment of power must be dealt with correctly in the design of development mode: the relation between 'rule by people and governance by laws'; between 'equity and efficiency'; between 'government and market', as well as the relation between 'centralization and decentralization' (Siwei Cheng 2008). And it also needs to be guaranteed through creating the appropriate soft environment, especially system, mechanism, laws and regulations, basic policies, etc.

2. A survey carried out by IMD showed that Japanese national competitiveness ranked fourth in the world in 1996, but that it had declined to the eighteenth by 1998 and seventeenth by 2000, even though Japanese scientific and technological competitiveness still ranked second from 1996 to 2000. The main reason for this sharp decline lay in Japan's financial and management systems (Hirasawa 1999). This decline indicated that the institutional environment that supported Japan in achieving its economic miracles had become an obstacle for further development in the age of economic globalization and was facing the inevitable challenge of institutional innovation. Take the IT industry in Japan as an example. The VI&P plan produced by NTT in 1991 was the first IT strategy in the world aimed at bringing broadband information networks to every family within the next twenty years. It was produced three years earlier than 'The United States Information Base Action Guide' but it failed to become a national strategy. Five years later, the computer adoption rate in Japan ranked only nineteenth in the world and Japanese users accounted for 49 per cent of those located in the United States, which occupied the first rank. In view of this situation, the Japanese government set up its 'IT Strategy Headquarter' in 2000 and proposed catching up and surpassing the United States within five years. The key here was the sensitivity of the prospect of science and technology development and the decision-making process of the leaders in the US and Japan. Since then, all successive presidents of the United States have attached great importance to the national development strategy by integrating the science and technology development strategy, the national comprehensive competitiveness strategy and the global strategy. The problems of Japanese government strategy are mirrored in the case of the Japanese venture capital industry. Besides the effect of the economic collapse following the bursting of the economic bubble, the incompetence of venture capital policies and mistakes of the Japanese venture capital industry were also core reasons for the loss of the 'glory days' of Japan's high technology industries.

Without doubt, the technology discussed here is technology as traditionally understood; scientific and technological competitiveness refers to that of hard technology. This conceptual bias results in great differences between the results for technological competitiveness and national competitiveness in the IMD study.

The following ideas are presently widely embraced in Japanese society: a defining trend of the twenty-first century is the merging of various types of knowledge, and the competitiveness of a country is embodied by new initiatives, new strategies, new management modes and its implementation capacity. Japan is therefore now seeking to practice a comprehensive type of total-systems reform. For example, when the government began implementing its twenty-first century strategy of 'nation-building on the basis of creation', its first step was to pursue a strategy of overall reform of the organs of government. The main content of this strategy included: reorganizing all ministries and their functions; making government information available to the public; improving policy evaluation functions; increasing the transparency of administrative management; and setting up independent administrative legal personnel systems, etc. The strategy also involves a special step for science and technology institutional reform. For example, the original MONBUSHO (the Ministry of Education) and Science and Technology Agency have been incorporated into the new MONBUSHO, involving education, culture, science and technology (including social sciences and natural science and technology), sports, etc., in order to vitalize science and technology on a full scale. Another example is the installation of an independent administrative legal personnel system in state-run R&D institutes and state-funded universities, which is one of the major reforms in Japanese history. Within the new cabinet system, the former 'science and technology convention' will be replaced by the 'comprehensive convention of science and technology', which exercises the function of coordinating the relationships to science and technology of different government departments. These reforms contribute to the making of a comprehensive strategy by taking science and technology as their object while including a comprehensive approach to the humanities, society and natural science.

3. The number of Nobel Prize laureates should be a powerful symbol of scientific competitiveness of a country. Since 1985 – except Nobel Prize in Literature and for Peace – among the 128 scientists who are the Nobel laureates in physics, chemistry, physiology or medicine, etc., 65 per cent of laureates have been Americans or have been engaged in the most of the research in the United States, followed by the United Kingdom and Germany. Among 64 Nobel laureates in economics, 44 of the laureates are Americans and 52 have been worked in the United States during the period that they obtained the award. Here, besides the abundant funds support for basic research in US, the soft environment, which is in favour of innovation, is laudable. There is a unique culture in the United States with the free movement of talents; the educational method which emphasizes 'width, elasticity, agility, independent, self-confidence'; the research culture of being courageous in challenging authority; competitiveness and innovativeness in the education sector and academia; being generally tolerant towards those who dare to

attempt to do something innovative but do not necessarily succeed; and they are also willing to listen to the opinions of young people, etc. This is very rare in the world. Moreover, US immigration policy and the talent system have created conditions to construct an innovative environment and attract a large number of scientifically and technologically talented people from around the world to their country. According to a report by *USA Today*, globalization has entered the office of top layer of management in the US companies; the CEOs of American enterprises come from nearly 100 countries (*USA Today* 2001). During the past six years, the number of CEOs of US companies of foreign descent has increased three or four times over. In addition, among the top Fortune 100 companies in the US, there were nine CEOs born in foreign countries in 1996, and the number has increased to fifteen CEOs born in foreign countries in 2007.

4. South Korea, a country severely afflicted by the Asian financial crisis in 1997–1998, recovered within only one year and managed an economic growth rate of 10.5 per cent in 1999. This should also be attributed to the Korean government's strong resolution to carry out institutional reform, especially financial reforms such as the reorganization of banks. After the financial crisis, the Korean venture capital industry grew vigorously. The resolute and effective policy input of the Korean government played a key role. In a short time, the government finished a unified planning system, established a complete system of organization setups, laws and regulations, research system reforms, capital support, incubator centres and favourable taxes (including the stock exchange system), and strictly defined the role and limitation of the government venture capital and market operations. They also implemented priority policies for innovation and start-ups ('venture firms') and they even implemented temporary policies, such as exempting young people engaged in start-ups from military service and releasing university professors engaged in new ventures from their duties for three years or permitting them to work only part-time. All these measures stimulated and encouraged the enthusiasm of Korean people for innovation and start-ups. Many talents who left the big corporations during the financial crisis and those who were fired from their companies tried to start new venture companies. It is reported that there are more than one thousand start-up enterprises within the 'Teheran Valley' district of southern Seoul.

5. It is difficult for SMEs to arrange the needed funds; the root cause lies in the lack of appropriate mechanisms and systems. China needs to carry out a comprehensive reform and institutional innovation in all directions for SMEs, such as: the legal system; the financial system, including the regulation system of entire banking sector; the social security system; SME credit system; guarantee system; risk compensation mechanism; socialized services system, etc., so as to optimize the growth circumstances for SMEs. Making them account for 99 per cent of the total number of enterprises will become an important engine of China's economic growth, and main position for addressing the employment and the transfer of rural labour forces.

6. Comparison and analysis of Japan's successful transfer of US technology after World War II, the experiences and lessons of China's 156 project (which involved transferring

technology from the former Soviet Union), and the phenomenon of the huge gap in foreign investment between the eastern and western area of China, reveals that the key success factor for technology transfer is whether or not the soft environment is sound.

4. Exceeding the Power of Institutions – Cultural Innovation

The institution is indeed the core content of the soft environment. However, the institution is not a panacea. Legal changes must be realized by changing people's fundamental attitudes. Culture and custom exert in-depth impacts on innovation, frequently going beyond the role of the market and government. That is why we have said that culture and values sometimes exceeding the power of institutions.

Many frustrated attempts to replicate Silicon Valley in the past few years illustrate the point. Silicon Valley, California, is a centre of innovation for America and the whole world. Almost every country has tried to create innovation centres in their own territory similar to that of Silicon Valley but the world has thus far failed to do so. In general, people have cited the visible factors of Silicon Valley, e.g. high-tech industries, venture-capital funding mechanisms, excellent universities, research institutes, good infrastructure, etc., as explanatory factors. However, it is easy to ignore the most fundamental factor that has supported Silicon Valley – the culture. It is the culture of Silicon Valley more than any of the above factors which provides the support system for its success in innovation. The culture of Silicon Valley encourages innovation and cooperation, allows entrepreneurial failures (with the concomitant learning they entail) without punishing the protagonists and emphasizes investing in talented people.

Take the present credit crisis of the world economy as an example. The basis for the normal operation of the market economy is a sincere and trusting relationship between two sides of a transaction. In this sense, the market economy is a credit economy. During recent years in China, various cases of commercial fraud, such as fake and shoddy goods flooding the market, economic swindling, non-fulfilment of contracts, breaking of faith and fraud in the market for bills and delays in arrears have become a 'malignant tumour' for China's economic development. According to official reports there were 7419 criminal cases related to the counterfeiting of currency and illegal acceptance of counterfeit currency investigated by the public security organs of China in 2000, and the amount of money involved in those cases reached 5.2 billion RMB. The credit crises occurred frequently in fields such as service, manufacturing, finance, stock-trading and tax, indicating the great need to establish and perfect a credit system that includes credit laws and regulations (in 1998 alone, 180,000 cases of malicious overdraft were handled by the authorities). Accordingly, Shanghai has therefore taken measures to monitor personal credit, and has established an inter-bank record system for personal credit; Beijing has launched a pilot project for enterprise credit in the Zhongguancun Science Park, involving establishing a business credit service system; Shenzhen has been developing laws to deal with these problems, and will promulgate

'Administrative Measures of Shenzhen Municipality on Individual Credit Information Collecting and Credit Rank Evaluating'.

Cases of such financial fraud and other criminal behaviour associated with the financial system still happen despite prohibition and repeated attempt by the authorities to prevent their spread. For example, in Tianjin City alone in recent years, the amount of money each year involved in these criminal financial cases has been as much as several billion dollars, seriously endangering the healthy functioning of the financial system. In spite of the fact that more than 80 per cent of medium and large-sized companies and a large number of small companies in China have been successfully certified as having passed ISO9000 standards, fake and shoddy products are everywhere. Shouldn't the granting of an ISO9000 certificate to an enterprise signify a guarantee of quality and symbol of the corporate culture of social responsibility for that enterprise?

Certainly, the credit crisis is a phenomenon that is not unique to China. Data released by United States Securities and Exchange Commission (SEC) show that in 2001, US government agencies investigated 112 cases of suspected fraudulent corporate financial reports, an increase of 41 per cent compared to 79 cases in 1998, including the case of Enron Corp. and Waste Management Inc. in which Arthur Andersen LLP has a responsibility for failed audit (W. Zhang 2002). Investors have been shocked because a historical record was created by the number of large companies among the Fortune 500 involved in these fraud cases, and because of the significant rise in the sum involved. Let's put aside reflection on market rules and regulations triggered by Enron's collapse aside for the moment and concentrate on the financial crisis that began in late 2008, which exposed cases of financial fraud in which hundreds of United States banks were involved. Today, the crisis of credit, ethics and corruption has even spread to academia. In an effort to mitigate such problems and risks, the US National Institutes of Health (NIH), for example, announced a far-reaching reform in 2005, which prohibited employees from making deals with pharmaceutical companies, forbad all the scientists from receiving any consulting fees or other receipts from the pharmaceutical companies, and forbad employees from holding stocks of pharmaceutical companies.

It seems that although it is no doubt important to continuously improve institutions and establish sound mechanisms for supervision, monitoring and feedback, as well as maintain the authority and seriousness of the institutions, the most fundamental thing is to raise the corporate and public awareness of credit and build the culture of credibility. Because the credit system must take sound ethics as the ideological foundation – namely, constraining force of credit needs no doubt to be guaranteed by a strict institution – but without a solid foundation based on a commitment to sound ethics, self-discipline, trustworthiness and honesty, the role of institution is limited. It is not possible to build up a successful credit system for a society that lacks ethics and morality as well as credibility.

Examples of positive efforts in this area may be found with Juhani Pekkola's research on business ethics, and Hazel Henderson's founding of the 'Ethical Markets Media' and launch of a TV programme on ethical market worldwide.[1] Moral construction is much more difficult than institutional construction. To build a society which generally takes ethical values, ethical compliance and ethical advocacy as important for good credit standing – and in which telling the truth and engaging in ethical daily practice is the culture basis of national life – we must start from children, from education in the family and school, and then extend this to ethical education in workplace. Additionally, the working style of government officials, and also the decisions and policies that emanate from the country's political life, must be treated as the touchstone of leadership, thereby testing the good faith of political leaders in the eyes of the domestic and foreign populace. The ethical (or unethical) behaviour of government leaders will become the teaching material for educating (or misleading, as the case may be) the people.

1. Conceptual Changes

Guangdong Province, China, enjoys the advantages of technological innovation that have been derived from conceptual innovation. Up to 2008, for fourteen consecutive years, the volume of patent grants, taken as one of important indicators for technological innovation in Guangdong province, have ranked first in China. In 1999, over 70% of the total investments of science and technology in the entire Guangdong province were invested by companies rather than government organizations, and this percentage rose to about 90% by 2004, causing enterprises to become the main force in technology innovation. Enterprises founded in Guangdong Province such as Kelong, Huawei and KONKA have established research institutes in foreign countries like the United States and Japan. Guangdong's exports of high-tech products in 1999 amounted to 11.8 billion US dollars, accounting for 48% of China's total exports, ranking them number one. This figure increased to 66.5 billion US dollars in 2004, accounting for 40% of China's total exports, still ranking Guandong as number one. Compared with Beijing, Shanghai and Xi'an Guangdong is a province without first class universities or institutes and it does not enjoy the obvious advantage of a strong scientific and technological base from which to draw for innovation. During the last few years, furthermore, foreign investments have been decreasing compared with the previous several years. However, Guangdong still retains the leading position in China vis-à-vis innovation. Some explanations for Guangdong's extraordinary position, despite its handicaps, are as follows:

- Guangdong was the first province in China to benefit from the preferential policy of having a special economic development zone. After thirty years of rapid development under the aegis of that policy, Guangdong possesses great economic strength. The province contains only one thirteenth of China's entire population yet it accounts for one eighth of the national GDP in 2007, while tax revenue accounts for one seventh.
- Guangdong attracts many qualified personnel with pioneering abilities and forethought from across the entire country because of its relaxed commercial environment. The GDP per capita of Shenzhen (a city of four million people and a history of only twenty years) reached the extraordinary high level of 39,700 RMB (which can be converted into more

than 4800 US dollars at that time), while the national GDP per capita was 7078 RMB in 2000. In 2007, its per capita GDP was 10,628 US dollars, becoming the first city which per capita GDP exceeded 10,000 US dollars in the Mainland of China. There are 74.5 personal computers per 100 people in Shenzhen, ranking the province first in the world in 2004. In 1999, the output value of its IT industry was 15% of that of the entire country, and this proportion had risen to one sixth by 2006! In Shenzhen there are at least 150 transnational companies investing in high-tech industries. Shenzhen contains a large-scale IT industry; it is the production centre for China's computer hardware and Shenzhen also produces 30% of the hard-disc drives and 10% of recording heads worldwide.

- The leaders of Shenzhen are aware of the weaknesses in their city's science and technology base, in qualified personnel and in industrial foundations, and they could only make breakthroughs in institutions. Therefore, Shenzhen cancelled the government authoritative department for enterprises in 1995 and has set up multi-channel information services to provide services for private enterprises. More than 40% of scientific and technological R&D personnel in Shenzhen are concentrated in the private enterprises. The gross industrial output value, sales and profits of SMEs accounted respectively for 65%, 62% and 51% of that of Shenzhen city. Shenzhen launched a series of new policies in 2001, for example, the 'Guideline on system reform of investment and financing of Shenzhen', announced in September 2001, indicated that, apart from the projects involving national and regional security, other areas are open to social capital. Its principle is to break monopoly and broaden market access; only those who invest may continue to receive benefits under the auspices of the project, and those who gains profit are also expected to undertake risks. The purpose of the programme is to achieve diversification of investment, commercialization of financing channels, transparency of government regulation and control and socialization of intermediary services, while the government protects various rights and interests of investors and the public. A new policy labelled 'Temporary provisions about certain registration problems in promoting the development of high-tech enterprises' has reduced the entry threshold for Shenzhen high-tech enterprises reduced to the minimum; namely, the required registered capital of high-tech enterprises has been reduced to 30,000 RMB, with the first payment reduced to only 15,000 RMB, the operating project is optional, and the ratio of investment by achievement can be decided freely. Such provisions are not available in other places within China at that time. Accordingly, the leaders of Shenzhen chose to focus on institutional innovation. They also knew, however, that breakthroughs in perception were pre-conditions for breakthroughs in institution. During the financial crisis of 2009 Shenzhen stood once again at the forefront of administrative reform.
- Open-mindedness and flexible commercial environments are important reasons for Guangdong's success in attracting talented people, new technology and investments. In China, the meaning of the names 'Guangdong' and 'Shenzhen' are now synonymous with 'open'. Government officials and the public in Shenzhen are more open-minded than their counterparts in Beijing and Shanghai. There are also fewer government restrictions for

high-tech enterprises in Guangdong than those in Beijing and Shanghai, and the province no longer allows the establishment of new state-owned enterprises, instead encouraging private companies to unite to compete with foreign companies.

To sum up, we should thank the policy environment of Guangzhou because it helps to incubate these excellent private companies, not only making these companies the backbone of technological innovation, but also enabling Guangdong to become the richest region in China. According to a report in *Guangzhou Daily* on 30 May 2001, the savings deposits of Guangzhou city's residents amounted to almost 500 billion RMB, which equals one-twentieth of the total deposits of the entire country, with only one two-hundredth of the total population. Guangzhou has become the largest financing centre across the country.

2. Trust, Cooperation, Share and Innovation

Another extremely important cultural factor affecting business and innovation is the existence of a system-wide ethos of trust and cooperation.

Zhongguancun spent nearly twenty years developing before it was able to experience the changes that it is now undergoing. It is worthy of praise. However, when compared with the international situation, or when judged against the investments it has absorbed, Zhongguancun's performance would be deemed unsatisfactory. For example, the Founder Group in Peking University monopolized relevant markets with their own key technology, realizing that they could be more successful than at present. For the past many years, most of the successful people in Zhongguancun have experienced the difficulties of learning new things and of doing everything themselves. Why do they have to do this? Some say it is because only enterprises and not entrepreneurs have been created in China during the past thirty years, and that there are no suitable models of entrepreneurship to emulate; some say that most of the Chinese state-owned companies lack the internal pressure or demand for technological change and that it is therefore hard to cooperate with them, and that scientists in Zhongguancun therefore have to do everything themselves. As the result, although the government encourages, and sometimes even forces, the combination of research institutes and enterprises, there are few successful examples of collaborative spin-offs of high-tech companies.

The fact that Zhongguancun has been largely ineffective in producing cooperative start-ups can be traced back to the following factors.

1. For a long time, China has organized economic activities along the lines of government ministries, departments and administrative regions. Under the mechanism of 'cut up the links among department and regions', the components of the R&D system and the production system have also become isolated from each other and great barriers have been built between scientist-technologist circles and business-industrial circles, causing misconceptions and distrust between each group. In addition, China's distinctive 'Unit System' has been an umbrella that legally protects the interests of each component of the system.

2. People have been known to blame every fault on the old system. However, such criticisms are not always justified. People should reflect on the influence of the traditional thinking and behavioural mode that is encapsulated in the phrase, 'It is better to be the head of a dog than the tail of a lion', 'although able to hear the dogs' barking and cocks' crowing from neighbours, but there is no need for intensive interaction with them till death – live close to each other but never be in contact with each other'. This kind of thinking is hazardous to the healthy establishment of joint ventures and partnerships for start-ups. In addition, it runs counter to the need for development of our era. All the old culture, such as not to advocate to do something unconventional or unorthodox ('Gun shots out head bird – the outstanding usually bear the brunt of an attack'); individual innovation is regarded as showing off or individualism; the old culture of being content with things as they are in the small peasant economic culture like 'being content with one's lot', 'live contentedly as a poor scholar and happy to lead a simple virtuous life', etc., are not conducive to innovation.
3. To lean how to cooperate, interest concessions and share. In the modern society the mix of resources required for success in business alters with each stage in the development of an enterprise. In particular, each stage requires a different approach to technology transfer corresponding to the unique mix of problems and risks that occur at each stage, and to be integrated at different levels to maintain competitiveness. This is the reason for no sign of simmering down in mergers boom across the world during the last couple of decades. Successful high-tech enterprises must learn to cooperate, sometimes on a grand scale, by various resources. However, managers of enterprises must know that it is indivisible to optimize the combination of resources and wisely remit appropriate interest to their partners. In other words, they must know that sometimes 'without 80 per cent of partner's interests, will not meet their own 20 per cent interests'. They need to avoid the trend of over-stressing their own interests.

China has the longest history of any country and possesses a rich cultural heritage that is an inexhaustible resource for future generations. However, in the process of developing modern science and technology as well as economy, China must overcome some salient attitudes and traits such as blind arrogance and looking backwards rather than forwards with regard to cultural issues; and China should consciously avoid factors that are not favourable to the opening-up of the economy and to innovation, such as continental culture, peasant culture and human-ruled culture, and establish a culture conducive to innovation, cooperation, sharing, credit and the toleration of failure. The cultural environment is just like soil for innovation, which is of great significance in retaining qualified personnel and attracting funds and technology.

Concerning cultural innovation, China may learn something from Japan's experience. Japan has managed to maintain its unique cultural heritage while achieving modernization.

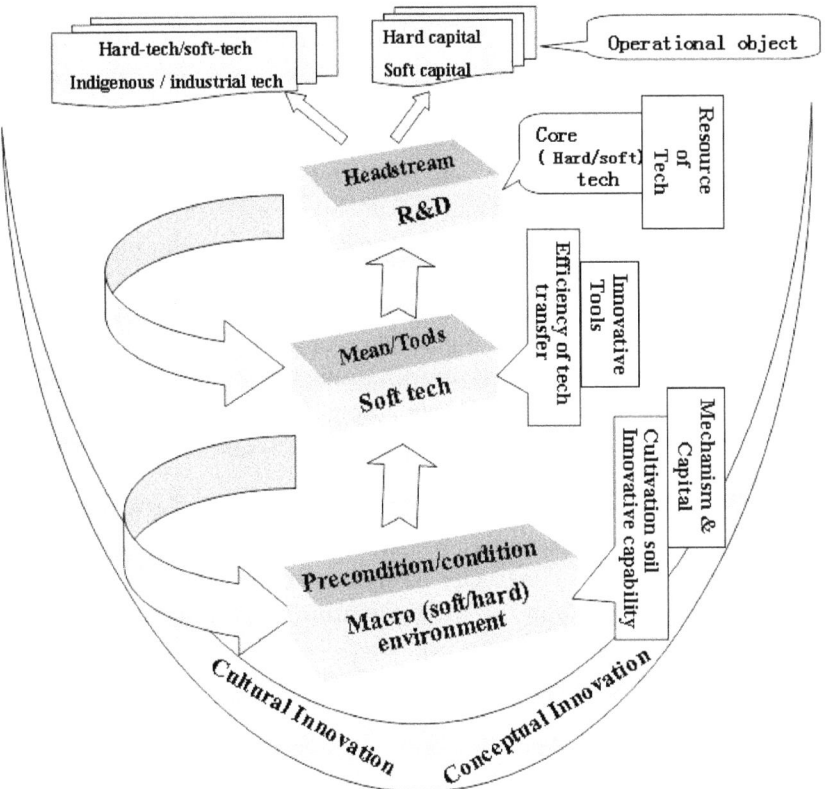

Figure 12: Three conditions for creating technology competitiveness.

5. The Three Key Factors of Competitiveness

In summary, it is necessary to foster competitiveness in three dimensions simultaneously: 1) sources of technology; 2) the means for innovation; and, 3) the innovation environment. Namely, the core of comprehensive competitiveness is the design capability for a solution (hard-tech and soft-tech), the capability of implementing the solution (soft-tech innovative ability) and the implemental environment for solution (the capability of environmental innovation).

Therefore, soft technology is an 'activating' means for the potential competitiveness of knowledge and technology; and the hard and soft environment is like soil for cultivating capability for innovation, and prerequisites for enhancing competitiveness (see Figure 12).

Let's take the Samsung Group, a world-leading high-tech enterprise, as an example. In 2005, Samsung ranked twentieth among the Top 100 Largest Global Brands with a brand value of USD 15 billion and a turnover of approximately USD 133 billion, and became one of the five fastest-growing brands during consecutive five years. Samsung Group has nearly twenty kinds of products with world market share ranked first among global enterprises. In 2003, three companies of Samsung were listed in the Fortune 500, with Samsung Electronics ranked fifty-ninth. Six key factors have helped the Samsung Group maintain and improve its competitiveness: 1) nurturing talents and putting them into important positions; 2) good company culture; 3) active investment in R&D; 4) constant operation innovation and structure adjustment; 5) operation at fast speed; and 6) leadership of the CEO (Strategy Research Group on National Innovation System 2004). Among the six factors, the third belongs to the first condition of competitiveness, strong R&D power; the second factor belongs to the third condition, good soft environment; and the other four fall under the scope of the second condition, constant innovation of soft technology for which Samsung laid strong emphasis on a flexible structure adjustment mechanism. During the financial crisis, the Samsung group reduced its organizational structure from over 40 enterprises with different characteristics into 28 enterprises. Thanks to this move, the Samsung Group withstood and overcame the financial crisis of 1998.

The Samsung Group builds upon a rich history that extends back to 1938 when it was still a small company called Samsung Corporation. Today it is a multi-national company with electronics and finance as the main business, with 285 offices with legal status in 69 countries throughout the world. Clearly, the growth of Samsung was not the result of advanced technology from the very beginning. In 1950 Samsung started producing products relating to people's livelihood. In 1960 it set foot in the electronics and finance industry. In 1970 it expanded its business into the heavy chemistry industry. In 1980 it entered the semi-conductor industry and in 1990 it established its presence in high technology sectors such as TFT-LCD and CDMA cell phone. However, in the process of enlarging its business scope and realizing a diversified operation, as well as in light of the real operation environment at home and abroad, Samsung did not stake its future only on the innovation of hard technology. It also focused on a coordinated operation of the three key factors of improving competitiveness, which are constant innovation of the organization culture, active investment in R&D in new technology and full utilization of various business technologies in the meantime. For example, in its different stages of development, the company used soft technologies including organization technology (structure adjustment), alliance technology, merge and acquisition technology, management technology (operation innovation) and virtual technology, etc. to seek further progress in the hard technology sector on one hand, and, at the same time, formed soft industries related to the company's logistics, trade, insurance, security, investment trust, venture capital, etc. The details of the utilization and innovation of these soft technologies always become the business secret of a company. This is why technology-oriented companies can only be fully developed after they are incorporated into large groups. 'The Status Quo of the Total Market Value of the Top 10 Groups', released by Korea Stock Exchange on 20 January 2004, shows that the total market value of Samsung Group accounted for 29.1 per cent of the total market value of Korea, with Samsung Electronics alone taking 22.5 per cent.

In summary, the expansion of Samsung integrated hard-tech innovation, soft-tech innovation, cultural innovation and institutional innovation to realize the success of industrial innovation. Conversely, the needs of industrial innovation promoted innovation in hard and soft technology and, moreover, the innovation environment provided by South Korean government and their society contributed to the improvement of overall corporate competitiveness.

C. Comprehensive Competitiveness and Soft Power

Our discussion above focused on technological competitiveness. In fact, the principles we enunciated can be applied for comprehensive competitiveness at either the national level or enterprise level.

The comprehensive competitiveness has always involved the integration and complementary combination of hard power and soft power. Over time we have also learnt that a large number of issues in the world today, such as sustainable development, globalization, and the challenges of new economy – including the negative impact brought by high-speed acceleration of industrialization – may be almost impossible to resolve if we rely solely on hard power. Therefore, the relative proportion of the contributions of hard technology and soft technology has been changing, which in turn has forced people to recognize and understand the existence and importance of soft power. This can be studied in terms of soft power, soft competitiveness, soft control and soft factors, etc (Jin 2008).

1. How to Study Soft Power

Professor Joseph Nye has pointed out that 'soft power is a country's ability to get what it wants by attracting rather than coercing others, or that to manipulate another country's political agenda' (Nye 1990). Nye defines a country's soft power as having three main sources: its culture, its political values and its diplomatic policies. However, if promoting and applying soft power is for the sake of controlling or manipulating other countries or other people, then it will never achieve the objective of truly improving and accumulating soft power. This is the reason why American soft power has often been frustrated in recent years.

It is well known that soft power is correlative with hard power and, moreover, that hard power and soft power are mutually dependent upon each other. Hard power is expressed as tangible strength, which is an ability to attain one's object by physical means and tools, and a country's hard power consists of economic power, military force, hard technology strength, hard environment, hard capital, etc.; while soft power is expressed as intangible ability, taking hard power as the base to attain one's aim but without using hard power directly.

It is noteworthy that the 'objectives' mentioned above include not only the motive to promote competitiveness but also, more importantly, to develop one's own exuberance,

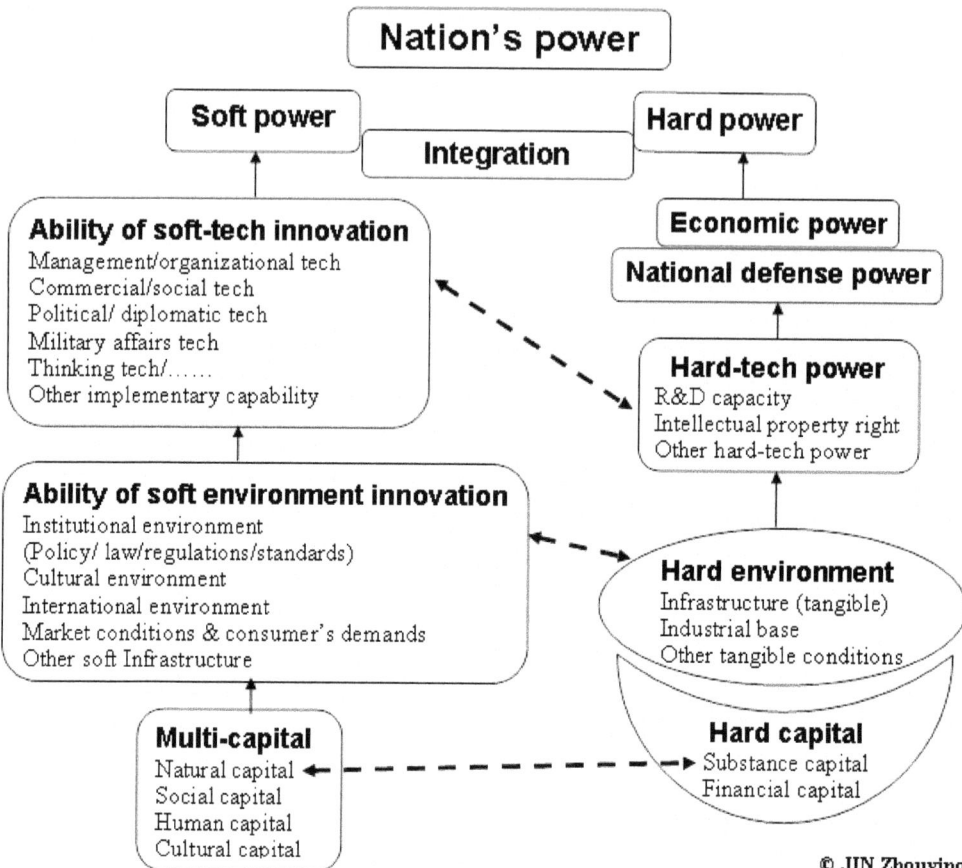

Figure 13: Comprehensive competitiveness.

self-improvement and long-term sustainability, and, furthermore, it is also necessary to learn to respect other countries and other people. Without internal satisfaction and success it is impossible to influence and attract others. Currently there is a kind of misunderstanding about overemphasizing soft power as a means to influence, attract and compete with other countries, enterprises or individuals, while neglecting that soft power is first of all about upgrading the sustainability of one's own country or enterprises. Therefore, it's crucial to study the meaning and method of creating soft power so as to make clear the direction, approach and measures for enhancing soft power.

I believe that it is most reasonable to study the formation of soft power from the viewpoint of the integration of soft capital, soft technology and soft environment (Jin, Jiang & Gong 2006); namely, the potential of soft capital, the innovative ability of soft technology, and the creativity, adaptability and innovative ability of soft environment, which helps develop the three core elements of soft power (or three resources of the soft power), and their integration and the capability of integration coming into being through the soft power of a country or an enterprise.

As shown in Figure 13, soft capital is correlative with hard capital, such as physical capital and monetary capital, etc (ibid.). In general, physical capital includes land, machines, material wealth and other artificial capital; monetary capital refers to the ability to support, acquire or invest in other forms of capital; natural capital refers to fresh air, water, soil, ecological environment, geographical conditions, mineral and other natural resources, which are the bases of emerging ecological economics research; and, moreover, soil and minerals, which belong to natural capital, can be regarded as hard capital. Human capital consists of individual knowledge, skills and abilities; social capital includes social network, norms and beliefs, etc. (See the section of 'Social Resources' in Chapter 2); cultural capital represents the ability to make acquaintance with social culture, understand and use the educated language, cultural resources, historical heritage, cultural relic, landscape, traditional culture, food, values and so on. Certainly, these different types of capital are not completely independent but interdependent and intertwined. For instance, social capital and cultural capital play a key role in the formation of human capital, while soft capital leads the formation of physical capital including financial investment. Individuals, consortiums or companies often face the choice of which the forms of capital mentioned above they should invest in.

Soft power is manifested differently in different dimensions; namely, it is different at the national, regional, corporate and individual levels, but is inseparable from the above-mentioned three cores. A country's soft power is a system constituted in the relationship amongst those three elements. Only with better innovative capability in soft technology and the soft environment could the potential of multi-capital, including soft capital, be turned into strength, enabling hard power to get better play (see the section of 'Revealing the Essence of Creating Value for Intellectual Capital' in Chapter 1).

Human capital can be classified as a part of soft capital, but in order to be expressed as soft power, human capital must be exploited and 'used'. For example, if top students of Chinese universities whose education has been financed by China cultivate their careers abroad after completing their basic education in China, they will fail to make a due contribution directly to the development of China's soft power. Certainly, this is a ubiquitous experience of developing countries. To make this part of human resources create value for China, we must strengthen the construction of soft environment; we need to pay more attention to creating an environment to facilitate innovative opportunities for using their talents (from the angle of institutions, policies and cultural atmosphere) and develop some 'solutions' which enable the technologies that these students apprehend abroad to be brought back to be effectively transferred or commercialized as they wish. Then some of them will not go abroad after returning to China for a second time. For another example, in order to transform large quantities of internal and external social resources in China into social capital, and to develop social industries, it also needs to create and innovate the soft environment (Jin & Ren 2004).

In the case of China, the most fundamental and long-term goal to enhance soft power is the coordinated and sustainable development of China's society, economy, environment and resources, benefiting the 1.3 billion people and their future generations. That is, to put own house in order – the nation is prosperous, society is stable and friendly in its coexistence with neighbours; the citizens live and work in peace and are content, the majority of them feel free, happy and proud of their own country. Only by achieving this goal could the national culture, values and institutions be convincing at home country and abroad, thereby enabling its image in the international community to have charm, so as to show its powerful soft power.

In order to achieve this goal, the Government must give priority and continue to invest (both financial and policy investments) in the sustainability of education, culture, health care, eco-environment, etc., and constantly foster, accumulate, use and optimize soft capital (as a resource of value creation); seize the opportunity to carry out political, economic and social reforms, making their development mode and code of conduct adapt to the needs of sustainable development; keep up with the times to promote the top-down and bottom-up innovation on culture and values; increase personal quality from ordinary citizen to policy-makers, including politicians, so as to enhance the ability to implement national strategic goals; and ensure the conditions where the varieties of reform and innovation can be realized, varieties of capacity can be brought into play by the institutions (laws, regulations and policies).

As the result, internally, their own potential and the ability to solve internal and external problems will be strengthened, and will be manifested in the national development model in line with the strategic objectives of building a harmonious society and sustainable development; the strategic management and implementation capacity of leaders at all levels are strengthened; and the level of material, social and ecological civilization in the whole society continues to increase. And consequently, the institutional attraction, the impulse power of values, cultural appeal, diplomatic persuasion, and the charm and the affinity of the leaders, as well as the national image, will also be involuntary enhanced. In other words, the attractiveness for other countries or other individuals is the result other than the primary objective.

When Japan was defeated in World War II, the war destroyed almost all tangible things. The reason why they can create an 'economic miracle' in the midst of the ruins within just twenty years is that the human capital built and accumulated during eighty years of Meiji Restoration have not been destroyed. This includes the experience and implementation capacity of developing science and technology and constructing industrialized countries; the required institutional environment, especially the experiences and lessons written in blood learned from the path moving from a complete imitation of the West to the integration of western culture and Japanese culture; as well as the fighting spirit (culture). In other words, the soft capital, the innovation ability of soft technology and the innovation ability of the soft environment, which forms soft power, persist. It is also the case that hard power rises rapidly based on the supports of soft power.

2. From National Soft Power to Corporate Soft Power

In the twenty-first century, enterprises are facing various challenges never encountered before. Taking corporate China as an example, in order to play the role of the engine for transforming the national development model, become the backbone of the a harmonious society and also compete in the international market, corporate China must succeed in meeting with eight big challenges including: how to adapt to the new economy and sustainable development; their role change in the international community; globalized operation; the new corporate culture and values; changes in corporate positioning; and others. These challenges may not be met only through the application of hard power; it is necessary to address them through improving soft power. Competition of enterprises must shift from exclusive competition in capital, products and hard technologies to a win-win competition business model expressed in culture and values, cooperation, innovation, and implementation of the 'triple bottom line' approach, etc. For instance, the Chinese government has a strategic goal of implementing a scientific approach to development, to building a harmonious society with national well-being that is friendly to the environment and to the conservation of natural resources. This requires that China will phase out its emphasis on economic growth imperatives alone and simply take economic construction as the central task in favour of a strategy that stresses a 'comprehensive, coordinated and sustainable' mode of development. To implement the new development model, renewing concepts and changes of government functions are certainly important but must ultimately implement the transformation of the corporate business model to achieve sustainability – the competition of enterprises will shift from hard power competition to competition based on comprehensive strength.

As is the case at the national level, corporate comprehensive strength is the integration of corporate hard power and soft power, and corporate soft power is correlative with hard power, such as products, outputs, financial strength, hard technological patents and financial capital, etc. Soft power embodies various forms, such as business model and operation principles, codes of conduct, core values and corporate culture, charm of corporate leaders, ability of strategic planning, management capability, ability of technological innovation, ability to meet an emergence in the domestic and international business environment, degree of fulfilling social and environmental responsibility, stakeholder relations, internal and external credibility, and so on. However, the above elements providing essential and strategic significance for healthy and sustainable development of enterprise need a long-term accumulating process, and can not be fostered in one day!

Some of China's big companies are struggling to establish a responsible corporate image in the crisis, but they do not know how to deal with and explain the negative information to stakeholders and the public. In fact, the corporate image does not depend on propaganda, but 'doing'. The responsibility for some large-scale security incidents cannot be brushed off on to others; they are due essentially to failures in corporate governance. For another example, as the concentrated expression of soft power, the enterprise's brand image can not be established only by advertising with a big budget. Fulfilling social responsibilities represents an expression of excellent corporate behaviour. However, a responsible corporate image can not be established by a few social donations; enterprises must implement their social and environmental responsibilities at all levels of management and gain recognition from the whole society, especially from stakeholders, through management practice. So far, most of Chinese social responsibility reports pay more attention to promoting the corporate image, but the companies devote very little attention to reforming their actual practices as a platform for enterprises to open doors to the outside world, achieve self-discipline and enable stakeholders and even the whole society to supervise them. Certainly, this kind of report itself has admitted social and environmental responsibility as part of the overall performance, which is a big step forward.

Corporate soft power is the core content and basic concept of national soft power. Enhancing corporate soft power is not only for the purpose of improving competitiveness at the present stage; more importantly, it is for maintaining an ability to sustain survival and development, and to flourish in the constantly changing business environments. Corporate soft power is more influential and durable than hard power for enduring survival and for the sustainable development of enterprises. Therefore, the long-term view needs to be emphasized along with implementation capacity (capability of problem solution), as well as the conditions (the environment in which its ability and potential can play a role).

1. **To increase and accumulate the potential of soft capital.** For enterprises, outstanding leadership and management team are the most critical human capital. The charm of corporate leaders in the twenty-first century has a profound meaning: they show great foresight and are good at long-term strategic thinking; they have proper understanding of the enterprise positioning, incorporating the enterprise strategies into the track of sustainable development; they should not only have the pioneering spirit, but also encourage innovation which does not stick to one pattern, and, moreover, be good at providing the development opportunities for staff; have a correct understanding of the crisis, challenges and failures; attach importance to soft technology talents, and do not regard themselves as scientists or senior engineers; they are good at capturing market opportunities in sustainable development (multinational enterprises have already made lots of money from China's pollution market, while some Chinese companies complain about the increase in costs associated with worrying about sustainability), etc. Corporate leaders must improve or consummate themselves with the above criteria so as to influence and cultivate their employees, human capital, relational capital, organizational capital and cultural capital in the organizational level effectively.

2. **The innovation ability of soft technology.** This refers the ability to enhance the implementation of the long-term strategy and the medium and short term business plan by creating, inventing and using flexibly a variety of soft technologies. It embodies strategic planning ability, strategic management capacity, innovation efficiency and innovative ability of business models suitable for corporate strategic objectives, etc. For instance,

access to market opportunities in advance, winning customers, maintaining talents (Novozymes China has recognized that the greatest cost for multinational corporations operating in China is the loss of senior talents), being in a superior position to access financing opportunities, the inclusion of various social investments, etc., will rest with the design and operation of various schemes in which it is propitious to improve and exert the efficiency of multiple types of capital. Other companies will find it to be difficult to imitate and reproduce such capital, and these conditions are also prerequisites to increase hard power. For another example, the social relation of enterprises, especially relation with stakeholders, is an important part of social capital. Moreover, how to develop and apply the social capital is an important issue of soft-tech innovation. In the development and application of social capital, there are orientation problems (only for the economic interests leading towards irregularities or corruption, or sharing in the legal framework), as well as issue in the design for multi-win-win solutions.

3. **Creativity, adaptability and innovative ability of soft environment.** The soft environment of an enterprise embodies the leadership system, the management mechanism, rules and regulations (codes of conduct), the innovative mechanism, core values and corporate culture, etc. and it must be created and innovated based on changes in the external environment of enterprises. Examples include internally creating a good corporate culture and promoting good global corporate citizenship, while creating good conditions for employees' work and life with innovative activities to stimulate their creativity to form a cohesive force. The soft environment is the basic guarantee for transforming the potential of multi-capital into the strength so as to exert the innovative ability of soft technology. For enterprises, good corporate governance and keeping pace with the times is the core content of the soft environment, while the corporate core values and culture enable the implementation of good corporate governance to be guaranteed.

In short, the enhancement of soft power requires systematic planning, implementation plans and operational deployment, manifesting the measures of enhancement soft power in all levels of enterprise management.

3. Future 500: The Successful Case of Assisting Enterprises to Upgrade their Soft Power

Aiming at maximizing profits, the market economy and globalization have created the Fortune 500 over fifty years. But their demerits of simplistically pursuing fortune, combined with the limits of the industrial economy, have rung the alarm bell.

Therefore, in the unstoppable huge wave of pursuing and creating wealth, persons with a keen awareness of their social responsibilities in academia and industrial circles have dared to challenge the old concept (but still very powerful model, the Fortune 500) and founded the Future 500 in 1995, which is an international non-profit green corporate alliance. It represents a new business model, and its mission is to promote enterprises to fulfil the triple bottom line; namely, economic benefits, business ethic and market credit, as well as eco-environmental protection and construction, and its management principle is not only to maximize the profits for shareholders, but also to deliver maximum returns for stakeholders (Kiuchi & Shireman 2002). Future 500 does not advocate the simplistic ranking of the top 500 enterprises on a scale, but instead encourages and celebrates examples of 'good' companies responsible for the next generation and the Earth.

The naissance of the Future 500 implies the reflection of the industrial circles on resources-consuming and highly polluting economy, as well as on the negative effects brought by rapid development of industry, which is the reflection of world's outstanding entrepreneurs on business objectives and profit models.

In the era when the stakeholders' interest in social and environmental issues are growing day by day, the Future 500 carried out a comprehensive integration based on the twenty-four standards and norms that are widely recognized worldwide, and developed an audit tool, the GC360 (Global Citizenship Audit System). It clued on the approaches and means to implement triple responsibility from key functional areas such as ethics, human resources, leadership, planning, operations, finance, marketing and sales, procurement/supply chain, public utilities/public relations, safety, health and environment, etc., helping businesses to integrate triple responsibility into all levels of management activities and making it to be part of corporate core values and culture. At the same time, from the angle of governance, workplace, community, marketplace and environment it pointed out approaches in how to increase corporate soft capital, improve the ability of soft environment to meet an emergence, and enhance the implementation capacity. In other words, to come up to the path to increase corporate soft power so as to promote a new profit model and operating principle adapting to new economy, actualizing the transformation from a mechanical model to industrial eco-business model in order to improve corporate comprehensive competitiveness.

Foremost among the members of Fortune 500 are companies determined to prosper in the future. Big names in this group include Coca-Cola, Hewlett-Packard, Nike, Boeing, Bank of America, Adobe Systems, among the hundreds of excellent companies who are now in the Future 500 fold. In view of China's importance in global sustainable development, Future 500 (China) was launched formally in 2004. Chinese outstanding enterprises such as Guodian Nanzi, oil giant Sinopec, DHV China, Wenzhou Tengxu Garments Group, Senluo Ecology Group, Haier Group and Shenzhen Water Group have become the members of the Future 500 through GC360 audit, and intend to be pioneers of new economy.[2]

D. The Essence of the Gap Between Developed and Developing Countries

Through the above analyses and examples it is safe to say, therefore, that the combined insufficiency in soft-tech capacity and weaknesses in the soft environment causes the inequality of opportunities for new technologies and new products, and generates the gap between developing and developed countries, as well as between different countries and regions. Soft technology and the soft environment are factors that have for a long time been neglected in developing countries.

For the last thirty years, every frustration experienced by Chinese enterprises as they have entered the market economy and as they have competed with foreign enterprises, has been related to operational mistakes in soft technology and to unfavourable soft environments.

1. Keeping a Clear Head and Pinpointing of Their Own Strength and Weakness

It is important for developing countries to track the cutting-edge technology at an international level. However, if a developing country develops only whatever is popular internationally, the result will only be 'more haste, less speed'. As stated above, just by supplying more hard technology R&D cannot shorten the gap between developing and developed countries in core technology. Developing countries should consider imperturbably and choose the right way for enhancing the competitiveness without ignoring their own unique advantages: to focus on the full development and application of indigenous cultural resources, intellectual resources and social resources which have a vista of limitless promise and may open a new path for 'leap frog' development. And, moreover, making the focus of investment lean towards the R&D of soft technology and soft environment in some way will greatly help to improve the efficiency of technology transfer, so as to reverse the inferior position in hard technology as quickly as possible.

The experiences of Japanese enterprises and Korean enterprises referred to above are worth using for reference. Through industrializing and refining technologies that originated in the United States, such as television technology and videocassette-recording technology, Japan has not only reaped rich profits but also applied for a number of the relevant patents. Similarly, the Korean company Samsung enjoys an advantageous position through its leading position in the development of CDMA wireless communications technology, which also originated in the United States. These examples show that while less developed countries try to strengthen their own endogenous R&D capacity, there is also ample scope for them to exercise their ability in applying core technologies that come from other countries, of which the key is the efficiency of technology transfer.

2. To Actualize 'Leap Frog' Development and Boost Institutional Innovation

Developing countries always advocate 'leap frog' strategies for development. This is a laudable wish but it is important to keep a clear mind about exactly which aspects of development are amenable to this desired leap frog approach. For instance, the general education levels of a country and cultivation of its citizens cannot be skipped over; neither can industrialization, and to some extent infrastructure, be skipped over because there are rules of economic, social and technological development that must be followed. However, leap frog development can be achieved in terms of the speed of economic development and the change of economic structure and technologies in some fields. Nevertheless, even in these fields, 'leap frog' development must be based on conceptual change and must start with institutional innovation. At the same time, more important is the need to invest in predictive research pertaining to soft technology in advance according to the prospect of the development of technology (both soft technology and hard technology). Only by doing this can developing countries avoid playing by the rules set by other countries in the international market competition.

Take China as an example. The high growth of the Chinese economy after the year of 1980 should be attributed to the institutional innovation, namely, the reform and opening up policy. The next step of development is still dependent on the environment innovation, especially on the innovation of institutional environment. We first need a suitable legal environment and to reform those rule of law based on the philosophy of the planned economic system. For example, various government approval systems which are not in accordance with the philosophies of rule of law in the market economy formed an implicit regulation. Under this condition, only the things which are recognized, permitted and approved by government should enterprises and individuals do; people should not do those which are not recognized or ratified, otherwise they have always the possibility and risk to be judged illegal 'after the event'. This kind of governance principle restricted the ability of corporate and individual innovation, while, as the Government's monopoly on power resources, it will inevitably become an institutional basis of a breeding ground for corruption (S. Gao 2000).

After entering the WTO, the biggest barrier facing China is the challenge of merging into the world market economy system and carrying out relevant institutional innovation. According to an announcement by national legal department, there will be a need to revise and eliminate more than 1100 departmental regulations following WTO rules, consistent with commitments made by the Chinese government. In addition, a great deal of effort is needed to work out the new laws, administrative regulations, departmental regulations and other policy measures consistent with membership of the WTO.

3. *The Serious Shortage of Soft-Tech Experts is the Core of This Gap Regardless of the National Level, Industrial Level or Enterprise Level*

Take the Zhongguancun Science Park in Beijing as an example. The correct operation of property rights has already become the bottleneck for the further development of its enterprises. After thirty years of ups and downs, Zhongguancun Science Park cultivated a number of entrepreneurs such as the founders of Lenovo, Stone, Founder and other companies, but they are still very limited in number.

It is claimed that up to 2008 there were more than one thousand intermediary organizations engaged in technical consultation, more than 150 correlative industrial associations, and about 500 professional service centres in Beijing. Nonetheless, a great proportion of scientific achievements cannot find a place to which they may be transferred, and this is true despite the fact that lots of enterprises are thirsty for projects and that there is venture capital worth several million RMB in this area looking for 'good projects' every day. The reason is lacking in technical brokers with technical backgrounds who can track certain technology over a long time, who are very familiar with both the market prospects and the technical complexity of that technology, who understand market environments and who also know how to choose appropriate experts, including relevant laws experts. It is obvious, therefore, they desperately needed a group of soft-technology experts who had not only mastered management technology and a meso-level of market-operation technology but who were also familiar with domestic and international soft environments. They are key figures in the Industry-University-Research cooperation. Such experts cannot be cultivated completely within schools. They must learn through experience and must have opportunities to experience successes and failures in the market, including facing the fierce competition of the international market. Certainly, Chinese companies must develop vigorously their own corporate technology centres or R&D institutes, studying the technologies the company needs so as to fundamentally alleviate the 'project famine'.

The orientation of personnel training.
Even though China is currently in dire need of senior management experts, incurs great expense in recruiting management experts from abroad and the whole society is in need of management innovation and institutional innovation, the training of personnel in China is still oriented almost entirely towards hard technology.

China's *Science and Technology Daily* carried an article entitled 'What will the popular major be ten years from now?' (Wen 2001). According to this study, China's requirements for qualified personnel will be centralized in six types of technology (biological technology, information technology, new materials technology, new energy technology, space technology, oceanography technology) and nine high-tech industries (biological engineering, biological pharmacy, photo-electronic information technology, intelligent machinery, software, superconductors, solar energy, space industry and oceanic industry). Furthermore, according to a forecast from the State Personnel Bureau, the fields in which the strongest demand for qualified personnel will be made in the future are similar to those reflected by the above-mentioned survey. They fall into eight categories: high-tech qualified personnel in the fields represented by electronic technology; biological engineering; space technology; the application of marine resources; new energy and new materials technology; IT experts, electro-mechanical integration professional technical experts; agriculture technology experts; environmental protection technology experts; biological engineering R&D experts; international trade experts; and lawyers.

Unfortunately, the above studies and forecasts have not focused enough attention on the need for qualified personnel from outside the fields of natural science. This oversight is troublesome for a country like China where the development of soft technology is more backwards than the development of hard technology. Shanghai has taken the lead in China in appreciating the value of soft technology experts. A comparison of demand for personnel recently identified in Shanghai and Beijing is instructive. Shanghai has identified that it needs experts in fields such as information technology and microelectronics; finance and insurance; biological pharmacy; petrochemicals and fine chemistry; automobiles and automotive systems equipment; urban agriculture; modern physical distribution; city construction and management; new materials; social services; investment and administration; and culture and sports. Among the twelve specialized professions listed by Shanghai, six are related to soft technology industries as defined in this book, namely, experts needed for the development of intellectual service industries, cultural industries and social industries. Beijing, on the other hand, does not include such professions in its list of priority personnel fields. In spite of being the national cultural centre, Beijing does not necessarily accommodate more soft technology experts than Shanghai does. The importance of soft industry and the popular high-tech industries are not stressed equally.

The main reasons for these short-sighted behaviours are, firstly, that most enterprises have not moved away from the established way of thinking that places hope on certain high technologies being the primary vehicle for improving competitiveness. Secondly, the public education still attaches more importance to knowledge education and makes light of intellectual education or skill education; an imbalance of attention in favour of natural sciences and hard technology; and against that in non-natural sciences and soft technology. The consequences are that students educated in school are generally not equipped to meet the needs of most enterprises.

The *International Herald Tribune* of the US has reported the difficulty of employment for graduates of less prestigious universities in India (Giridharadas 2006). They concluded that the reason why these graduates were in an inferior position vis-à-vis finding a job was not lack of knowledge but a lack of 'soft skills' for which a new generation of employers were looking. However, those colleges and universities that are unwilling to reform are still reluctant to teach these skills, including fluent and pure presentation ability in English, capability to write a decent article, ability to conduct teamwork, as well as catch in depth the point of leaders' intentions. In India, on the one hand, the unemployment rate of college graduates is higher than that of secondary school graduates, while on the other hand, Indian companies have keenly felt the lack of available technical talents. A lack of 'soft skills' becomes the 'soft underbelly' of college students. The challenge facing the old educational system is that the university has produced a large number of graduates that nobody wants to hire.

Although for different reasons, China also faces a difficult situation which a large number of college graduates can not find jobs.

What did developed countries do?

Soft-tech industries are high added-value industries. Many of the 'rules of the game' that control the hard-tech development are produced, developed and updated within soft-tech industry. This is one of the reasons why an increasing number of local students in developed countries are shifting towards the realm of soft technology and why such a large proportion of engineers working in high-tech fields in America originate in Asia or other developing countries. For example, there is a severe worldwide shortage of talented IT personnel, and many talented hard-tech experts educated in China work as senior employees in American companies. In a sense, famous Chinese universities serve as preparatory schools for these people. A metaphor can be applied here: in the enterprise or in the international market, the role of hard-tech experts may be likened to that of 'manufacturing', while the role of soft-tech experts may be likened to that of 'product design and marketing' – with higher added-value. According to a survey on wages of CEOs and employees in global big enterprises conducted by TowersPerrin, an international management consultancy, the US bosses received the highest annual salary of 1.06 million US dollars; the average annual salary of British president was about 0.7 million US dollars; French CEOs ranked third with an annual salary of 0.6 million US dollars; that of German bosses was 0.41 million; and Swedish presidents had 0.35 million US dollars a year (*China Economic Times* 2001). This is part of the reason why MBA is so popular in developed countries. Because of the technological gap associated with the current international division of industrial structure, China may have to play the role of the 'manufacturer' for quite some time and reluctantly allow high added value to be accrued by developed countries. Therefore, China's strategy for training and producing qualified personnel must be altered as soon as possible to redress this unfortunate situation, and to thereby position itself from a 'manufacture giant' to become a 'design giant'. In order to actualize this goal, we must first change the structure and direction of personnel training.

Leaders should be soft-tech experts
Due to a lack of awareness of soft technology, we have been mired in the misunderstanding of 'talent'. Since China began to implement its policy of reform and opening-up, it has promoted many engineers and technicians to leading positions with the idea of making cadres younger and more knowledgeable. However, by not paying enough attention to the education of soft technology experts, many painful experiences have occurred. For example, some excellent engineers have been promoted to executive directorship of enterprises; outstanding scientists have been chosen to be administrative leaders of academic institutes; some excellent teachers or professors have been promoted to head positions of schools and universities; some physicians and surgeons with excellent medical skills have been promoted to being chief administrators of hospitals; and some technicians without a financial education have been appointed presidents of banks. These kinds of promotions of intellectuals to important positions generally produce situations where neither side gains, i.e. neither the promoted expert nor the organization under his charge are able to thrive optimally. Many examples of these kinds of organizational mismatches have produced bitter experiences that have forced enlightened observers in China to note that excellent engineers are not necessarily capable of being executive directors, and that excellent scientists may not necessarily be competent entrepreneurs. Moreover, entrepreneurs and venture capitalists cannot be nominated or appointed by government.

We do not deny that a number of successful entrepreneurs were the scientific and technological personnel. However, in general, scientists need to explore the laws of the objective world; engineers and technicians need to address the technical problems in production practice. Besides well-knit professional knowledge, these people also need to not be captivated by money in the dazzling market environment, to be able to endure 'loneliness' and 'a cold reception', to have the courage to hang upon and not to be afraid of failures and setbacks in order to find a solution, as well as the truth. As a manager or leader, their operational object is the human psychology and human behaviours, and they need to focus on various 'relationships' (human-human, human-nature) to coordinate, control and balance; their mission is to actualize the strategic objectives of the organization (enterprise, department or even state) which they are taking charge of so as to improve overall performance and competitiveness, and they are willing to put their lifetime energy to do it. From the angle of speciality, they should be soft-tech experts. However, due to the limitation of institution and evaluation system, those who are in the leading positions with scientists and technical backgrounds are still pursuing their own scientific achievements or academic titles in the field of hard technology. As a result, their level of leadership had not been enhanced, while it has led to the exaggeration and even corruption in the academic atmosphere and discipline.

The misunderstanding of soft-tech talent training

We may take management experts as an example. As explained earlier in this book, management expertise requires at least three elements: moral standing or quality, knowledge and technical background, and intelligence and practical ability. As to knowledge and technical background, training and education in soft science and soft technology is absolutely necessary. However, because our understanding of knowledge and technology stresses hard technology and the natural sciences, soft technology has not been regarded as a special field of technology. Thus, people with neither social science knowledge nor sufficient specialized education and training relevant to soft technology have been promoted to management positions and even to extremely important senior management positions. This practice seems to have developed in the belief that managerial capability can be easily gained through experience or be 'compelled' through practice.

Many organizations have paid a high price following poor decisions that arose as part of a 'trial-and-error' approach to management, based upon the so-called 'compelled-into-practice' approach to management. The seriousness of the mistakes that have happened is related to the managerial levels of the decision-makers. In general, the higher the position of the unqualified manager, the heavier will be the losses flowing from decision-making mistakes. It is widely understood that mistakes of entrepreneurs are responsible for 85 per cent of enterprise bankruptcies worldwide.

Education is a great cause for the future

Hence, a significant proportion of the efforts should be directed towards the needs of talented personnel during the next one or two decades or more long term. If the plans for educating and training qualified personnel are designed to meet the current, short-term profile of market needs, or if students are educated only in professions popular in recent years, it will never be able to change its fate as a follower of developed countries and as a passive supplier of casual labour in the global market.

For example, although in order to restore its status as a treaty power in the General Agreement on Tariffs and Trade (GATT) and join the World Trade Organization (WTO) China carried through the negotiations for as long as fifteen years, it failed to pay attention to fostering qualified personnel for the WTO in advance. Monterey Institute of International Studies is the first training base in the world for a masters degree related specifically to the WTO. It has trained more than 80 WTO experts from over ten countries worldwide, including the United States. The Academic Director of the Ron Brown Center for Politics and Commercial Diplomacy of this institute, Geza Feketekuty, believes that WTO personnel should be familiar with the WTO rules, with high quality in analysis, advocacy, alliance, decision-making, negotiation and settlement of disputes, and also be able to handle with ease the phenomena brought by economic globalization after the open door (*Beijing Youth Daily* 2001). He has classified WTO experts into two types: one is commercial diplomats, who have knowledge not only about business, economy, politics, domestic policy-making, trade, work processes of organizations, negotiation skills, etc., but also maintain good relations with the political circles, media and business community; the other type of expert is the professional in finance, telecommunications and laws. After China's accession to the WTO, it is not only the Ministry of Foreign Trade and Economic Cooperation of China (MOFTEC) which needs people talented in WTO matters, but also other relevant ministries, regulatory agencies, large enterprises (including state-owned enterprises), associations, law firms and trade-related organizations. It is widely estimated that China needs to have at least 350,000 management experts. China has been involved in MBA education since 1991 (Ma 2008). The number of schools which have set up MBA education increased from nine in 1991 to 127 in 2008, recruiting more than 30,000 students each year compared with less a hundred students in 1991. The cumulative enrolments have exceeded 212,000 MBA students in China up to September 2008, in which over 100,000 students have been awarded the MBA degree. In contrast, more than 70,000 students graduate with an MBA degree every year in America.

Chapter 4

Soft Technology and Innovation

Innovation is the process of creating value through new ideas, new ways and new means; and the value referred to here includes not only economic values, but also social values and eco-environmental values. From the perspective of a broad understanding of technology, the process of creating value is actually a process of inventing, creating and applying soft technology, focusing on the operational resources of soft technology, or new application processes of existing soft technologies. In other words, innovation is an operational process of soft technology aimed at creating value.

We will further understand the relationship between soft technology and innovation below through analyzing the functions of soft-tech, innovation space, institutional innovation and the structural framework of innovation.

A. The Functions of Soft Technology

In this book the functions of soft technology are classified into three categories, as summarized in Table 7: to act as tool of innovation, to act as core technology for independent industries, to provide content and basis for institutional innovation.

1. Soft Technology Provides Innovation Ability of Hard Technology

As the process technology for technology transfer, technology commercialization and industrialization, the role of soft technology is to act as a servant for the transfer, and even industrialization, of other technologies. Therefore, soft technology is both the tool and the content of technological innovation in a broad sense.

The relationship between soft-tech innovation and hard-tech innovation consists in soft-tech being a tool for hard-tech innovation while, simultaneously, the efficiency of soft-tech innovation, namely the efficiency of creating added value, depends on whether or not the integrated innovation of soft-tech and hard-tech is successful. In the process of soft-tech innovation, hard-tech, as well as all indispensable natural objects and products, are instruments for soft-tech innovation.

Table 7: Functions of Soft Technology.

Function of soft technology	Tool of innovation (technological innovation)	Key technology for independent industries (industrial innovation)	Content and basis of institutional innovation (institutional innovation)
Examples	• process technology of technology transfer • tools and contents of innovation(including hard-tech, soft-tech and non-tech)	• intellectual service industries • cultural industries • social industries • soft-life industries	• venture capital technology and Nasdaq stock market • stock technology and stock company act • patent technology and exclusive law, as well as intellective property rights • life tech and the relevant laws and regulations on ethics • nuclear technology and relevant laws

Source: Zhouying Jin, 2000.

It was said that technology from one context should not normally be copied 'mechanically' in other contexts. Hard technology, however, can actually be 'copied' and it should be standardized. It is soft technology that cannot be applied 'mechanically'.

It is widely understood that most knowledge may only be converted into actual products and services through the medium of technology. In order to insert technology into products (so-called materialization), to expand the presence of products in the market and to enable enterprises to make profits a series of 'insert' technologies and 'expand' technologies is needed. These technologies include the technology for protecting the inventor's interests, the technology of fund raising for high-tech industries, the technology for improving the efficiency and quality of materialization, and the technology for delivering more goods and better services to customers. Only when the above 'intermediate' technologies are actively developed and applied is it possible to speed up the technology transfer; namely, to accelerate the process of embodying market value of knowledge and technology, and to exert competitiveness in hard technology. That is the reason why managerial innovation and technological innovation are often mentioned in the same breath even though it is not appropriate to describe the above activities as 'management'. We could say, in short, that a series of soft technologies is the means and tools for technological innovation. Namely, soft technologies provide the means and tools to inaugurate new markets, and to develop new products and new services creatively (as understood in the Schumpeterian view of innovation). Soft technologies also act as the medium and the bridge for adapting hard technologies to contemporary circumstances. In order for a community to improve its overall ability in technological innovation, it must grasp and take full advantage of soft technology.

Soft technology is, in fact, a vehicle for moving technology from the outside (including 'foreign' sources from both the home country and abroad) to the inside of an enterprise or industry. In other words, it is the means for conducting technology transfer.

An example from China, Zhongguancun, the High-Technology Development Zone in Beijing, serves as an excellent illustration of the above principles. For some time arguments have raged about whether or not the development of Zhongguancun during the first decade was really based on technology, let alone high technology. Some say that it developed, instead, through '*trade → manufacturing → technology*'.[1] Some people even say that Zhongguancun's 'electronic street' is a 'cheater's camp'. However, from the soft-tech perspective, 'trade' is a process of utilizing commodity exchange technology and market technology to create value-added; 'manufacturing' is a process of utilizing manufacturing technologies, organization technologies and management technologies comprehensively. Trade and manufacturing in Zhongguancun during the last decades has primarily been about applying processes of soft technology in a special environment within China.

Before China's reform and opening up, Zhongguancun had already become a well-known educational centre and a national venue for research in natural sciences. With 68 universities and over 200 research institutes, over 30 per cent of the academicians of the Chinese Academy of Science and Chinese Academy of Engineering are gathered in this region of Beijing. However, after China's reform and opening up, especially at present, this region was no longer a pure research centre. It became a centre of a new type of economic activity, an incubator of high-tech industries and a centre for the commercialization of knowledge and technology in China, as well as a pilot and demonstration base for new things in a variety of senses. Zhongguancun's unique charm helps it attract many talented people from across the country and from abroad.

Since 1990 the number of registered Zhongguancun 'residents' has increased by 37% annually. The growth rate of these enterprises is significantly greater than that of the national average: from 1988 to 1998 the total technical, industrial and trade income of the Zone recorded an annual growth rate of 42.58% and the industrial output had a value of 48.66%; in 2000, the total technical, industrial and trade income amounted to 154.03 billion Yen, increasing 46.8% compared to the same period in the previous year (M. Zhao 1999). As of 2007, the Zone's industrial scale of new and high-tech industries reached 859.5 billion Yuan, forming an industrial structure based on the electronic information industry, and the two other industries, the clean-tech industry and the biopharmaceutical industry. The proportion of the total economic output of the Zone which accounted for high-tech service industries was 50% by 2007. There were nearly 20,000 enterprises in total, of which 4000 were launched by Chinese citizens returned from abroad. Nearly 10,000 practitioners who work in the Zone are Chinese citizens returned from abroad, while there are 2000–3000 new start-up enterprises every year. By 2008, about seventy multinational corporations had established R&D institutions in Zhongguancun and there were more than 4000 foreign technicians employed there ('Zhongguancun is on the rise' 2009). 'Lonesome' Zhongguancun has changed, and the zone is now with brimming vitality.

The reason why even greater, more fundamental, changes did not occur in Zhongguancun during the past several decades is partly due to the lack of 'trade and manufacturing' links. Under the planned economy, most enterprises had less inner demand for technology progress and they were weak in technology-transfer ability. In a sense, there were enterprises but no entrepreneurs. Those who mastered natural science knowledge and technology could not raise funds to transfer knowledge and technology to products or to endow them with market value because of the limitations of the environment. Nor could they turn products into commodities in the market or accumulate money for further innovations.

Therefore, during the era of reform and opening up in China, the road of *trade* → *manufacturing* → *technology* has been the common solution for people who have had no market economic knowledge and experience under the planned economic system. This gave them the opportunity to experience the economic laws of the market for themselves through the processes of trading, acting as agents, conducting assembly activities, copying and imitating, etc. for the purpose of independent innovation. In this way, they became familiar with markets and by following these steps they carried out the design and production of their own products. In other words, they learned soft technology. If a sound macro environment for exploring and doing business did not exist, or there were no masters of soft technology available, innovation ability would surely not have exerted and improved. The achievements of science and technology would have been doomed to remain as samples and as display items in a showroom. We can say that the original hard technologies in Zhongguancun are only potential competitiveness. Without the support of suitable soft environment and better operable ability of soft technology even the best high-tech could not have become products and services with market value.

Thus, we can now understand that the preceding decade at Zhongguancun has involved the process of Chinese scientists and engineers mastering and making use of soft technology to foster their innovation capability under the new environment presented by Chinese government policy reforms. This capability could not have grown from textbook knowledge; it had to be learned the hard way. Who could seriously claim that during the preceding decade Zhongguancun has not created technology of its own?

In fact, both soft-tech and hard-tech innovation provide each other with the tools of innovation. Because no matter whether we are concerned with soft-tech or hard-tech, they are both either the objects of innovation or tools for innovation, and their relationship in the innovation is that both of them are the means of innovation for the other.

2. Soft Technology is to Act as the Core Technology to Create Independent Industries – Providing Core Technology for Soft Industries

An industry is an enterprise cluster that engages and links identical or similar businesses. If there is a technology that underlies and links all the enterprises in the cluster, thereby adding value to their business, that technology may be considered to be a core technology of that industry. According to the traditional understanding, the so-called 'core technology' of the industry refers to the hard-tech dominating the secondary industry such as the technologies of chemistry, electronics, textiles, information, and construction. Interestingly, although the tertiary industry has on average accounted for nearly 70 per cent of the world GDP, and while this level is higher in the developed countries, it is not generally agreed that the service industries have core technologies. The driving forces in these industries have therefore been described as 'non-technological factors'. This is a great obstacle to service innovation. In fact, in the same way that hard-tech may form the basis of 'hard' industries, soft-tech can form the basis for new, independent industries or act as the core technology of those industries. Cultural industries, social industries, soft-life industries, as well as intellectual service industries with unlimited prospects, have embraced omnifarious soft-tech as their core technology. Simply because of our 'stubborn' old notion of technology, we turn a blind eye to it now. As long as we change the thinking mode, we can find business opportunities everywhere and will develop more new creative industries (see Chapter 5 – 'Soft Industry').

Take the contribution of the soft industries in Silicon Valley as an example. The University of California's Professor Martin Kenney points out in his *Institutions for New Firm Formation in Silicon Valley* that Silicon Valley has the ability to periodically generate new industries and clusters (Kenney 2000). This is the point where Silicon Valley differs from ordinary industrial clusters and cannot be explained by common industrial-cluster theories. Silicon Valley is 'not only a productive site for the creation of new firms, technologies and industries, but it also has fostered the development of innovative, new business models', some of which have been successfully used throughout the world. To be regarded as an electronic industrial cluster is just part of the picture of Silicon Valley; it's key to success lies in the 'emergence of those non-technical institutions', which acted as interactive supplements to high-tech industry. Martin Kenney believes that two kinds of economies exist in the Silicon Valley. The first economy consists of those 'firms, research laboratories and universities that are the constituents of the existing economy, which are not unusual for any industrial cluster'. Their goals are, 'in the case of private firms, profitability and growth; in the case of universities and non-profit research institutions success is measured in terms of research and education'.

In addition, there are a series of another type of organization in Silicon Valley that constitute the second economy, which set Silicon Valley apart from most other regional clusters and which are combined to create an 'economy' that is 'predicated on facilitating entrepreneurs in the creation of new firms in some foreseeable areas'. This is the reason why the second economy is so attractive for entrepreneurs, and why entire start-up companies from regions around the world locate in Silicon Valley. In fact, the second economy is generated from those organizations whose nature is institutional rather than corporate or whose nature belongs to what we might call 'social industry'. Their activities are engaged mainly in 'nurturing new firms and the transformation of quite informal business arrangements in the stage of the beginning, then into the delivery of specific services and institutional arrangement for new firms'.

Though the firms and universities that belong to the first economy do also initiate new independent firms or act as a source of entrepreneurs, in themselves these kinds of activities tend to lack continuity and are prone to miss opportunities since they tend to be rather fixed in the set of businesses and clients with which they deal. In Silicon Valley, venture capital investors, investment agencies, venture banks, recruitment companies, law offices, accounting offices and sales companies combine to form a strong network for the second economy. Of course, it is difficult to distinguish clearly between the first and second economies because they are interdependent and interwoven. Nevertheless, for second economy organizations, entrepreneurs are the initial and essential investment (their ideas and contributions), and venture capital is the second most important investment, as well as the main source of capital.

The process, by which the second economy has emerged, as proposed by Martin Kenney, is actually the process of forming soft industry. Kenney's Second Economy is the soft industry that serves the First Economy; and those activities that create new business models and new firms are the innovative activities of soft technology. Silicon Valley's success also reveals that the integration of hard-tech innovation and soft-tech innovation, as well as interactive innovation, is the only way for successfully building a high-tech development zone.

Now Zhongguancun has also begun to form a second economy. The proportion of the total economic output of the Zone accounted for by high-tech service industries is 50 per cent, and most of them belong to the 'second economy'. Recognizing the essence of the soft industries will help us to form an interdependent organic ecosystem.

3. Soft Technology Provides Content and a Basis for Institutional Innovation

'Institutional innovation' means those processes of substituting new institutions for the old ones, or the process of changing, breaking or establishing institutions according to the requirements of social development and technological progress. Technological progress (whether hard or soft technology), the new application of technology and industrial innovation typically associated with rapid change constantly generates the need for institutional innovation, and institutional innovation, in turn, generates a favourable environment for innovation, with the strengthening of the soft-tech implementation capacity thereby improving the efficiency of innovation. In contrast, the basis and content of institutional innovation is provided by the constant development of soft technology. For example, intellectual capital management (in particular the development of patent technology) gave birth to a variety of patent systems and their continuous innovations; the development of corporate merger technology stimulated the emergence of anti-monopoly laws during different development phases, and so on. (See the section of 'Technological Institutions' in this chapter for details).

B. Soft Technology and Innovative Space

An awareness of technology, in the broad sense, enables us to move the concept and activities of innovation from its narrow form to its broad form.

According to Joseph Schumpeter, the path-setting economist who pioneered the development of innovation theory, innovation is about establishing new production functions; namely, it is about introducing a new combination of production factors and production conditions into the production system. The new combination, as described by Schumpeter, includes the following five cases (Schumpeter [1912] 1934):

1. The introduction of a new good – that is, one with which consumers are not yet familiar – or of a new quality of a good.
2. The introduction of a new method of production – this need not be founded upon a scientifically new discovery and can also be a new way of handling a commodity commercially.
3. The opening of a new market – that is a market into which the country in question has not previously entered, whether or not this market has existed before.
4. The procuring or control of a new source of supply of raw materials or half-manufactured goods, again irrespective of whether this source already exists or whether it has first to be created.
5. The creation of a new organization in any industry, such as the creation of a monopoly position (for example, by setting up a trust) or the breaking up of a monopoly position. Schumpeter also referred to these 'new combinations', or 'innovations', as 'economic development'.

David Sawers further stated that technological innovation is 'the application of new technology to some practical purpose, or the new application of the existing technology for some practical purpose [...] Innovation is characterized by its variety [...] Innovation is a technological process, as well as a commercial process, military process, or social process' (in Williams [1979] 1989: 37).

Alexander King has defined technological innovation as 'the first successful application of science and technology in commerce and military affairs' (ibid: 147). He divided technology development into three stages: invention, innovation and dissemination. 'Invention' is the emergence of a new concept of how to apply science and technology to a particular purpose; 'innovation' is the process in which invention is transferred to commodities and services;

'dissemination' is applying innovation widely to an industry, or worldwide, thus contributing to the growth of industrial productivity.

Following Alexander King, we also need to recognize that invention and innovation are not the same thing: they are two stages of technology development. Invention is the emergence of an idea, concept or thought of some new technology; while innovation is the process of embodying its value through which end-users utilize new technology.

It is obvious that the aim of innovation is to create added value while the process of innovation is the process of creating added value.

The innovation process, as enumerated by Schumpeter, could be interpreted involving the following elements: 1) adopting the new design or strict management to develop new products or improve product quality; 2) operating a variety of tools for technology transfer such as productive technology, management technology and organizational technology; 3) the technology of creating markets; 4) the creation of new market channels or physical circulation and distribution systems around raw materials and products; and, 5) organizational innovation inside and outside of enterprises. From the view of soft technology, the above elements of innovation are part of soft technology activities. David Sawers emphasizes that innovation is the application process of technology; namely, the activity of process technology; Alexander King stresses that technology innovation comprises both technological activities and commercial activities and that the commercial activities constitute the technology transfer process of hard technology; namely, the process of value creation. King's definition of innovation comes closer to the definition of soft technology in this book.

From the perspective of soft technology, the various combinations of ideas of innovation that have emerged from the work of Schumpeter, Sawers, King's or others have tended to place particular stress on tangible products and hard-tech industries or on innovation in the production sectors. However, the objects or means for creating value are not only tangible products or hard technologies. If the Schumpeterian 'new combinations' can be extended to the soft-tech sphere, then the Shumpeterian view of innovation will be able to embrace new concepts and new contents.

In fact, awareness of technology in the broad sense has expanded the space of innovation to five dimensions.

First, in the micro- and meso- dimensions, innovation space should be expanded from hard-tech innovation to soft-tech innovation, including service innovation; from hard industry innovation to soft industry innovation; from hard environment innovation to soft environment innovation, as well as integrated innovation incorporating all of the above factors. We can think of these as the first level of the innovation space. Figure 14 provides an illustrated comparison of Schumpeter's innovation concept and the first level of innovation space.

Secondly, in the macro- dimension, innovation space should be expanded from focusing on the sectors of material production to the sectors of non-material production; from economic activities to social and cultural activities; from innovation aimed solely at

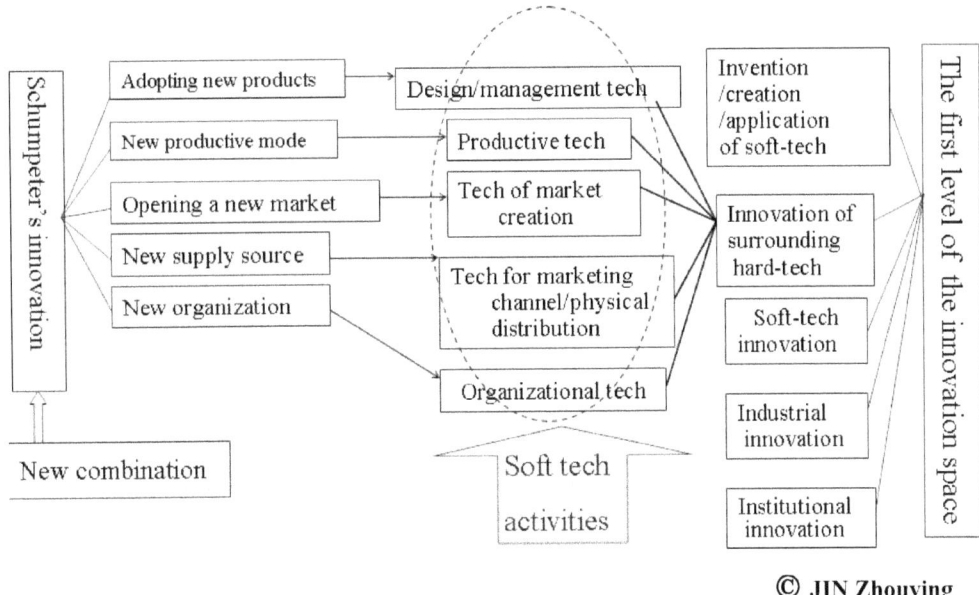

Figure 14: The first level of the innovation space.

economic benefits or economic value to innovation under the social, economic and eco-environmental framework of sustainable development; from technological innovation to institutional innovation; cultural innovation; as well as the integrated innovation of all above dimensions. We can name these as the 'second level of innovation space'.

Third, from the viewpoint of the innovative resources of soft technology (see Figure 15) innovation should not only include technological innovation, industrial innovation and environmental innovation, but should also include innovation in multiple types of capital. That is, innovation focusing on hard capital and soft capital. We may think of these as the third level of innovation space.

Fourth, from the perspective of corporate management, innovative space is expanded from the product, technology and internal management of enterprise to the operation process of the entire enterprise, including strategy, supply chain management and business model innovation. IBM provides a typical case how an enterprise may expand its space of innovation in its publications, *Soft manufacturing* (IBM Global Service 2008). Figure 16 is IBM's understanding for the transformation of innovation. We can call the innovation space involving a specific enterprise management as the fourth level of innovation space.

Fifth, the innovation space formed from the main body of innovation can be thought of as the fifth level of innovation space. For more details, please see the section of 'The Main Body of Innovative Activities and the System of Technological Innovation' in this chapter.

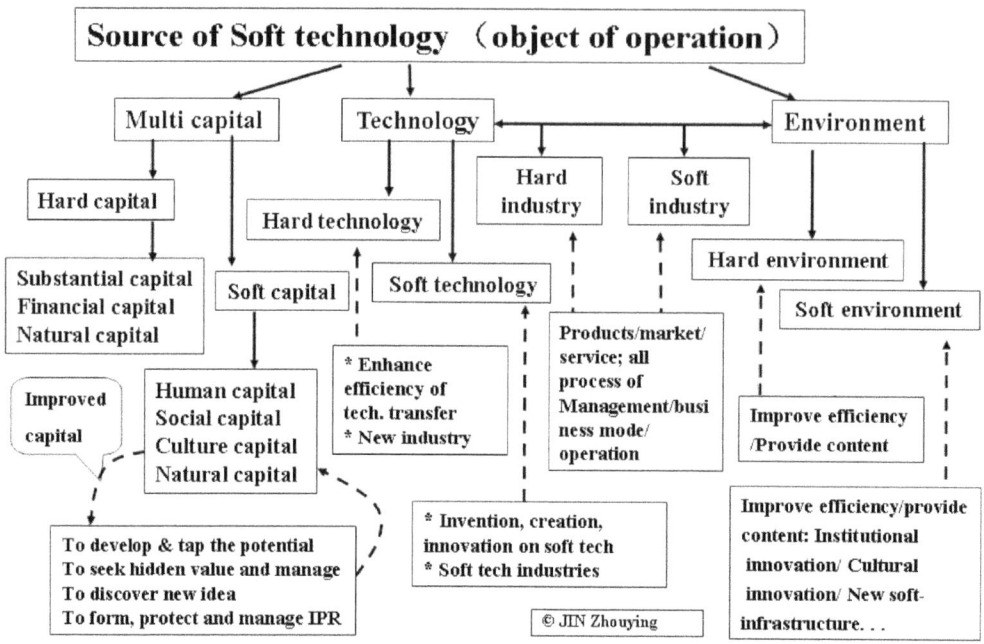

Figure 15: Main source of soft technology.

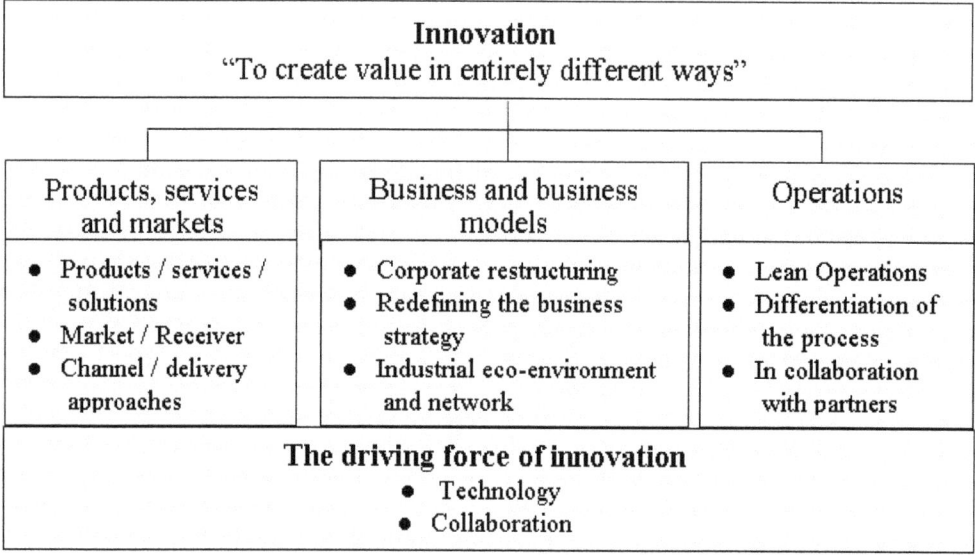

Figure 16: IBM's understanding of the transformation of innovation.
Source: IBM Global Service (© Copyright IBM Corporation 2008)

There is no doubt that innovation is the process of creating new added-value through all resources, means, tools and approaches; the essence of it is the operational process of soft technology aiming at creating value. In other words, the essence of innovation is the process of invention, innovation and dissemination of soft technology, or the process of new application of existing soft technology.

That is to say, innovation theory should be endowed with a deeper and broader meaning than has typically been the case up until now. Correctly understanding the space and essence of innovation will help us to move beyond the faulty but currently dominant focus on hard-tech innovation and product innovation.

C. Soft Technology and Institutional Innovation

For a long time, economists, scientists and sociologists have studied the institution from a variety of perspectives. The subject of institutional economics was very popular in the US during the 1920s and 1930s. New institutional economics emerged in the 1970s, causing economists to study the contribution of institutions to the development of the economy from different points of view, such as property rights theories, transaction costs, trust agents and contract theories. For example, Douglass North, the primary representative of the new institutional economics school of thought, believed that a series of institutional changes paved the road for the Industrial Revolution – a fundamental change for human society. Institutions are also studied in science and technology fields because they are among the most significant driving forces of technology; they not only set the basic conditions of technological innovation but they also determine the allocation mechanisms of the income generated through innovation. History is replete with evidence that the social and economic development of nations involves continuous reformation, and that institutional innovation is at the core of social and economic reformation.

David S. Landes discusses the importance of institutions in his book, *The Wealth and Poverty of Nations*, in which he analyses the resistance to technological progress that has been observable in China and other ancient civilized countries (Landes [1999] 2001: 278). He points out that this resistance comes not from lack of innovation but from institutional ossification.

What, then, is an institution? What is the essence of institutional innovation? For the institution, there are different understandings and interpretations by different schools of thought or from different points of view.

The *Comprehensive Dictionary of Chinese Language* (*CDCL*) offers three definitions of institutions: 1) 'rules requiring all members of a community or organization to stand by and do things according to certain procedures'; 2) 'political, economic and cultural systems that are formed under certain historical conditions'; and 3) 'political course and policies' (*CDCL* 1989: 210).

From the perspective of institutional economics, Douglass North proposes that institutions are a series of formulated rules, law-abiding procedures and moral criteria of conduct, which aim to restrain the behaviours of individuals and are directed towards seeking aggregate social welfare or the greatest marginal benefit (North 1981: 325), and that institutions are the game rules of a society (North 1994). V. W. Ruttan, another representative of the school of new institutional economics, points out that an institution is a code of conduct that is used to control a particular behaviour pattern and mutual relationship (Coase, Alchain & North 2000: 329). Wolfgang Kasper and Manfred E. Streit of Germany think that an institution is a code of conduct, and that it thus becomes a means to guide people's behaviours. It usually excludes some behaviour and limits possible reactions. Therefore, institutions make people's actions foreseeable. Institutions are regulations made by people, that restrain random and opportunist actions in interpersonal relationships. Institutions decide to a large extent how people should realize their goals and whether or not they can realize their basic values (Kasper & Streit 1998: 32; 37; 112; 142).

To summarize, an institution is an aggregate of a series of codes of conduct and standards, which is used to limit and regulate people's mutual behaviours.

As institutions establish the behavioural rules of people, they can be classified according to what behaviours these rules concern (encourage or restrict) into religious institutions, political institutions, social institutions, economic institutions, technological institutions and others.

Religious institutions restrict people's behaviour regarding religion and they offer rules that restrict people's beliefs; political institutions are the rules that govern political authority and the allocation of social resources and income; social institutions restrict and formalize people's behaviours in social activities and social lives; economic institutions restrict and formalize all kinds of relations in economic activities, including property rights delimitation, relations of business and competitions, economic organization and property allocation, etc.; technological institutions encourage or restrict all kinds of behaviours in scientific research, technological invention, technological innovation and technology expansion.

Familiar behavioural rules include regulations, mechanisms, institutions, policies and all kinds of standards. Generally speaking, laws are behaviour rules made by legislative bodies that are guaranteed to be carried out by state power; policies are the regulated criteria for making decisions about actions and are established by government for accomplishing tasks in a certain historical period. There are a variety types of definition on regulations but regulations normally refer to 'the government behaviour of controlling the actions of citizens, juridical persons and subordinate organizations of the government' (JETRO 1995) (Some commentators also portray regulations as rules authorized by law through government administrative organizations that restrict and supervise behaviours of enterprises active in the market. Generally, they refer to the Government's micro-management tools). And, moreover, systems are created from a set of institutions that exhibit special relations among themselves. The main body through which an institution is established may be either a governmental or non-governmental organization according to the situation.

This book takes economic and technological institutions as examples. It probes into how institutions come into being (requirement and motivation), what the content of institutions is and what the relationship is between institutional innovation and technological innovation from the perspective of a broad understanding of technology.

For convenience, the term 'institution' will be used here to cover regulations, institutional mechanisms, institutions, policies, the rule of law, laws and standards, etc.

1. Essence of Institutional Innovation

Now let us review the process from soft-tech invention or innovation to institutional design and institutional innovation (Figure 17).

From the perspective of inventing, producing or innovating soft technology, a creative concept or an idea is first needed. However, the idea or the concept is not actually soft technology because it is not in itself operable in a practical sense.

After having the creative concept, idea or experience, and addressing the 'objective' at which the idea aims, it is first necessary to identify whether it can be achieved by existing soft technology; if not, it needs to invent new soft technologies. Namely, it needs to design new approaches and methods of achieving the objective. If, however, through numerous experimentation or experience, a common way is found to convert the concept or idea into a practical operational system (or operational mode, operational procedures, operational solution) for solving problems, then we may legitimately say that soft technology has been created. Namely, a new soft technology can emerge depending on the different 'objective', while if the original soft technologies can be applied then it enters into the innovative stage of the soft technology.

However, as soft technology becomes popularized and implemented on a broad scale in society – because soft technology always involves the three variables of human factors,

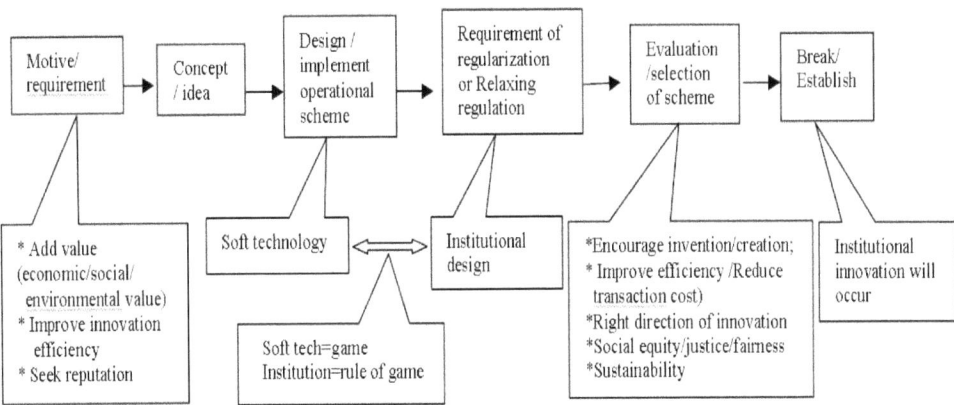

Figure 17: From soft-tech innovation to institutional innovation.

social factors and cultural factors – any changes in soft technology always require changes in organizational structure, patterns of economic operation, modes of social activity, values and the relative roles, positions and interests of individuals and groups in the community.

For instance, the application of commercial technology is usually accompanied by competition under the market economy. To become more competitive, companies rely not only on improving the performance of their enterprise internally (e.g. by increasing efficiency, improving the quality of products and services, or acquiring intelligence and talent), but also employ new external means to improve competitiveness (e.g. adopting new corporate strategies, employing competitive intelligence or pursuing merger techniques, virtual technology, advertising technology and PR technology, or even using head-hunters to 'steal' talent from rivals).

Competition can be an effective way to promote industrial development but sometimes companies engage in immoral, divisive or criminal tactics as part of competition, thereby damaging normal market order. If free competition is not restrained, unscrupulous application of above technologies may lead to monopoly and unfair competition, and even furthering activities of enriching oneself at the expense of others or serious criminal activities, e.g., the dishonest financial reporting or accounting scandals as recently revealed in the cases of Enron, Arthur Anderson, WorldCom and others in the United States.

Hence, when new soft technology suitable beyond the organizations, regions and domains is disseminated, or when it needs to apply soft technology which serves new hard-tech innovation, there come into being requirements of the standardization of soft technology or deregulation. Namely, it becomes necessary to innovate institutions, design new institutions or repeal the institution related to specific activities (including soft and hard technologies). That is why soft technology is described as the content and basis for formulating appropriate institutions and mechanisms, and the relevant soft-tech innovation provides content and basis for institutional innovation. For example, the Antitrust Act was established in the USA in 1914 to address the need to standardize merger technology; and in 1933 and 1934, the Stock Act and the Stock Exchange Act were established to address the need for standardizing stock technology and the stock market at the climax of enterprise mergers during that period. It can be said that in order to be able to ultimately implement soft technology, it should be supported by institutions.

The institutional design is in fact to regularize and maintain routine in the soft technology used in the process of value creation (such as economic, social and eco-environmental value) and accessing to benefits, etc., thereby becoming the implementation criterion or the game rule of operating soft technology, and then making it into the publicly recognized code of conduct and social rule so as to encourage, regulate or inhibit or even prohibit some organizational and individual behaviours. However, in view of the characteristics of soft technology, institutional scheme – changing or creating new work routines or code of conduct to replace the old ones – must incorporate international, national and regional conditions (international circumstances, economic and technological levels of development, values, ethics, social norms, religious beliefs, language, culture, social organizations, administrative systems, government's role and the natural environment are important factors for institutional innovation).

Once a new of institutional scheme has been designed, it is important for the relative merits of alternative institutional designs to be properly evaluated. Common evaluation principles for economic or technological institutions, for example, include whether or not the institution favours creation, invention and innovation; favours defined property rights; improves cooperative or competitive relationships among individuals and groups; increases innovation efficiency; expands economical scale; and reduces the transaction costs or restrains those unfair competitions which violate the above principles.

The principle described above can be regarded as the precondition of institutional change. However, these types of rules tend to change according to the progress of human society and the changes of values. Entering the twenty-first century, in view of the challenges facing human beings and the need to generate examples of 'good' companies, institutional design must conform to high standards of 'good' institutions. That is, it should be consistent with sustainability principles from the points of view of economy, society, environment, natural resources, etc.; comply with either social progress and civilized moral standards, or the principles of equity, fairness, justice and transparency; and consider either innovation efficiency or innovative direction and institutional ethics, as well as whether it is conducive to social harmony. Certainly, it must satisfy the mission and functions with which the soft technology designers endow it.

Basically, if the new institution, supported by a new soft technology, conforms to the above benchmarks it will be a successful institutional innovation, as long as the advantages or (economic, social and environmental) benefits of institutional change or overall return rate it obtained are sufficient to cover its costs, and if most of people can share the costs and benefits of institutional change. If it is successful then a new cycle of institutional innovation will take place: soft technology invention → innovation → diffusion → application → regularization → establishment of institution → institutional change → breaking → replacing with the new.

Through the above vehicles (the relationships between soft technology and institutions), the second attribute (the relations of production) of soft technology is manifested. Conversely, institutions (the relations of production) are dualistic (the productivity and the relations of production). Namely, institutions have two basic technological attributes (see the section of 'What is Soft Technology?' in Chapter 1): first, institutions are not goals in themselves, they are only means to dominate given behaviour modes to restrict interactive behaviour and to realize basic values; second, they provide services of many kinds, in the economy, in politics, in society and in technology. For example, institutions serve economic activities. Such institutions include market economy institutions, proprietary institutions, enterprise institutions, accounting institutions, interior by-laws and regulations of enterprises. Institutions also serve societies, e.g. various kinds of social security institutions, the laws of leagues, association institutions, the laws of non-profit organizations, the laws of publications, etc.

Obviously, the requirements of different soft-tech applications have provided the content and basis for the corresponding institutions. In this context, we can say that the institutions are not only the tools of regulating soft technology and operational regulation for soft technology, but also a kind of 'product' of soft technology: institutional design technology is one of the soft technologies (see details in the section of 'The Classifications of Soft Technology' in Chapter 1).

However, we should distinguish between institutions and soft technology. Soft technology is in fact a 'game' in economic terms, i.e. soft technology is the technology of making 'games'. The invention and creation of new soft technology is therefore the invention and development of new 'content of games' and institutions are the game rules that correspond to the operation rules of soft technology.

It is common for people to talk about technological innovation, managerial innovation and institutional innovation as if they were three discrete phenomena, and argue which is more important. In talking this way, however, they do not quite understand the essence and content of institutional innovation; and, in another sense, it is a misunderstanding of soft technology to simply classify it in the category of management. Moreover, the innovation about which they talk is aimed mainly at hard technology.

There are two opposite schools of thought in the field of institutional economics: one emphasizes that institutional change relies on technological change, while the other emphasizes that technological change relies on institutional change (Coase, Alchain & North 2000: 329). The root causes of the opposition of these two points of view lies in ignoring the existence of soft technology. Without soft technology, innovation in technology (understood here as hard technology) is impossible to realize: namely, hard technology cannot 'automatically' create added-value in the economy. When the need for the application of new hard technology arises, or the need for new applications of existing hard technology arises, then the need for invention and creation of soft technology will also arise; and questions will also arise about how those existing soft technologies should be applied in new fields. Continuous R&D and innovation in soft technology result in the necessity of constant institutional innovation. In this way, new technology, or the new application of existing technology, provides bases and content for the birth of new institutions. Technological innovation will continue to impel or require institutional innovation, and without the safeguard of institutional innovation, technological innovation is also unlikely to succeed (see details in the next section).

The study of the process of soft-tech innovation and institutional innovation also clarifies what institutional economists have said regarding the correlation between technological change and institutional change. We may quote V. W. Ruttan, as follows: 'The conversion of supply for technological change and institutional change is formed by a similar force, the progress of scientific and technological knowledge lowers the cost of the new income flow caused by technological change; the progress of social science and related professional knowledge lowers the cost of the new income flow caused by institutional efficiency income (including improved skill in settling conflicts)' (ibid.). 'The progress of social science and related professional knowledge' that he mentions here is, in fact, a part of soft-tech progress; the so-called 'skill in settling conflicts' is soft technology.

2. Making Soft-Tech Institutionalization and 'Mechanism-ization' Keep Pace with Technological Innovation as well as Socio-Economic Development – The Enlightenment from Thousands of Years of Technology History and Hundreds of the History of Industrial Revolution

From the initial concept to the operational solution – the emergence of technology from the technological invention to the commoditization, then from new product to the market expansion – commercialization is a complicated process.

Historically, it has taken almost a hundred years or even several hundred years for a revolutionary technological invention to become widely used.

The steam engine is an example. Denis Papin, a Frenchman, designed the first steam engine in 1687. The first equipment for driving water pumps by using steam to make a vacuum was invented by Thomas Savary of Britain, and he gained the patent for the steam pump (the Miners Friend) in 1698. In 1707, Papin published a book on steam engines. In 1786, Jonathan Hulls of Britain received the patent for the steam engine boat. In 1765, James Watt of Britain invented the condenser, revolutionized the steam engine and obtained a British patent in 1769. In 1786, John Fitch of the United States built the first recorded oar-propelled steamboat (Ito et. al. [1983] 1984). The first steam-powered ship was used for waterpower transportation in the early nineteenth century (1807). The end of the nineteenth century witnessed the replacement of sailing vessels with the steam engine; however, even until 1880, most of the large cargo shipments worldwide were transported by sail vessels (Landes [1999] 2001). It is obvious that it had taken almost 200 years to actually apply the steam engine in transportation from the initial invention.

Take the application of steel and iron in the field of transportation as another example. Benjamin Huntsman of Britain invented the crucible steel-making method in 1740; R. Reynolds of Britain designed the cast iron railway in 1767; in 1784, Henry Cort of Britain invented the puddle method of steel-making technology; in 1804, Richard Trevithick's first train came out; in 1811, Germany established the Kelubo Iron factory. In 1822, Britain launched the first iron ship. In 1825, the first railway in Britain was built and put to use; in 1830, the American railway was open to traffic; and railways were put to use in France in 1833 and in Germany and Belgium in 1835. It had also taken almost 100 years from the invention of steel to its application in the railways.

The slow pace of technological innovation stemmed mainly from the fact that the incentives for developing new technology were only incidental. Under those circumstances where, as Douglas North has pointed out, systematic property rights in the field of innovation did not really exist (North 1981: 185), technological inventions could be imitated at very little cost and the inventors would often go without reward. However, I believe that property rights were not the only reason. The deep-seated causes lie with weakness in the system, culture and institutionalization of soft technology. It speaks volumes to analyze why the first Industrial Revolution was centred in Britain rather than in Portugal and Spain, which were the overlords of that time. The Portuguese and Spanish accumulated great wealth by exploiting their colonies by monopolizing trade with the Orient, pirate robbery and the slave trade; and they were also the owners of shipbuilding technology, navigation technology and compass technology in the sixteenth century. It is also very interesting that the first Industrial Revolution did not happen in either the Netherlands, which was the overlord during the first half of the seventeenth century, nor in India, which was the number one textile industry giant of that time.

The pre-conditions for the Industrial Revolution in Britain can be traced back to the thirteenth century. David S. Landes, who has analyzed many aspects of the background of the Industrial Revolution, has identified a number of features of the political and social system as well as culture of Britain over several hundred years that played a role, including: the elimination of slavery in the fifteenth century; the economic management mechanisms – for instance, the road and canal systems that were all built and managed by private enterprises; businessmen involved in the production process (purchasing, shipping, inventory, looking for auction timing); the introduction of the industry and trade in rural areas promoted the commercialization of planting and its marketing, and accelerated the whole progress of agriculture in rural areas, reducing the gap between urban and rural life; the initial formation of industrial clusters; loose immigration policies (see the section of 'Patent Technology' in Chapter 2); a culture and values which supported invention and creation, emphasized mechanization, gave emphasis to transport speed rather than comfort, and favoured punctuality and time-saving (both in rural and urban areas – British people of the eighteenth century were the manufacturers and consumers of the world's leading timer); the mercantilism policies of the authoritarian dynasty of Great Britain (Elizabethan Era beginning from 1558). All of these factors encouraged the British pioneering spirit and formed an indispensable prerequisite for the subsequent Industrial Revolution.

It is noteworthy that although Britain did accumulate a great amount of wealth through colonization – just as did Portugal, Spain, the Netherlands and France – the key reason why Britain was located at the centre of the first Industrial Revolution was that Britain had *consciously* implemented institutional innovation. In the long history of commercial technology, many modern soft technologies in the field of commerce were first invented and institutionalized in Britain.

For example, the first overseas trade chartered company, Moskel, was set up in Britain in 1553 in the form of a joint stock company; in 1581, the first formal stock company in overseas trade was established in Britain; in 1657, a comparatively stable stock exchange organization first appeared in Britain; in 1624, the first invention patent law appeared in Britain; in 1610, the first advertising agency appeared in Britain and in 1812, the first formally and professionally organized advertising company in the world was set up in London; during the seventeenth century, paper currency, the European contemporary currency, first appeared in Britain (although the modern financial industry emanated from Italy, the Bank of England, established in 1694, was the beginning of modern banks) and during the beginning of the nineteenth century, London became the money house of the world; the Charter Law and the Royal Charter, issued by the British king in 1694, were the first bank laws; founded in 1710, the Sun Insurance Company was the first share-holding fire insurance company; in 1762, the first life insurance company in the world, London Fair Insurance Company, was set up in Britain; the first Anti-Monopoly Law in the world was enacted in Britain in 1642. Furthermore, the Company Law of 1844 in Britain, together with all the other innovations just mentioned, helped to create a good institutional environment for the emergence of Industrial Revolution in Britain.

At the same time, institutional innovation further created a good condition for the scientific development and technological innovation in Britain. As a result, from 1642 to 1764, the technological achievements of Britain were much higher than Italy, Spain, Portugal and the Netherlands (see Table 8). During that period, Britain was responsible for 59 of the 220 crucial events that occurred in the history of western science, and for 93 of the 137 most important events in the development of western technology history. Simultaneously, by one account Britain was responsible for one quarter of the 93 important events that took place in the history of western social culture development (Ito et. al. [1983] 1984). In addition, from the beginning of the seventeenth century to the end of the eighteenth century, there were 53 important economic events that happened worldwide, eighteen of which took place in England (*Encyclopaedia China – Economy, Finance, Agriculture* 1994: 1476–95).

After the first Industrial Revolution, people consciously expedited the development of modern soft technology, especially commercial technology innovation and institutional innovation. This resulted in the second Industrial Revolution, which occurred less than a hundred years after the first. This time, the centre of innovative industrial activity shifted from England to Germany and the United States, further proving the argumentation mentioned above.

In the United States, a large number of phenomena created the institutional environment for the flowering of the second Industrial Revolution. These included: the great innovation

Table 8: The technological achievements of Britain from 1642 to 1764

	Whole western countries (event)	England (event)	Proportion
Important events in western science history	220	59	
Important events in western technological science history	137	49	
Important events in western social & cultural history	93		25%
Important events in the world economic history from 1600 to the end of eighteenth century	53	18	

Source: Data rearranged by Zhouying Jin according to Chronology of *Concise World History of Science and Technology*

of the patent system; the quick expansion of the research institute system; the theorization and normalization of 'scientific management' technology; the creation of batch production techniques; the popularization of the stock market; the development of monopoly enterprises; and the sanction of the antitrust laws established during the first enterprise annexation tide. All of these things together created the environment for the emergence of the second Industrial Revolution. They also elucidate the development and entry of commercial technology into the stage of institutionalization.

Centred in America, a climax in the development of commercial technology was reached in the 1950s and 1960s and again in the 1980s and 1990s (see the section of 'Soft Technology and Thrice Industrial Revolutions' in Chapter 2). As a result, the centre of the third Industrial Revolution was without doubt located in the United States. This was an age of ubiquitous creativity in soft technology, pushing the world economy to enter into the new age of the intellectual service economy. For many decades now the United States has continually occupied first place in the world in competitiveness in science and technology. This is

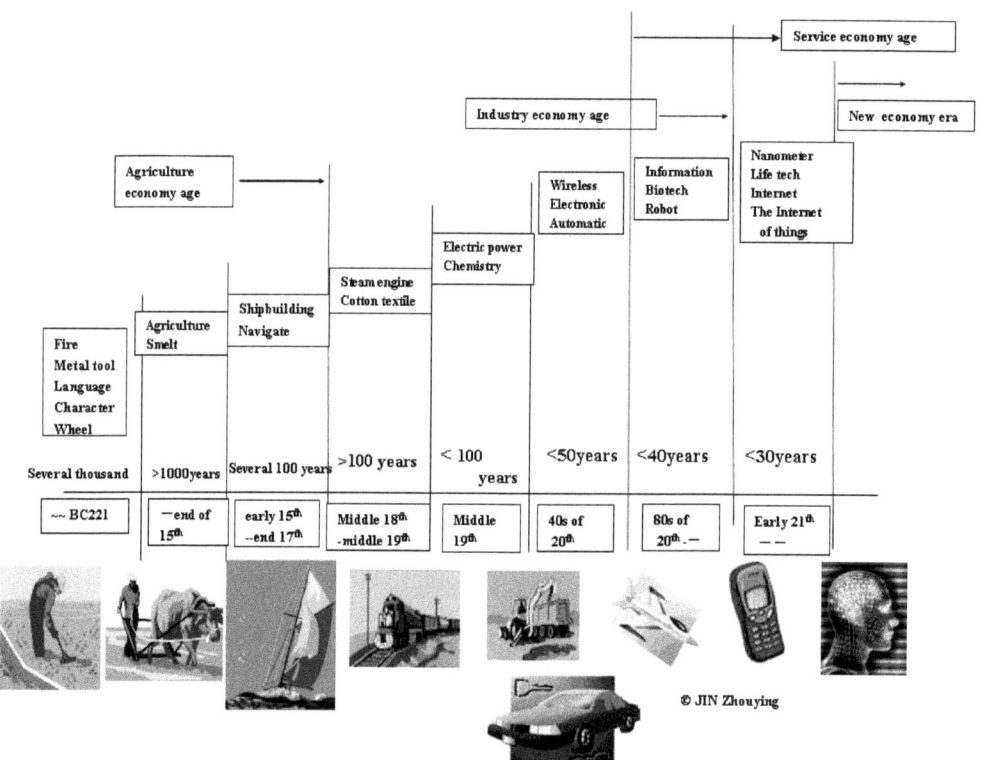

Figure 18: The frequency of technology revolution is quickening.

mostly because the United States attaches importance to the development of soft technology, consciously pushing institutional innovation and ensuring that it is synchronously matched with innovation in hard technology. The United States even carries out proactive research on institutional innovation.

The great achievements of scientific discovery and technological invention during the last several hundred years have been transferred into operable products with remarkable success. The result was the formation of great industries during the twentieth century, the success of which stemmed from the large amount of modern soft technology created during the course of successive Industrial Revolutions, especially the second and third. The majority of soft technology applications involved institutional innovation and, together with innovation in countless traditional soft technologies, the speed of hard technology innovation was thereby increased.

A survey of thousands of years of technological development history, as shown in Figure 18, reveals the shortening of the cycle of the Technological Revolution. Reducing the time between invention and commercialization for hard technologies – namely the time from invention to innovation, and even market-expansion (forming of market scale) – rested with the efficiency of invention and innovation in soft technology as well as that of soft-tech institutionalization. It has been the wave upon wave of soft-tech innovation which has impelled and accelerated the pace of hard-tech revolution.

3. The Relative Rigidity of Institutions and the Difficulty of Institutional Innovation

The above paragraphs explicate the necessity and contribution of institutions. However, institutional innovation is a difficult process; we must fully understand the relative rigidity of institutions, and new soft technologies need to be continuously supplied, demonstrated and assessed until the institutional innovation is ultimately accepted.

Once soft technologies have been regularized in the form of governing tools (of government, market and social organizations) such as institutions, policies, rules, laws and formal assessment criteria, they enrich the soft environment for a certain period and provide the essential conditions for (soft and hard) technological innovation. However, once soft technology forms the governing tools – institutions that are fully recognized and accepted by society and government – it usually becomes solidified. Firstly, because it will take time to change people's ideas, behaviour patterns and habits the formation process of a new institution (or the destruction process) must undergo much experimentation, receive public recognition and eventually be approved by governments or other institutions in order to be implemented as a form of institution. For example, the World Trade Organization must pass any formulation of major international trade rules before they can be implemented effectively as institutions. Secondly, any further institutional reform always affects the interests of groups or individuals with vested interests related to the old institutions and, hence, it always encounters resistance and the risk of rejection. In addition, sometimes the institutional changes tend to involve vast commercial profits and even political power, marking the formulation of the game rules: institutional innovation technique is no longer a purely technical issue. Thus, the task of replacing old systems, laws, regulations and policies can often be daunting. Thirdly, it is subject to the level of productivity increases likely to ensue, as well as the costs and benefits of reforms (i.e. only if the benefits of institutional reform and innovation are greater than expected costs can it be implemented).

In China's case, there still remains the influence of project management systems, which overemphasize affiliation among departments; and sectoral interests are placed above everything else so that policy contradictions and conflicts still exist, which are brought by departmental and regional barriers, as well as fragmentation caused by above situation. Because those institutions intend to be conducive to efficiency and to sustainable development, which in turn tend to undermine the interests of groups or individuals with vested interests related to the old systems, the majority of people and groups with vested interests create barriers to protect their existing 'territory' and go their own way, impeding the process of institutional innovation. For instance, the integration of 'Industry-University-Research' has been advocated for many years. However, it has been extraordinarily difficult to accomplish this goal – as reflected in the Chinese metaphor 'keep two skins between technology and economy'. Besides the faultiness of the interest distribution mechanism, there is another reason why the status quo has been largely maintained. In the orthodox system of centralized planning, which has lasted for a long time in China, scientific and technological activities and economic activities have been separated in to two completely different administrative systems involving different behaviour patterns, different value preferences and different sets of interpersonal relationships. Together, these factors created a kind of organizational and social culture that has prevented the two systems from transmogrifying and infiltrating each other. Moreover, the selfish interests of each system have usually been justified under the banner of the wider interests of the community and state, thereby increasing the overall rigidity of the established institutions.

Good institutions will encourage creative spirits and promote technological innovation, but if institutional innovation cannot keep pace with the progress of (soft and hard) technology and the needs of economic and social development, it will turn into a great obstacle to the development of enterprises or even of a country as a whole. For example, many industries in China were not open to private capital and private enterprise, while in the coastal regions (especially in Shenzhen and Zhejiang), where such restrictions were relaxed, the economy developed much faster. Before the entrance of the WTO, service industries such as finance and insurance industries were not accessible to domestic enterprises, but now that access is permitted, domestic enterprises are handicapped by inexperience and are placed at a disadvantageous position in competition with foreign enterprises service areas. Industrial and business innovation in Chinese enterprises has been limited by too many restrictions and old institutions and policies.

Therefore, the 'old' institution is often the biggest stumbling block of technological innovation, and we must first break it in order to implement the new one. As North said,

'[I]nstitutions provide a framework for people to exert mutual influence on each other, and they establish the relationships which compose the society, or rather the relationships, in which economic orders cooperate and compete' (North 1981: 132–225). The environmental changes in society, the economy and technology all require further adjustments to the balance between cooperative and competitive relationships, and those behaviours that are illegal or inimical to the new kinds of relationships of cooperation and competition need to be restricted by new mechanisms (in other words, the new institutions need to be established). Conversely, it is also necessary to relax or abandon restrictions that do not accord with the new environment and new technology (in other words, the old institutions need to be abandoned).

In conclusion, institutional innovation must keep pace with economic, social and technological developments; only when soft-tech R&D progresses at least as fast as the R&D of related hard technologies can appropriate institutional innovation be guaranteed. Only through innovation in related soft technologies, consciously and with clear objectives (rather than accidental or individual incentive), may an adequate theoretical basis and experience be provided for future institutional and systematic innovation. When people are aware of this, they may boldly create the soft technology needed by their country or community. They may also flexibly absorb and introduce selected advanced foreign soft technology, taking into account the actual situation of their country, and promote related institutional innovations (this is different from seeking to clone foreign institutions which, in any case, should only ever be treated as references rather than as globally replicable models). Just as Robert M. Solow, the Nobel Prize winner for economics, has indicated, only those individuals, enterprises and countries that perceive the rules of new games in advance can occupy an advantageous position in the new globally competitive environment.

Kazunori Ishiguro, a law professor at Tokyo University, criticized the liberalization of the service trade and the intellectual property regime in his article 'Globalization and Law', pointing out that 'the theory of the neoclassic school, which put all emphasis on the market, is academically insufficient (Ishiguro 2000). Now that Japan's background is different from that of the US and Europe, why should it follow the US in so-called multi-international laws concerning aviation, finance, insurance, intellectual property and the environment?' I believe that, apart from the fact that the US is an advanced economy, the other reason why the US occupied such advantageous position is that it consistently and continuously develops soft technologies, regularizes them, makes them into various institutions and laws during the course of its application, and then requires others who use them later to abide by them. That is, the US has created new games (namely, the soft technology that has been independently innovated) in many fields and has made relevant rules for these games. Developing countries should work harder at creating new games and then they will have the chance and right to make their own game rules, therefore, assuming the advantageous position in the competition.

4. Establishing an Institutional Audit and Evaluation System

The institutionalization of soft technology must avoid the institutionalization of bad practices or irregular operations.

An article in the US newspaper *The Washington Post* contained an article with the following evocative heading, 'Lobbyists Emergence Reflects Shift in Capital Culture' (Edsall 2006). In the United States, access to power on Capitol Hill is lubricated by lobbyist-funded meals, travel and campaign contributions. The relationships between lobbyists and politicians have become institutionalized in recent years. Since 1998, lobbyists have served as treasurers on 79 lawmakers' campaign committees and leadership political action committees. At election time, many lobbyists put on a new hat and become political consultants, guiding incumbents to re-election. Afterward, these lobbyists return to their traditional roles, being able now to ask for votes from those they helped put in office. The House Republican leadership has also brought them directly into the legislating process, institutionalizing the practice of using large networks of lobbyists to help 'whip' bills. Recent years have even shown how the traditional lobbyist-lawmaker financial pipeline can flow in both directions. In addition to helping legislators raise re-election money, lobbyists have received large sums from politicians' political action committees for vaguely defined services. L. Sandy Maisel, director of the Goldfarb Center for Public Affairs and Civic Engagement at Colby College, said the issue now is less a matter of criminality or overt corruption than a change in standards: a degradation of ethics in the Congress.

In China, there are many examples of institutionalizing bad practices or irregular operations. The large number of 'jerry-built' projects, the high frequency of mining accidents and especially organizational corruption are expressions of long-tolerated bad practices or irregular operations, and even their institutionalization. The institutionalization of bad practice or profit-oriented operation processes or loopholes in the institutions are the cradle enabling the social backlash, breeding corruption and widening the gap between the rich and the poor. Thus, the loss of high standards of ethics, and the institutionalization of that loss, will be far greater than the degree of damage that individuals are able to cause.

In order to maintain the sound standards of institutional design, each country must establish various institutional audit and evaluation systems according to the influence scope and extent of the institutions, and a periodic audit should be carried out by government departments, together with social organizations, associations or trade organizations.

5. The Limitations of Institutions

We have fully affirmed the importance of the institutions; however, special attention should be placed on the fact that institutional innovation is not a magic elixir for development. Across the broad spectrum of business activities, we can see thousands of instances where the limitations of institutions are revealed. The institution is insufficient to completely restrain all human behaviour and psychological activities, and using policies and laws can only be part of the solution. In addition to those factors, such as culture, concepts and awareness that are discussed in the section on 'Exceeding the Power of Institutions – Cultural Innovation' in Chapter 3, there are at least two other reasons why the power of institutions may be undermined.

The first reason may be found in situations where the institution itself is unfair and unreasonable. For example, the principles of publicly funded expenditure systems in government departments and state-owned enterprises are basically framed by the officials of relevant departments or business executives. If there is a little flaw in designing such an institution, it may be very easy for departmentalization or personalization of public interests to take place by using 'legal' authority as the justification. All power owners tend to abuse it; without supervision, power will certainly lead to corruption.

Another example may be found with the British 'reimbursement gate' scandal of 2009, in which British Government officials and members of Parliament were found to abuse the privilege of their subsidy system. More than 200 members of the three major political parties in the British Parliament were shown to have abused the expenditure of public funds by such means as multiple reimbursements, false expense claims and other types of fraudulence, etc. This case revealed the irregularities which exist within the British parliamentary system and the vulnerabilities of systems in which public funds are used to reimburse public officials for expenses ostensibly incurred in the public service.

In China, it is an open secret that many local officials go on foreign tours for the ostensible purposes of official inspection or training, but who instead engage in recreational activities using public funds. Although the 'Notice of the General Office of the CCCPC and the General Office of the State Council on Firmly Preventing Overseas Travel with Public Money' has been issued and dispatched everywhere in China, violations of the rules have happened repeatedly. During 2008–2009, there were a number of 'travel gate' cases exposed in Nanjing, Wenzhou and Guangdong. Some travel companies, or the facilitating agencies, played the role of 'accomplice' in publicly funded sightseeing tours abroad. There are over 100 travel companies licensed as an official travel agent, and several agencies are known even to have counterfeited the foreign invitation letters in order to make a profit.

The second reason may be found in the resistance to new institutions by various interest groups. For instance, the difficulties of enforcing patent rights or preventing damage to the eco-environment are immense. In China, although the number of laws and regulations concerning water is increasing, water pollution incidents have recently become increasingly severe. Since the Songhua River incident at the end of 2005, China has experienced more than 140 incidents of water pollution in total; namely, on average, a water-related pollution incident occurs every two or three days. The outbreaks of toxic cyanobacteria (blue algae) that have occurred in Taihu Lake, Dianchi Lake and Chaohu since the summer of 2007 reveal that the water pollution in China has approached a critically dangerous point.

From the point of view of institutional analysis, the reason why implementation of environmental regulations of the local governments have repeatedly failed is that local protectionism is rampant and the supervision of enterprises by local government is insufficient. However, the roots of this problem may be ascribed to the central government (Yi & Ma 2009). In order to implement national environmental regulations, a series of complementary regulations and policies must be developed in harmony with each other. These policies should include the implementation of targeted assessment systems (implemented by central authorities to mitigate against the risk of GDP centred policy) for the performance of local government officials during their tenure, financial monitoring mechanism for environmental protection, and compensation mechanism for the regional eco-environment protection.

In China, the central government places more and more emphasis on environmental protection and has taken drastic measures to punish polluting enterprises now. For example, from 1996 to 2005, in order to implement national environmental protection laws and administrative regulations, China has formulated and promulgated more than 660 rules and local statutes, and promulgated more than 800 national environmental protection standards (State Council Information Office 2006). The State Environmental Protection Administration (SEPA) has already launched three 'Environmental Protection Storm' actions in succession, including 30 illegal construction projects halted in early 2005, 56 projects ruled inadmissible by SEPA in 2006, and in 2007, SEPA notified 82 projects in steel, electricity, metallurgy and so on with the investment of 112.3 billion Yuan, which had serious violations of environmental-impact-assessment laws. The 'Three Simultaneous' System is for cases when the equipment and facilities for environment protection must be designed, constructed and put in to use simultaneously with whole projects, and the first adoption of the 'Regional Limited Approval' approach imposed sanctions against Tangshan City, Luliang City, Laiwu City and Liupanshui City, as well as four electric power enterprises, including China GuoDian Group Corp., in order to contain the blind expansion of high-polluting industries. Despite these initiatives by the central government aimed at environmental protection, during this period environmental disasters continued apace and, in fact, increased in frequency: a sudden environmental accident has occurred every two days, countrywide; public environmental complaints have increased by 30 per cent; and the central leaders' directives on environmental issues increased by 52 per cent over the previous year (Tang 2007a).

In February 2008, the 'Guide to Strengthen Environmental Regulation on Listed Companies' was released by SEPA, taking the verification system of environmental protection and environmental information disclosure system of listed companies as the core for containing the overexpansion of the 'double high' industries; namely, the ones with high pollution and high environmental risk. The guidance concerning 'green' negotiable securities is the third environmental economic policy that has been promulgated following the 'green' credit and 'green' insurance. It indicates that the Government is seeking a long-term mechanism to mitigate the ecological crisis at the level of institutional construction.

However, in order to enable the above-mentioned institutions to work, and to avoid the risk of the 'Green Storm' launched by government's environmental protection agency becoming no more than 'a gust of wind', it is insufficient to depend only on environmental investigation and assessment, environmental monitoring, pollution penalties, raising the entry barriers for public listing and other forms of government authority and economic leverage. It also requires changes in corporate culture and values, the effective participation of the public, and the appropriate education of government officials, entrepreneurs, and even the citizens as a whole. Moreover, these three requirements all involve deeper institutional reform and innovation in China's management system of non-profit organizations.

Various 'anti-environmental interest communities' have already formed in China to resist the implementation of environmental protection initiatives of the central Government (ibid.). We will not be effective in pursuing the environment-friendly goals of the civil society unless those 'anti-environmental interest communities' (some of which have backing from local authorities) are opposed. For that to succeed it will be necessary to provide organizational help to those citizens engaged in positive efforts to protect the environment; to turn 'social appeals' into a kind of organized action; to ensure sustained pressure; and to form an environmental community which is comprised of environmental protection organizations, community-based organizations, media, local governments and environmental protection departments. If the main body of environmental protection undertaking lacks the involvement of organized environmental NGOs, the slogan of public participation – 'environmental protection undertaking must have participations and concerns of whole community' – will be pale, and neither stable nor lasting.

Moreover, the goals and principles of the environmental movement must ultimately be reflected in the practices of business enterprises, which are the most important vehicles for environmental protection. Here, the key point is to take environmental protection as the core of corporate culture and values, and as a precondition for gaining economic profit. Accomplishing this will require a long and difficult journey all over the world. Thus, many social welfare organizations and NGOs in the international community are active in promoting the principle of the 'triple bottom line' (i.e. the economic, social and environmental bottom-line) for application in decision-making and management in corporations. This requires educating enterprises and helping them to implement responsibility for corporate citizenship at all levels of business operations and, ultimately, to embody the triple bottom-line approach in normal business practice through the transformation of the corporate business model. Thus, it may be possible to gradually move towards creating 'firm ground' of protection for humans and the environment in business through a tripartite effort of government, the market (enterprises) and civil society jointly.

Due to the value dualism of the development value in social resources (Jin & Ren 2004), China has adopted a policy of tight restrictions of the development of social groups, setting a very high entrance threshold for permission for non-governmental organizations to operate. For those which form associations or academic societies in the discipline or industries, China adopted a system with dual approval (namely to get the approval from both the registration authority and business governing unit), affiliating oneself with a unit (the executive organ must have sponsoring unit which is admitted by government) and hierarchical management (by setting up the association in national and local level) (W. Yang 2003). And, moreover, regardless of to which areas the social organizations belong, the system has come in 'one size fits all'. This is advantageous in reducing to a minimum the risks of developing social resources, and it is conducive to social stability and facilitating monitoring and management by the government. However, it has also created a serious obstacle to social groups playing various positive roles as social resources in China, thus affecting socio-economic development. Namely, in the area of social organizations, China faces the problem of the so-called 'refusing to eat for fear of choking' (refraining from doing something for fear of a slight risk). As China's reform progresses, and as there is a sharp increase in demands of the social market, the emergence of a large number of non-governmental, non-profit organizations is difficult to avoid. Now, many government departments have also recognized that this is part of the urgent need for social transformation in China; so many new organizations (including consulting organizations) with all kinds of names, under the direct control of the government, appear on behalf of NGOs and compete with civil NGOs, resulting in more chaotic management so that the entrance threshold for a real NGO has been even higher. In recent years, it has become almost impossible to register a new civil social organization. In such circumstances, the emergence and survival of an NGO engaged in social welfare undertakings, such as environmental protection, or corporate social responsibility, etc. is extremely difficult, and even its legitimacy may be questioned.

Obviously, this is no time for China to delay in carrying out institutional reforms and innovation in non-governmental organizations or in non-profit organizations and, moreover, it must become an important component of the institutional framework of China's environmental protection.

However, as mentioned above, these institutional reforms depend mainly on the transformation of the Chinese Government's understanding of, and attitude towards, the strength of 'third parties' – the social resources.

To sum up, we should not expect institutions to totally restrict the behaviour and psychological activity of enterprises and individuals. Therefore, it is important to, on the one hand, improve the quality of government decision-making and enable the timely institutional innovation to be possible; on the other hand, the most fundamental measures will include raising public awareness, enhancing public trust and credit consciousness, and strengthening business-ethics education to foster a corporate culture adapted to sustainable development. It will also be necessary to encourage self-discipline in the behaviour of enterprises under the general framework of pursuing a strong 'triple bottom line'.

6. *Technological Institutions*

If the aim of technological innovation is to increase the added-value of technology, then the most critical soft environment is technological institution in the process that technology becomes products or services and delivers to demanders (consumers); that is, the process of industrialization. Owing to lack of interdisciplinary cooperation between experts in the social sciences and experts in the natural sciences, scholars who have studied institutions have neglected technological institutions. Once the concept of soft technology has been understood, however, it is possible and necessary to conceptually separate technological institutions from economic ones. This, in turn, makes it possible to conduct systematic research on the design of optimal technological institutions (including systems of laws, standards and policies that pertain to technology) for the purpose of further promoting technological innovation in the broad sense.

While it may be difficult to distinguish between technological institutions and economic institutions, the two types of institutions nevertheless differ quite significantly. Furthermore, with the rapid development of technology, the systematic study of technological institutions bears more and more significance.

1. Technological institution research should go ahead of the R&D.

Technological R&D activities, which should be regulated by technological institution, have become increasingly linked to scientific research activities directly. Scientific research activities themselves also face institutional requirements when their results are applied in practice; in other words, as they get translated into technology. However, some contemporary scientific fields such as genetics are perceived as having potentially significant implications for the structure of society, ethics, lifestyles and human civilization even before they reach the stage of applied technology. Laws are therefore needed to direct and control such research from its beginnings and sometimes such laws even need to be international in scope (e.g. the strict prohibition of experiments involving cloning of human beings by most countries). In such cases, institutions need to be developed earlier than is normal in natural science research. Moreover, the motivations of some contemporary scientific research activities, in addition to the objective of understanding the world and exploring uncharted territory, involve also the pursuit of long-term social and economic benefits. That is, apart from the pursuit of scientific value, they aspire after the social, economic, resource and environmental value (such as exploration and research on the Moon and other planets). Therefore, as the innovation of technological institution continues, a new type of legal system – a series of Science and Technology Acts – are put on the agenda.

2. Although technological activities differ from scientific research activities in the sense that they are intended to directly contribute to economic and social development, they are nevertheless not, strictly speaking, economic activities themselves.

The motivations of a great many scientists and technologists, which are included in the main bodies of implementation, are not primarily economic in nature. They are often more about gaining a sense of achievement by way of invention and innovation, about self-fulfilment and social fame, or even simply about personal interests; namely, non-economic motivations. As a consequence, the normal kinds of economic institutions may not be very suitable for controlling the behaviours of scientists and technologists.

3. **As the influence of technology in society has increased, normal people have become more aware of the significance of research and development. The era when large-scale technological research and development projects were controlled and funded mainly by governments has passed.**

The great temptation and profits in the future market have increased the role and influence of private enterprises, private individuals and non-government organizations investing in technological research, technological innovation and technology commercialization. This in turn has increased the importance of strengthening (or relaxing) the regulations pertaining to science and technology.

The Japanese Trade Promotion Committee, asked by the Japanese Academy of Industrial Technology, has conducted research on the institutional requirements of science and technology by surveying many institutions that influence science and technology activity (JETRO 1995). Take the United States as an example:

Institutions influencing R&D activities and the development of infrastructure
- Regulations concerning research projects (related to political purposes, environmental pollution, economic profits and moral values).
- Regulations concerning methods and agreements of research (related to human health and security, protection of animals used in experiments and environmental protection).
- Regulations concerning the spread of scientific knowledge (keeping certain levels or fields of science available, protecting economic interests, personal health, privacy and security, and national military and economic security).

Institutions that indirectly restrict R&D activities
- Regulations for federal research organizations and their research activities (the restriction on individual researchers, such as incentives and payments and restrictions on federal researchers, restrictions on research organizations of the federal government or those funded by the federal government).
- Regulations for cooperative research and development (related R&D laws, including Steven-Wydler Technology Innovation Act of 1980, Federal Technology Transfer Act of 1986, National Competitiveness Technology Transfer Act of 1989, Bayh-Dole Act of 1980, policies and Anti-monopolization Law pertaining to cooperative development).
- Regulations for technology transfer and intellectual property rights (intellectual property rights for the inventions of federal employees, property rights of patents owned by the federal government, central management of the supply of patents and property rights, cooperative development agreements and technology commercialization, etc.).
- Privatization of public research organizations.

Other institutions and regulations that influence R&D activities

4. In modern society, new hard technologies are emerging every day and soft technologies are also changing quickly. The rapid pace of technological innovation is creating pressure for research on technological institutions and also on innovation in technological institutions.

People are aware that technology can bring human beings blessings as well as various disasters. For example, nuclear technology can provide efficient and clean energy as well as the most dreadful weapons of mass destruction. However, as discussed earlier under the heading of 'Characteristics of Soft Technology', it is unfair to blame technology for all these situations. The blessings or disasters produced by hard technology are dependent upon the behaviour of humans who manipulate the technology or are dependent upon the soft technology controlling hard technological innovation. Therefore, it is necessary to adequately develop the positive aspects of new technologies while taking measures beforehand against their potentially negative or even disastrous impacts. This, of course, places even more importance on the task of developing specific institutional innovations for the emergence and development of new technologies and new industries. In other words, the process of the application and industrialization of any new technology, such as biological technology, robotic technology, nuclear energy technology, gene technology, cultural technology and emerging intelligent service technology, all need the support of related soft technologies. As a result, the need for regularizing soft technology also arises. Therefore, the joint effort of hard and soft technology experts is needed to discuss, conclude and summarize the related requirements of technological institutions in the process of operations.

Take the development of cultural industry as an example. Owing to the dual value of cultural technology, regulation of cultural industries in the form of company law, anti-monopoly law, etc., is needed, just as is the case for all other industries. In addition, copyright law is especially pertinent to cultural industries. The publishing and printing of all publications, the production and playing of movies, television and broadcasting programmes and the creation and the performing of art programmes – in other words, the 'products' of cultural industries – are all afforded protection under copyright law. Furthermore, many countries have laws aimed at creating moral conditions in cultural industries. For instance, there is an Illicit Publications Law in Singapore that restricts and controls the content of publications, and many countries ban the production and spread of books and programmes with pornographic and violent content (Le 2000).

Japan is sometimes called the 'Kingdom of Robots,' owing to the fact that it owns almost 60 per cent of the world's industrial robots. More than twenty institutions, rules of laws and projects concerning the development of robots have contributed to their striking development in Japan. Listed below are some of these examples:

- Loan system enhancing short-term investment (started in 1992) is used to encourage investments in labour-saving equipment. Investors can get long-term cheap credit from the Japanese Development Bank and the Northeast Development organizations in Hokkaido.
- Temporary Measure Law promotes special enterprises' innovations (Business Innovation Law), stipulating industrial robot users who meet the specified conditions in this law, which can allow the users to receive special compensation and extremely soft loans.
- Non-interest loans promote the popularization of industrial robots and their applications, and encourage the development of high-level industrial robots and their applications by providing necessary non-interest loans to manufacturers for the expense of machinery manufacturers and corresponding software developments.
- Registration system of robots and engineering technological enterprises, with formal enterprise members and sponsoring enterprise members of the Japanese Robot Association as its object, provides consultancy and technical guidance to help those industrial robot users solve all kinds of economic and technical problems.
- Technology development projects improve the working environments of SMEs, loans for equipment of modernized SMEs, interest compensation systems for loans used for industrial robot equipment, and subsidy systems of industrial robots, etc. are all used to encourage SMEs to introduce robotics systems, improving their working environments, preventing work injuries and enhancing the competitiveness of enterprises.
- Credit insurance systems of instalment and loans in the field of machinery employ instalment and credit insurance systems to equip SMEs with modern equipment and reasonable management so as to achieve the goal of vitalizing mechanical industry.
- Robot insurance and damage insurance companies (twenty-one domestic and twenty-four foreign insurance companies provide robot insurance in Japan). The insurance covers all parts of the industrial robot body, software, robot peripheral equipment and semi-manufactured products.
- The Manufacture Science and Technology Centre and the Mini Machinery Centre are set up to promote the research and development of robot technology and FA technology.

Another example may be found in the financial collapse of the American company Enron and its relationship with the Anderson Consulting Company, a topic that was a front-page event in the business media during 2002. This incident not only illustrates the importance of institutional innovation for business enterprises and for innovation in accounting laws, it also poses a challenge to the fast developing intelligent service industry (i.e. the need for pertinent soft-tech innovation and institutional innovation).

5. The process of designing, evaluating, establishing and implementing technological institutions is very complicated.

There are institutional requirements associated with all aspects of scientific research, fundamental technological research, and applied research associated with technology development and commercialization. The complexity of technological institutions, especially the challenge of institutional ethics, is at least as great as that of other kinds of institutional design. In addition, the boundaries of institutions at different stages of development may overlap or diverge. For example, gene technology creates a bright future for medical developments and human health; however, at the same time, its impact on ethics and social security will make people shudder with fear and its challenge to the law is also unprecedented. Laws and judicial practices related to genetics are not isolated from international laws, different political systems and legal systems, economic development issues and the cultural backgrounds of different countries.

With recent breakthroughs in the life sciences, most countries have strengthened their research concerning life, ethics and the issues of human rights protection. The United States established its fundamental framework for the regulatory system pertaining to the life sciences and medical care in the early 1970s. Related research organizations and institutions were subsequently established, for example, the DNA Reconstruction Advisory Committee (RAC) of the National Institutes of Health (NIH); National Biological Ethics Advisory Committee (NBAC) under the leadership of the president; Ethic, Legal, Social Problems, Investigation Programme (ELSI), centred around research on human gene groups; the Institutional Research Board (IRB) system, etc. These groups actually conduct the advanced soft technology research and development that is needed for the future development of the life sciences and are undertaking the related institutional innovation. Accordingly, the budget in the US during 2001 for these kinds of activities was 18.8 billion US dollars, an increase of 5.6 per cent from 2000.

As to the security of transferred gene technology, all governments and international organizations currently attach importance to the creation of laws to govern the transfer and security of biological/genetic materials. The manner in which this is accomplished varies internationally according to differences in technological levels, economic interest and cultural backgrounds and public acceptance.

Although China has attached importance to research in this field in recent times (e.g. the Applied Ethics Research Centre was formed in the Chinese Academy of Social Science), there is still a great gap in soft technology research aimed at the application of gene technology, especially in the area of institutional arrangements. There is a lack of systematic laws, regulations and other related institutions concerning the research, development, application and industrialization of gene technology and the use and protection of genetic resources. Because it is lagging behind in lawmaking for the protection of genetic resources, China failed to protect part of the gene map of its own population. One foreign organization has defrauded China of more than 10,000 blood samples 'in ignorance and passion' under the guise of sponsoring a health project. The patent for the gene associated with asthma has consequently been applied for by people outside China, even though the resources for the work came from China. These are two examples of where China has lost some national advantages owing to a lack of advanced soft technology research. As a result, China has not only lost the opportunity of turning genetic resources into wealth but has also lost the above mentioned genetic research asset. This example shows that there are huge losses of the national interest caused by lack of soft technologies and ahead-research on technological institution.

6. Technical standards are among the technological institutions that affect the relative competitive advantage of firms and nations in industrial development.

The ones who control or own the process of establishing related technical standards will tend to dominate the market by manoeuvring for their standard to become the dominant one. Part of the great value of technical standards is that they enable generalization and interchangeability of product parts and technical components across organizations, regions and countries. Increasingly, however, technical standards will become the main form of non-tariff barrier.

Why does the traditional Chinese medical manufacturing industry, which has accumulated more than 2000 years of experience, has more than 300,000 classical prescriptions and 6500 factories producing medicines, account for only about 3 per cent of the international trade in Chinese traditional medicine? One of the most important explanations lies with the fact that there is a lack of standards for all the linkages between different aspects of the Chinese traditional medicine industry – from planting, to harvesting, to production and to distribution. Similar problems have also created the so-called 'green barrier' for the exportation from China of food, ceramics, leather, tobacco, vegetables, mechanical and electronic products and toys, thereby leading to many missed opportunities for the Chinese economy. Their core technologies are often composed according to hard-tech principles but then applied to soft targets.

The field of pharmaceuticals provides an excellent illustration of the salient issues. Technical standards are usually created by foreigners and executed by the Chinese ('GEP – Chinese made the rule of game' 2001). However, there is a challenge and an opportunity for China to do something to reverse its misfortunes in this area. An example of what might be done lie with the GEP standard ('Good Extracting Practice' standard) of the Chinese traditional medicine industry, which was first pushed forward by the Tianjin Tianshili Group. The GEP standard is designed to take into account the peculiarity of the Chinese traditional medical system. It is also the first occasion on which a Chinese company has set the rules of the standards game in the international pharmacy field. Hopefully, many more examples of this kind of institutional innovation will soon emerge in China and other developing countries.

D. Soft Technology and the Innovation System Framework

Why do we need to study the innovation system?

Firstly, since the process of innovation is the process of creating added-value, it is important for us to make clear what kind of value-adding creation activities are optimal, what kind of channels are appropriate, and what is the main body of innovation and its object. This is the main purpose of studying the structural framework of the innovation system.

Secondly, with reference to the above, the essence of innovation activity is the application of soft technology, which includes the application of new soft technology and the application of existing soft technology in a new environment or towards a new object. There are two reasons for this: 1) a single soft technology cannot function on its own and its successful application depends upon it being integrated and combined with other soft technologies; 2) as shown in Figure 19, the components of the innovative system – whether they are concerned with hard technological innovation, industrial innovation or institutional innovation – are all connected with each other, forming a system, through soft technology. Therefore, study of the innovation system is beneficial for elucidating how various elements operate and how they play complementary roles with each other.

However, there are a variety of ways in which the innovative system can be portrayed, depending upon the perspective from which it is viewed, or according to different criteria and different levels (national, industrial or corporate level). Here, we will consider only innovation at the enterprise-level, as an example of how to look at the whole innovation system.

1. Innovation Approach and Technology Innovation System Framework

We can view the approach and object of creating added-value from a variety of perspectives, such as the sources of competitiveness, the innovative space, the function of soft technology, comprehensive national power, the main body of innovation, and so on.

We may divide the broad sense of innovation system framework, considering both the approach of innovation and its objects, into six interactive innovation centres (Jin 2005), namely: hard-tech innovation (therein soft-tech is the tool of technology transfer and innovation); soft-tech innovation (therein hard-tech is used to enhance the efficiency of soft-tech innovation); hard industrial innovation; soft industrial innovation; environmental innovation (soft and hard environment); innovation on both of hard capital and soft capital. Together, these six plates of the technology innovation system framework, plus the integrated process of innovation among these plates, constitute the so-called '6+1 mode'.

The labels for the internal links of the technological innovation system framework shown in Figure 19 are as follows:

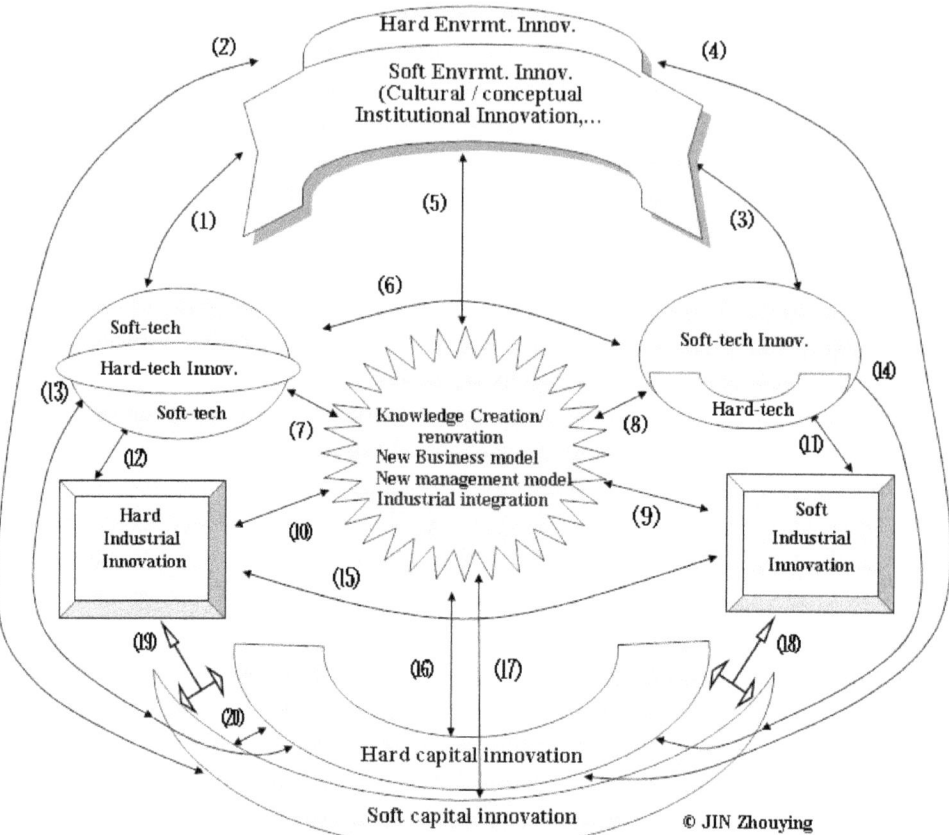

Figure 19: Technology Innovation System Framework (by innovation approach).

1. To bring forward the demands of environmental innovation; provide the bases and contents for the innovation of soft environment; create innovative environment for hard technologies.
2. To bring forward the demands, bases and contents of environmental innovation; provide the conditions for innovation in soft capital.
3. To bring forward the demands of soft-tech institutionalization and institutional innovation; provide the bases and contents for the innovation of soft environment; create innovative environment for soft technology.
4. To bring forward the demands, bases and contents of environmental innovation; provide for the environment for the hard capital innovation.
5. To construct the institutional and cultural environment needed by knowledge creation and updating; provide demand for environmental innovation.

6. To provide tools and means of innovation; improve the efficiency of innovation mutually.
7. To provide tangible solutions; provide the resources of integrated innovation.
8. To provide invisible solutions; provide the resources of integrated innovation.
9. To improve competitiveness; provide the resources of integrated innovation; bring forward the demands of innovation.
10. ibid.
11. To provide the core technology to improve productivity, structural upgrading and expansion; provide the resources of integrated innovation; bring forward the demands of innovation.
12. ibid.
13. To provide the resources of hard-tech innovation and the demands of innovation; provide the means and methods of capital innovation.
14. To provide the resources of soft-tech innovation and the demands of innovation; provide the means and methods of capital innovation.
15. To provide each other with the resources, needs and conditions of industrial integrated innovation.
16. To bring forward the demands of capital innovation; provide the resources of integrated innovation.
17. ibid.
18. To provide sources and conditions of innovation; provide the platform for capital innovation.
19. ibid.
20. To provide each other with the conditions and demands of innovation

The above figure also shows that, in addition to hard-tech innovation and to innovation in hard-tech industries, the other four plates belong to 'soft' areas; even for the two plates with a 'hard' main body, as mentioned previously, soft-tech is a means of innovation and the soft environment is the foundation and condition for innovation.

Since people are already reasonably familiar with hard-tech, the following discussion will be devoted mainly about several plates related to soft technology.

1. Innovation in soft and hard technology
If hard-tech innovation is considered as the technology transfer process around the hard technologies (that is, the soft-tech operations focusing on the tangible products), then soft-tech innovation refers to the invention, creation and application of new soft technology, which creates value focusing on the process of soft-tech transfer (that is, soft-tech operations focusing on the intangible products). What needs to be emphasized is that hard-tech innovation cannot be separated from soft technology and cannot exist on its own in isolation. That is to say, hard-tech innovation must depend on soft technology as its means and tool. Soft-tech innovation needs innovation in human thinking mode and behaviour

style, and then innovation in policy, rules, mechanisms and systems can be derived from it so that it creates the path for further hard-tech and soft-tech renovation. In the mean time, hard-tech is an important means for improving the efficiency of soft-tech innovation, just like computer technology provides visualization tools for strategic tracking management, generating a variety of solutions through digitization, visualization and flexibility. Integrated innovation in hard and soft-tech has an infinite space.

In our classification system enumerated earlier, six knowledge sources and eight major operable resources of soft technology were identified. Each category of sources and resources provides new objectives, new contents and new space for innovation, resulting in the endless generation of new soft technologies, new solutions, new industries, new markets and new jobs that may act as inexhaustible sources for economic and social development. Many Chinese companies suffer from 'project famine'. In China's western development, the economic strategic planning and industrial structure in the various provinces and cities are strikingly similar to each other, and most of them stress accelerating high-tech industry; they take tourism, biological resources development, electronic information industry and automobiles, etc. as leading industries. This indicates that regardless of social and economic development, or the business profile of the region, there are blind areas in the thinking mode for those leaders who believe that hard-tech projects (hopefully 'high-tech') are the only way to make profits – that is, their train of thought is still stuck in the stage of the industrial economy.

2. Industrial innovation

Industrial innovation here refers to the formation of a new industry or to the reform and upgrading of an old one. Increasingly, soft technologies function in a manner similar to that of many hard technologies, as core technologies in the formation of new industries

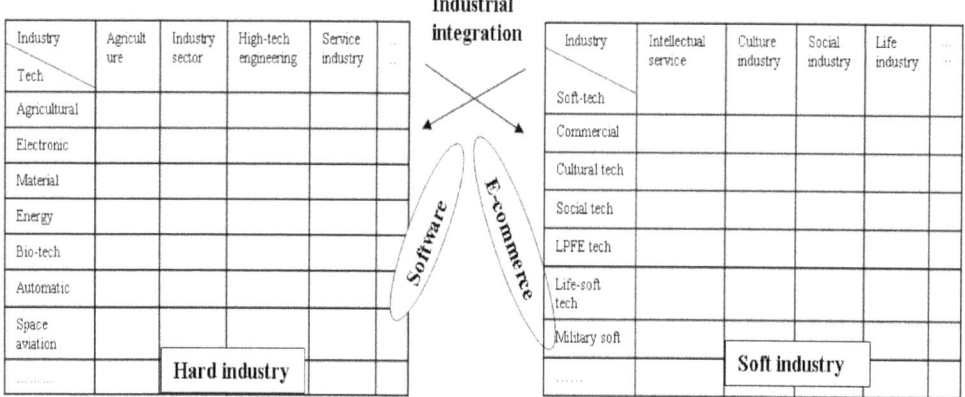

Figure 20: Industrial integration.

(see the section of 'The Functions of Soft Technology' in this chapter). In the same way, innovation in hard industry may not happen without soft technology because the process of the formation of new industries in which hard technology is the core is actually the process of applying soft technology.

It is noteworthy that with the development of information technology the Internet industry, the network industry, the e-commerce industry, the software industry, the eco-industry, the health industry and other industries the integration of hard technology and soft technology has emerged.

3. Environmental Innovation

Environmental innovation includes innovation in the soft environment and the hard environment. We are more familiar with the latter. Regarding innovation of soft environment, please see the section 'Hard Environments and Soft Environments are the Basic Conditions for Competitiveness' in Chapter 3, as well as the relevant part of the section of 'Soft Technology and Institutional Innovation' in this chapter for more details.

4. Integrated innovation

In Figure 19, the 'highlighting' plate represents knowledge updating/creation, business model innovation and innovation in management pattern, etc. In fact, integrated innovation exists among all plates, including between hard and soft technology, between the soft and hard environments and between soft and hard capital. For example:

- **Knowledge creation and updating**
  In the previous section, I enumerated six sources of knowledge. The continuous process of creation, production and updating of knowledge will constantly generate new technologies, and the invention of new technology and technological innovation in turn will enrich humanity's knowledge.
- **Business Model**
  This is the integrated result of various business technologies. See the section of 'Business Model and Management Pattern' in Chapter 2 and 'The Softening of Primary Industries and the Agriculture Service Industry' in Chapter 5 for more details.

2. *The Main Body of Innovative Activities and the System of Technological Innovation*

From the perspective of the main body of innovation, the technological innovation system consists of enterprises, universities and scientific research institutes, social enterprises, social communities, associations and other non-profit organizations, governments and individuals or consumers, etc. These main bodies have their own indispensable position and roles in technological innovation.

Enterprises

The enterprise is the core of technological innovation. Because technological innovation is the process in which technology is applied by users, the economic and social value of most technologies is realized through what we may call the 'economic body', i.e. enterprises. However, for enterprises to truly become the location for the main body of technological innovation, the key is to establish and improve an open, legal, equitable and orderly market economic environment and a reasonable competition mechanism.

Non-government organizations

Non-government organizations such as social communities, associations and social enterprises are important social resources because they coordinate government and the market and also embody the economic and social values of social technology in the social market to form the so-called 'social industry'. The social enterprises will play an increasingly important role in the flourishing social market, in steadying society, in encouraging the innovative enthusiasm of all people, and in strengthening the economic power of 'third' sectors. These are roles which governments or the traditional enterprises are unable to play.

Government

As a non-market association, the government has several roles to play in the innovation system. Firstly, it can create environments, nourish markets, map out relevant strategies, formulate relevant institutions, laws, regulations and policies (e.g. policies favourable to innovation or tariff barriers to protect new or weak technologies) and supervise the implementation of policies. Secondly, through education and training the government may improve civil quality, train experts and create a cultural atmosphere to fit the requirements of innovation. Thirdly, it can continue pushing and expanding institutional reforms, establishing the bridges between government, industry and the academic world and it can promote connections between the world of enterprises, the academic world and the financial world. Fourthly, it can direct innovation, provide technological foresight and support strategic technological innovation through direct investment (such as by investing in and carrying out long-term R&D projects or large-scale science and technology plans). Fifthly, government can shoulder the responsibility for protecting domestic technology (the military and civilian technology of its own country) in the face of international competition.

Universities and Research Institutes

Throughout most of history, technological innovation was dependent mainly upon individuals and entrepreneurs. However, the growth of universities and research institutes, especially industrial research institutes following the nineteenth century, changed the mixture of sources for innovation. As the main agencies of producing, accumulating and spreading knowledge and technology, they have become responsible for providing the dominant sources of inventions for the technological innovation system.

Entrepreneurs
It is not enough to emphasize universities and research institutes as sources of innovation. Research has revealed that although industrial laboratories and research institutes provided many technological sources for innovation during the first half of twentieth century, individual inventions were still the important source of innovation (Jewkes, Sawers & Stillerman 1968). More than half of the 70 important inventions during the first half of the twentieth century came from individual inventors. A similar pattern exists in the case of minor inventions. The development of the Internet has narrowed the distance between individuals and enterprises, and has tightened the relationship between customers and enterprises. In addition, an increasing number of technologically talented people are self-employed. People want to make full use of their own talents and demonstrate their own values to sell their own products in the market. Individual creativity, innovative desire and motivation become an important part of the driving force of technology. Individuals are becoming an important innovative resource. Some people even claim that the future is the time for self-employed workers. The numerous SMEs that exhibit vitality and strength in technological innovation have emerged for this reason. It is also for this reason that the present business world attaches such great importance to customer-based innovation. Individuals are playing an increasingly important role in the start-up of companies. Creative entrepreneurs and experts themselves are an important source of wealth to enterprises and they represent the innovative ability of enterprises.

The main elements of the innovation system do not function as discrete, stand-alone items; rather, they are mutually dependent on each other and require mutual promotion and mutual constraint. Moreover, to be effective, the national innovation system must be open and connected with the international innovation environment and should be coordinated with the process of globalization. A foreign expert of Chinese origin recently commented on the high-tech R&D strategies of some Chinese organizations saying that the biggest problem was that they were not internationally competitive. The reason given, apart from insufficient access to information in the R&D process, was the 'closed down thinking' pattern that allowed researchers to continue having their one-sided wishes granted. China should learn to face both domestic and overseas markets and make full use of and share domestic and overseas resources, including experts, technology, capital and natural resources. China also needs to engage in global institutional innovation.

3. The Advantages and Disadvantages of the Technological Innovation System in China

The following analysis will focus on the advantages, obstacles or challenges of China's technological innovation system. The main advantages of China's technological innovation system may be described as follows:

1. Ability to mobilize resources at the national level

The ability to achieve grand long-term goals reflecting the national will and interests is relatively strong. The essence is the superiority of the socialist system. Although the relatively centralized system is the key point for the next step of institutional reform, it nevertheless enables the central government to mobilize and control resources in order to concentrate resources which will accomplish great things, such as the creation of the aerospace industry. This capability is very important for China (a developing country with a large population), which has an advantage in total quantity but where all of the per capita resources are in a relatively weak position compared with the more developed countries (i.e. the implementation capacity of innovation goals at the macro level).

2. The policy of continued reform and opening-up

Continuous reform and opening up of China has released the productivity potential of the country by liberating the intelligence and wisdom of Chinese people and, moreover, continuous reform of various institutions has made the innovation environment mature day by day. Under the policy, talent can flow freely and almost all multinational companies in China are allowed to set up research institutions in China; a large number of young people study abroad, while more and more people would like to come back to China to start businesses – the innovation potential of the soft environment.

3. Rich human resources

In particular, China's abundant supply of low-cost but high-quality people with specialized abilities presents great potential for future innovation. Policy for foreign investment is shifting from an emphasis on the advantages of low cost labour to the advantages of access to qualified people of great ability – the potential of soft capital.

4. China's rich cultural resources

China's rich cultural resources are infinite treasures which need to be developed and innovated – the innovative potential of soft capital.

The obstacles or challenges of China's technological innovation system are embodied as follows:

1. Systemic problem

System reform and administrative reform lag behind economic reform. For example, the manifestation of the excessive concentration of power on technological innovation is to integrate the innovation into a top-down controlled administrative system, involving the top-down allocation of national resources and investment orientation; as for the main body of innovation, it attaches importance to those research institutions and universities which are directly under the control of government, but neglects social resources such as communities, individuals, associations and customers; it attaches great importance to the innovation of large enterprises, but ignores that of small and medium-sized enterprises, and so on.

2. Social reform lags behind economic reform

Social reform lags behind economic reform and various social institutions are far from perfect, thereby negatively affecting the creative atmosphere and innovative ability of the entire society. The lack of comprehensive awareness of social capital and lack of good solutions to deal with its negative impact so as to be unable to develop and utilize the potential value of China's rich social resources, is a particular problem.

3. The education system and the direction of education reform

The education system and the direction of reform in the education system place limits on the civil innovative capacity fundamentally.

4. Backwards thinking

Firstly, excessive emphasis is placed on tangible innovation sources and approaches and, as a result, innovation in soft technology and the soft environment (including institutions and culture), especially innovation in technological institutions, are neglected. This leads to a weak environment for hard-tech innovation, including the measures for encouraging and protecting innovation. Secondly, even though its importance has recently come to be understood, lack of full respect for intellectual capital persists and an imbalance of attention in favour of intellectual assets coming from natural sciences and hard technology, and against those coming from non-natural sciences and soft technology, persists. Thirdly, an imbalance of attention in favour of new ideas, new concepts and new technology coming from abroad and against those coming from domestic experts and indigenous technologies, persists.

5. Enterprise has not been taken as the main body of innovation

R&D investment in enterprises is extremely limited, and insufficient attention is paid to innovation that takes place in private enterprises (compared to state-owned enterprises) and small and medium-sized enterprises (compared to large enterprises).

6. Ignored and lacking in encouragement and supports for individual innovation

This is one of the important reasons for China's weak innovation capacity before its reform and opening-up was that the country, owing to the long-time influence of left-leaning thinking, let individual innovation become seen as the symbol of individualistic heroism and as seeking publicity in the eyes of many Chinese people. In the time of economic globalization and 'knowledge-ification', individuals have become the more important main body of innovation. As the most important resource of innovation, talented persons are footloose and can be easily attracted to those places worldwide with environments most conducive to innovation. In addition, even China's special 'units' mechanism is also weakening. Slogans and spiritual encouragement alone will be ineffective in the attraction and cultivation of talents if they are not complemented by mechanism of talents for nurturing the co-development of individuals and enterprises.

7. Lack of cultural atmosphere for innovation

Traditionally, China has lacked a culture of 'maverick' thinking (the bravery to advocate something unconventional or unorthodox) and the spirit of sharing. From the viewpoint of the social ethos, the culture of money is in vogue, and there is insufficient respect for knowledge and intellectual property rights. From the perspective of academic atmosphere and discipline, a 'hare-brained' style of studying is common, and the worship of business has encroached on many academic circles. Additionally, many scholars aim solely at a career as a government official as soon as they have been able to accumulate sufficient achievements, and so on.

Comparing above-mentioned advantages and disadvantages with the elements of the broad sense of technological innovation system framework, it's obvious that China's advantages lie in 'soft' innovation source and means, while its vulnerabilities lie also in 'soft' innovation source and means. It can be said that the key to enhance China's innovation ability lies in the innovation ability of soft technology, soft environment and soft capital.

E. Innovation and Corporate Competitiveness – Changing the Thinking Mode to Adjust to Corporate Strategy

In this section, we will broaden our scope of mind, from the perspective of soft technology, soft environment and soft capital as well as sustainable development, to renew the understanding of the issues most familiar in the traditional enterprise management, such as business strategy, R&D strategy, product strategy, organizational strategy, talent strategy, market strategy, corporate culture, etc.

Based on a new understanding of technology and the innovation system, we can greatly expand the approaches, space, objects and contents of corporate innovation mentioned above. Figure 21 shows that soft technology is like a nerve centre in a corporate innovation system and a variety of innovative activities are in fact soft-tech innovation with different objectives in different areas.

1. Strategy Innovation – Opening Up and Adjusting the Business Divisions to Expand the Life Cycle of Enterprises

The key to success for an enterprise lies in the correct business strategy. What is more important is that managers should adjust or transform their operating philosophy and strategy in a timely manner and according to the changes of epoch and the development of changes in the market environment.

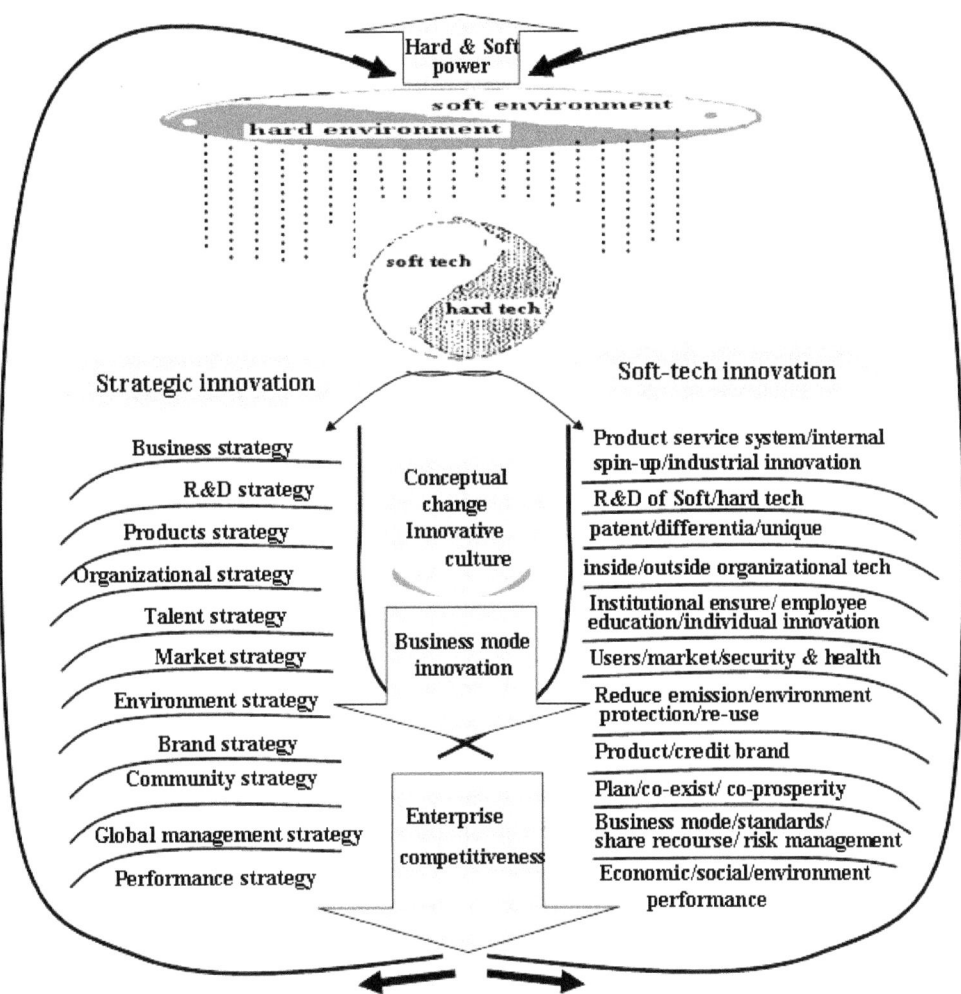

Figure 21: Soft technology and Enterprise strategy.

Strategy and future studies

Long-term enterprise strategy must be based on a global perspective and prospective study. Many fatal mistakes of decision-making are mostly due to lack of the research based on a long-term perspective. Years of cooperation between Institute for the Future (IFTF) in United States and Procter & Gamble (P&G), and the practices that have ensued from that cooperation, have proven that it is of great benefit for enterprises to actively participate in the study of 'the future' in related fields (Johansen 2007). In 1999, IFTF forecasted that biotech was becoming increasingly important and that it was mixing in very creative ways with information technologies. IFTF presented this foresight to the Global Leadership Council of P&G, and explained that biotech would become increasingly important for many P&G products. The top twelve people at P&G realized that none of them had the expertise needed to make good business decisions with regard to biotech. Thus, the action was to create the Biotech Reverse Mentoring Program for the top twelve people at P&G, who were willing to become reverse mentors for their senior executive colleagues (meeting about once a month for one year). The result was a considerable increase in the biotech expertise of the top executives: they did not become scientists, but they certainly knew a lot more about the business implications of this new area of science. At the end of the year, P&G had a biotech strategy, and you can now see the result of this strategy reflected in many P&G products, especially in detergents and hair care.

This example illustrates the full cycle from foresight to insight to action: the foresight was that biotech would have major impacts on P&G products; the insight was that the leaders did not have enough background to make good business decisions in this important emerging area of science; and the action was a reverse mentoring programme that paired young scientists with the top managers in the company. The follow-up action was a biotech strategy that has now become part of many P&G products strategies. Foresight provokes insight, insights spark action, and action reveals lessons that can only be learned in the field to avoid repeating old mistakes and to suggest new futures to explore. It is the art of getting there early for enterprise through the future study, which is summarized by Bob Johansen, the former president of IFTF.

Getting out of the office to find inspiration in the real world

The business operators and the administrative personnel should often go out of the office. Most Chinese company leaders are eager to attend conferences, new product exhibitions and project seminars. They think that these conferences will provide business opportunities or new projects for them and, accordingly, they feel obliged to attend. In contrast, if they receive an invitation to attend a world forum or a conference of an international future study association, they probably think it to be engaged in idle theorizing and be a waste of time and money. If their company is a large state-owned enterprise, the higher level units may not allow the firm's leaders to go abroad to attend such conferences. In contrast, foreign companies such as Sun Microsystems have sent their people to attend various forums of this kind over the past ten years to find inspiration for innovation. These forums may not offer ready-made remedies, concrete programmes and projects for enterprises. Instead, they provide a setting for different kinds of creative persons – successful entrepreneurs, Nobel Prize winners, poets, artists, scientists, sociologists and economists – to get together to voice their viewpoints and achievements, and to share their pioneering explorations and their views on the world and the future. Entrepreneurs must learn to communicate with those who have 'nothing' in common with them in order to discover inspiration regarding new markets, products, services, organizational methods and new business divisions. They should not merely depend on their subordinates' reports, but strive themselves to become corporate strategy experts in their own domain through training.

Moving past misunderstandings about high-tech and moving towards soft industry
Many people think that the supporting industry of the new economy is the high-tech industry or simply the information industry. However, they are quite mistaken. I have often met entrepreneurs who hope to be introduced to some promising high-tech projects. In the new market environment, every enterprise must orient itself in accordance with its business advantage.

Ever since China carried out its policy of reform and opening-up, foreign transnational corporations have occupied nearly all the Chinese high-tech markets with their advanced technology, abundant funds and advanced sales methods, and, in the name of cooperation or joint venture, they have swallowed up many of China's famous brands. For example, the Chinese car market has been divided up among joint ventures. Brands that have been famous for a long time, such as Red Flag and Shanghai, are now struggling or no long exist. While China was working hard at creating material production and striving to be the best manufacturing country in the world, enterprises from developed countries took advantage of emerging opportunities in China by utilizing various service technologies, especially finance, insurance and other high value-added industries.

Chinese people have now realized that many Chinese high-tech markets, such as the mobile communications market, have been forfeited to foreign enterprises. However, while Chinese people are mourning the loss of those markets and talents in the high-tech sphere that have been appropriated by foreign enterprises, these enterprises are occupying the 'last fort' in the name of the new commercial approaches, and China's market for services is being lost, without being noticed, to foreign corporations. Several of the largest consulting companies in the world have entered China's market and are making a fortune from Chinese firms. Transnational corporations in the retail field have broken into the Chinese market on a large scale through supermarkets, chain stores, shopping centres, convenience stores and discount shops. The French Supermarket Carrefour, for example, has entered 43 cities in China and has established 145 supermarkets in their chain up to September 2009. Carrefour's business is booming day by day.

Those entrepreneurs who are determined to open up new businesses may consider moving towards soft industries and in becoming entrepreneurs in the intellectual service industries, culture industries and social industries; or they should take the lead from the market and their customers and turn their existing businesses in the direction of the service industries. This approach may be particularly suitable for those private entrepreneurs who decide to undergo their 'second start-up' but who are not quite familiar with fields of hard technology. So long as they make full use of their special advantages, they are sure to find a new world of opportunities in the intellectual service industries.

Reflect calmly and 'get to the truth by verifying the facts', technological 'highness' is not the key to success

Chinese entrepreneurs should learn first to combine ambitious long-term goals with the reality. While entrepreneurs take into account how to make profits, they need to emphasize products that contain market value or that have the ability to further develop. An enterprise need not necessarily pursue new products that are 'high' in technological terms. It should pursue those products that are 'high and innovative' in terms of their application and their added-value. Eventually, in the long run, not all products of high added-value in the future will be derived from the high technology of the present. (See the section of 'Economic Softening and Soft Industry' in Chapter 5).

If China follows in the footsteps of economic globalization and 'knowledge-ification' along its present trajectory, based on its present level of industrial competitiveness, it will remain on a lower level in the international industrial division of knowledge production for a considerable period of time in the future. Therefore, the research and application of new and practical technology is a critically important job of most enterprises and most research institutes. It is imperative that during the upsurge of high-tech development nationwide, most enterprises in China, especially SMEs, do not ignore applicable and indigenous technology. All developing countries have their own advantages. For example, foreign enterprises will generally find it difficult to match Chinese corporations in the local market in matters that involve culture, social market, sales networks, relationship networks and customer information – at least in the short term. Hence, the enterprises of developing countries have a better chance than foreign enterprises to design and produce products that require sensitivity to the distinctive material, spiritual, cultural and service needs of local customers.

Many successful cultural companies have begun to emerge in China and, moreover, there are also many high technologies in the soft industries. Enterprises should not follow the academic tidal wave blindly. The Chinese company Lenovo, for example, has developed from its beginnings as a computer assembling and sales commission company for about ten years up to its present status as a sophisticated information technology company. Since Lenovo was familiar with the demands of the Chinese customers, it was able to control sales channels. By the late 1990s, Lenovo had become the largest enterprise producing Chinese-made computers, and now it has its own high-tech team with the ability to track the global technology trends. It was reported that the market value of Lenovo increased from 200 million US dollars in 1994 to 4.3 billion US dollars in 1999, and the global brand value of Lenovo has reached 14 billion US dollars by 2007. The Kelong Group, the biggest refrigerator manufacturer in China, started as a village and township company. Facing fierce competition caused by the overproduction of refrigerators, they conducted a design strategy that focused on consumers from villages and small towns. As a result, the products were cheaper in price and simpler in function. In addition, the Kelong Group provided more advanced refrigerators for consumers in the coastal developed areas, while turning higher profits.

Integrating Strategy with Globalization

Whether it is a network company, a consulting company, a materials company or an agriculture company, every company should adjust to the trend towards the 'informatization', globalization and 'knowledge-ification'. The strategies, business and product structures, as well as business models, need to address sustainable development, enhance the awareness of credit and follow international rules in order to maintain competitiveness. Thus, the enterprises should follow conditions and demands: 1) the characteristic of globalization is the coexistence of opportunities and risks, so it is therefore necessary not only to take full advantage of the benefits of globalization to achieve global optimal resource allocation, but also to prevent cross-border risks and pitfalls of globalization at full steam; 2) it is necessary to conform to new forms of global governance and to carry out corporate governance reforms in a timely manner so as to increase the transparency of the organization's management and business; 3) it is important to take initiatives to evaluate the progress or gap in attaining sustainability; 4) it is necessary to adapt to the next generation of accounting standards (evaluating intangible assets – human capital, environmental capital, alliances and partnerships, trademarks and goodwill); 5) it is imperative to comprehend the dynamic international criteria and make great efforts to be geared towards various international standards; 6) whereas globalized supply chains are facing the unceasing 'shuffling of cards', it is necessary to learn how to take the initiative to clean and leverage it proactively for competitive advantage.

The Davos annual conference of the World Economic Forum held in 1999 proposed three essential conditions for an enterprise in the twenty-first century: organizational structure adapted to external market changes; global brands; and the capacity for online sales. Haier has implemented three changes to comply with the above three conditions: the shift of organizational structure to a market chain involving reproducing the business process so as to shift the goal of the enterprise from profit maximization to serving clients; the shift of market direction from domestic to international; and the shift of industrial direction from manufacturing to service industry. Haier set up three promotion headquarters: physical circulation; commerce circulation; and capital circulation, in August 1999. These promotion headquarters provide a platform for the promotion of e-commerce to line up with the external market. It also set up the Haier Group E-commerce Co., Ltd. and began to cooperate with the Chinese Construction Bank in the e-commerce mode of payment, providing a means for achieving a fire-new trading mode.

2. Strengthening the R&D for Soft Technology

What should the focus of research and development in enterprises be? In the age of the industrial economy, ordinary enterprises conducted R&D according to the 'gender' of their products. Today, however, we understand that the carrier of products and services comes not only from hard technology but also from soft technology. Whether a company's main business lies with high-tech industry or with traditional industry, it may benefit from strengthening R&D in relevant soft technology and by constantly conducting R&D for new 'games' around its business fields in order to increase its added-value. In particular, the innovation in business model should be enhanced as the focus of R&D. Soft technology may be the endless source for future intangible products.

Several points need to be made with regard to R&D in soft technology. First of all, a forward-thinking approach is required for organizing or participating in pertinent future research and in order to understand future trends and obtain insights. It is also a necessary foundation if the people of a company are to persistently locate and apperceive knowledge systems with the potential of creating value in practice, or if they wish to collect, summarize, organize, integrate, upgrade and refine operable knowledge systems, and even protect the company's intellectual property rights.

Second, with regard to R&D in soft technology, the key is to expand horizons and locate sources of innovation along various pathways: via the process of industrial technology transfer and indigenous technology transfer; via the innovation process of hard capital and soft capital (human capital, organizational capital, network capital, customer capital and cultural capital); based on the strategic management, the management of whole industrial chain according to the overall process of intellectual capital management, or according to a company's business model or management flow. To take developing new forms of service mode, business model and patent as examples, the proprietary technology can be developed not only during the process of inventing new technologies and new processes. Rather, in all aspects of the enterprise's work – such as product design, market orientation, consumer orientation, service content and ways, production mode, both competitive and cooperative manner with competitors, the collaborative approach and content with customers, the contents and methods of cooperation with the suppliers, market services, etc. – there are the opportunities for strategic innovation and soft-tech innovation as well as patent 'generation'.

Third, a country's soft-tech research and innovation should embrace continuous institutional innovation because only then will it be able to build advantages in setting the rules of the game of global competition. Patenting business methods in the US and the recent operations of the United Sates Patent and Trademark Office are good examples.

Fourth, only with the soft technologies suited to local situations can soft industries be formed ultimately.

To take service technology as an example, in the past people did not conduct enough R&D on service industries because managers in both the government and companies tended to think that no high technology existed in service industries and that the service itself was not 'high' enough to warrant the effort. However, an increasing proportion of the added value in manufacturing industry is derived from services, and R&D in services is also put on the agenda. In developed countries, the percentage of investment accounted for by R&D has gradually become larger in the service industries than in the manufacturing industries. For example, in 1980, R&D investments in service industries in the US had reached only 4.1%

of total investment in those industries, while the percentage had reached 19.5% in 1996 and 26.1% in 2006. In 2006 that proportion in Canada is highest in the world, reaching 41.9%, followed by 33.6% in Ireland, 33.5% in Denmark and 32.7% in Singapore (*Journal of Taiwan Economy Forum* 2009). US business schools are now typically equipped with various service management and service market courses, and the number of doctoral students in the field of services has recently been increasing at a two-digit growth rate each year.

At present, many modern soft technologies emanate from western developed countries and, of course, the relevant rules of 'games' are formulated by key players in those countries. Thus, R&D in enterprises should stress both hard and soft technology and should emphasize R&D for both professional technology and industrial technology.

3. *Organizational Innovation*

Organizational innovation is the premise and basis for technology innovation. Chinese enterprises normally conduct organizational reforms when adjusting agencies and appointing new leaders. This is far from enough. Why do so many promising young people in China prefer to go out and start their own small companies rather than work for state-owned enterprises?

As for internal organizational innovation in enterprises, the system by which internal venture companies are formed, internal project organizations are formed and branch companies are established provides a good way to create an innovative environment, and to protect the enterprise against the loss of entrepreneurial talents and creative talents. Many large companies in Silicon Valley maintain an internal business incubation function, encouraging the 'intraentrepreneurship'.

External organizational innovation of enterprises involves the flexible operation of coordination technology, merger technology and virtual technology, all aimed at increasing enterprise competitiveness. Making use of virtual manufacturing methods to overcome China's traditional old concept of 'large and all-inclusive' and 'small and all-inclusive', gain access to external resources and to more effectively employ internal resources of the enterprise, are examples. In addition, those Chinese enterprises weak in R&D are not necessary for every company to establish its own research institute or adopt formal research institutes (the case of the failure in cooperation between an institute of the Chinese Academy of Sciences and a state-owned large company is illustrative). They should instead consider using virtual research institutes. For example, when the Haier Company, located in Shan Dong province, moved into the intelligent robot business, it established a common research institute for intelligent robots with the Robotics Research Center of Harbin Industry University, which is located in Heilongjang province, and has assumed an increasingly important position in the Chinese robot industry.

4. Talent Strategy

The global competition for talents is now emerging in full force. Manpower Inc. announced the results of its '2009 Talent Shortage Survey' in July 2009, revealing that the financial crisis intensified competition for senior managing personnel. Manpower surveyed nearly 39,000 employers across 33 countries and territories in late January 2009 to determine the extent in which talent shortages are impacting today's labour markets. Despite the context of the global economic downturn, 30 per cent of employers surveyed worldwide say they still face talented person shortages. Although a sluggish hiring pace is expected to continue in most of enterprises due to the financial crisis, there exists a mismatch between the type of individuals available for work and the specific skills that employers are looking for. It's this specificity of skills required in the individuals that employers are now seeking that is creating a sense of talent shortage amidst an overabundant pool of available workers.

What kind of talents do enterprises need?

Innovation is a continuous process of bringing out the full potential of people. In the age of attaching more importance to educational background, people are somewhat biased in their perception of talents. They put more emphasis on the number of those holding masters and doctorate degrees in enterprises.

Human abilities should be categorized into the possession of knowledge, skill or technique and intelligence respectively (Jin 2000a). From the perspective of ability development, ability can also be divided into the categories of congenital ability and acquired ability. For example, as for a person's intelligence, the judgment, insight, leadership ability, bargaining power, organizational capacity, charisma and charm of human beings may be perceived to be based on education and experience. However, numerous examples show that these qualities are closely related to innate factors such as genes, heredity and cultural background, and are developed by the combined force of congenital factors and acquired factors. Acquired abilities are determined by the education a person receives (the education provided by family, school, work and life) and the environment in which that person grows up (family, social and economic). Therefore, people who receive the same education may exhibit different abilities. Conversely, people with similar congenital abilities may exhibit different abilities in practice owing to the different environments in which they were educated, raised and employed. Thus, congenital ability and acquired ability are complementary but cannot be substituted for each other. It is important to realize this and admit it. It matters a great deal to employ talents correctly and to bring into play the abilities of different people appropriately, i.e. bring everyone's ability into full play.

The new character in Japanese business circles, Sun Zhengyi, has succeeded because of his extraordinary ability and charm. His courage and resolution in decision-making have put him ahead of other entrepreneurs, enabling him to make money by using time intervals; he is good at making use of wide interpersonal relationships in order to establish super partnerships all over the world; he is good at recruiting worthy talents and he practices the art of respecting and using talents. He has attracted senior management talents from Japan's successful enterprises, such as Mitsui & Co., Ltd. and Nomura Securities to form the 'Base of the Internet Revolution' (Yamada 2000). After starting up the companies, he entrusts executive operations to the experts in the relevant area. He does not micro-manage experts under his authority. Therefore, those tycoons in the financial and business circles were able to bring into play their creativity and ability, even under Sun, and he himself began to consider new businesses. Hence, it can be said that what pushes Softbank is the brain bank around Sun Zhengyi. Such ability to create new enterprise does not have much to do with possessing a doctoral degree. As it is easy to recite the theories in the complicated environment of contemporary enterprises, intelligence is more important than knowledge.

The gap between developing countries and developed countries lies, in the final analysis, in the shortage of soft-tech talents. In particular, far-sighted pioneers like Sun Zhengyi of Japan and Ruimin Zhang of China's Haier Group are rare finds. Good talents are the real wealth of their enterprises. With them, the necessary funds can be raised and technology and technological talents can be acquired. Thus, the key to success in the enterprise is to dig out, find, train and protect creative talents without any constraints.

The so-called 'innovative' environment should be the first of all mechanisms and environments to invest in people. As for the intangible assets of enterprises, its property rights can be established through original shares, operating shares and technological shares, and the institution can guarantee that the interest of creative talents and start-ups will be protected.

Many enterprises can accept and tolerate ordinary employees who are incompetent, but still engage in disputes and conflicts with their talented and creative employees when they accomplish outstanding achievements. Why doesn't the enterprise consider setting up a branch company to be managed by its talented employees and giving them stock shares in order to provide a foundation and motivation for them to run the company independently? Alternatively, their entrepreneurial effort could be adopted by the parent enterprise and the creative employees could be given the stocks or the right for dividends in return.

Flexible use of external think-tank and external resources
In general, the extent of using external intelligent services in enterprise reflects a country's level of corporate governance and that of economic development, which is the base of the booming development of world consulting industry. Now, along with the progress of globalization and 'knowledge-ification', external resources for enterprises are increasingly available. In addition to the traditional modes such as the external think-tank, the external professional management-consulting firms, etc., open innovation has recently become a powerful way of enhancing the competitiveness of enterprises. In open innovation, all the relevant knowledge, technologies, organizations and relations inside and outside of the enterprises are resources for innovation. Procter & Gamble is a good example in this regard. P&G fundamentally changed its model for proposing new ideas and for developing new products. As was the case with most enterprises in the world, their new products have

previously been developed with the company's R&D department as the centre in a closed or secret manner. Now, however, P&G encourages cooperation with universities, suppliers and external inventors and provides a certain portion of the return for these partners. Within less than a decade, P&G has increased the proportion of new product ideas acquired from the outside from less than one fifth to almost 50 per cent. In contrast, the majority of Chinese enterprises do not yet have the customs and culture to seek the aid of an 'external brain' when they encounter problems and, moreover, in most cases, even when help is sought, major consulting services in China are usually provided by foreign consulting companies. In contrast, for success in open innovation, people need first of all to be open-minded.

5. *Product Strategy*

If an enterprise wants to maintain its own characteristic product structure and technological structure, it must open up a unique market, service and operation mode, or possess its own proprietary technology, so as to seize a certain market in order to obtain the monopolized profits for a certain period of time.

For a long time, enterprises have paid more attention to innovation in their owned tangible physical capital and financial capital, and at the same time they have focused on research and development for new technologies, especially the technology amenable to so-called independent property rights. Studying soft technology can broaden our mind; we would no long seek to obtain individualized 'products' or so-called 'independent intellectual property rights' by focusing just on hard technology or on high-tech.

At present, many Chinese enterprises do not have a strong sense for intellectual property rights and they do not pay much attention to applying for patents or to the advantages of owning patents over time. Many important S&T projects in China are regarded as having produced advanced technology at national and international levels. Some projects have cost tens of millions, even billions, of Yuan but most companies do not really pay attention to studying how many of these projects have produced technologies suitable for patenting, let alone applying for patents for invention and creative work from their own small projects.

This cognition is extremely important for those SMEs that are regarded as the source of innovation. Chinese SMEs need not continue to dream of receiving small shares from a market that is already occupied by large enterprises or of only transferring technology from abroad. According to the concept of soft technology, SMEs may base themselves on creating their own innovative platforms centred in services, and they may also develop their unique market and product. They may especially develop new spheres in the intellectual service industries as a way of prolonging the life cycle of their enterprise – through such means as doing business in unconventional or unorthodox ways, 'small but specialized' ways, in 'small but peculiar' ways and in 'small but new' markets, products and services.

Indeed, enterprises not only need protection for existing intellectual property rights. More importantly, they badly need specialized department or personnel to conduct research on their own intellectual property and know-how development, including how to produce, generate and package, file and apply for patents. This is an important aspect of strategy that will enhance corporate competitiveness. They should learn from the experiences of outstanding enterprises known for their prowess in cultivating and exploiting intellectual property, both in the routine filing of patent applications, and in the artful management of trade-mark protection. Coca-Cola and McDonald come to mind as companies that have managed to maintain their business in the long term through the skilled management of trade-marks and other forms of intellectual property.

6. Market Innovation

Innovation of market technology is the sphere with the richest content and greatest potential.

Creating Customers, Markets and Industries
The market network of Kodak is a really wonderful example. Kodak came into China in 1994 and has since expanded significantly. In a short period of six years, it established itself in over 500 large and medium cities, the number of allied shops increased to 5000 retail chains networking in China (the 700 shops in the West created 5,000 jobs) and the company provided 40,000 jobs for the local labour force, forming China's largest retail chain network. According to Kodak's forecast then, the annual consumption of film per capita in China is 0.1 of a roll, and the growth rate of popular consumption of cameras is 15 per cent per year. Compared with America's 3.6 rolls per capita, there is still great potential. The annual increase in the Chinese photography market is around 10 per cent. Based on this estimation, Kodak introduced a 'start-up with 99,000 Yuan' project for medium and small investors. Therefore, as long as investors invest 99,000 Yuan (including buying one set of 955E colour developing equipment and shop decorating fees), one can immediately become the owner of a Kodak Express Printing Shop (KEX), after the investor installs the necessary equipment and finishes the required decoration (Chiwei Wang 2000). This programme met the local demands of small investors who are eagerly looking for chances to invest their money but are fearful of taking risks. It has also broadened the marketing channels and has increased the number of product agents. Now that the roll-film camera has been replaced by digital camera the adoption rate of household digital cameras in Beijing has reached 68 per cent. However, many of the lessons that may be learnt from Kodak's market expansion techniques and sales network technology for the chemical film business – as well as their experience in creating new consumer demand, new markets and new industries – may be applied to new categories of digital business and other businesses based on soft technology.

Innovation focusing on supply chain, delivery approach and marketing channel (see the section on 'Supply Chain Technology' in Chapter 2).

7. Changes in Corporate Culture and Values – Shaping 'Good' Enterprises of the Twenty-first Century

So-called 'corporate culture' should be built by integrating the values and codes of business leaders' conduct as well as those of the staff.

A culture encouraging innovation

The corporate culture that encourages innovation cannot be summarized in a few sentences of management psychology. Nevertheless, corporate culture creates the atmosphere and cultural environment that encourages innovation, respects knowledge, tolerates the failure and advocates cooperation.

For instance, does your enterprise respect knowledge? Is it willing to pay for concepts and advice? Does it use outside experts flexibly? Some companies are willing to pay millions of dollars for 'high' technology and for investments in universities and hard-tech research institutes, but not for seeking advices of a management consultant, strategic consultant or consulting firm, because it is 'soft,' and apparently they think anybody can create it. How a company deals with on-the-job inventions and individual innovations is a symbol of the company's culture.

Although we always say that failure is the mother of success, for decades China has been good at summarizing its successful experiences but very reluctant to admit its failures, summarize the lessons and learn from its failures. Whether those who failed in running businesses will be allowed to return to their original companies and be allowed a new opportunity is an indicator of whether or not a culture of tolerating failure exists.

The change in concept and values

On the path to sustainable development in the twenty-first century, we need to shape and advocate the 'good' company', rather than big company. Shaping a good enterprise, first of all, requires upgrading to a higher level of concepts and values. It then requires ascertaining: what an enterprise is; what corporate responsibility is; what a 'good company' is; how achieving corporate profits in the new economic era may be accomplished; and determining how enterprises may endeavour to shift their operation from the pursuit of financial performance to the pursuit of overall performance so as to create a green business pattern.

Thus, excellent entrepreneurs must go beyond the scope of a single enterprise when they consider corporate survival and sustainability. They should stand higher and take a broader view to shape corporate culture and values according to the historical mission of our generation (see the section of 'The Characteristics of Soft Technology are Suited for Coping with the Challenges in the Twenty-first Century' in Chapter 1).

What is an enterprise? Why should it fulfil the triple responsibility?

Along with several stages of the development of corporate positioning; namely, the tool or property of the owner → the individual or group's contract organization → the community of stakeholders → the agency of the legal person → the citizen in civil society, the theory of corporate responsibility comes into being as today's theory of global corporate citizenship. In keeping with their roles as global corporate citizens, enterprises enjoy the same rights and responsibilities as other citizens. The simplest explanation is that companies have the right to use their lawfully acquired resources to carry on their operations; while, in turn, they have the responsibility to repay society in kind for those various operational resources they obtained and upon which their enterprises depend.

How should we understand the 'operational resource'? In the capitalist market economy, enterprises first pursued profit fanatically and unscrupulously, according to the logic that maximizing returns to shareholders who provided financial capital was perfectly justified, while the employees who contributed their wisdom and work for the enterprise (and the related community providing the business environment, the consumers purchasing their products and services, the natural capital providing water, air, eco-environment, as well as the government and society providing soft environment for corporate operation) tended to be regarded as no more than cheap tools and resources and (except for the case of paying tax to the government),even free resources. As explained in *Das Kapital*, this approach was adopted because the fundamental purpose or motivation of capitalist production was the pursuit of the infinite proliferation of capital value and pursuing the maximization of profits through extracting surplus value (as determined the nature of capitalist production itself) (Ru 2009). Now, human society is facing serious challenges that it has never before faced (see the section of 'The Challenges in the Twenty-first Century and the Characteristics of Soft Technology' in Chapter 1), getting bogged down in the crisis of development mode. Human beings are therefore beginning to realize that the problems are caused by simplistic human pursuit of material civilization, and need to be addressed as such.

As an economic cell, enterprises must understand that their responsibility is to provide returns not only to shareholders, but also to the suppliers of the various resources, the so-called 'stakeholders'. That is, corporate citizens must simultaneously fulfil their economic responsibility (creating profit/surviving), their ethical and social responsibility, and their responsibility towards the natural environment and its ecology.

Creating green business model and shape the example of good enterprise
In order to create not only greater profits, but also to fulfil their social and environmental responsibilities, enterprises must actualize the new profit pattern: they need to integrate their triple responsibilities into their corporate core values, and to then seek commercial profits and business opportunities through sustainable development.

With the endeavours of men of insight and outstanding entrepreneurs worldwide during several decades, the above ideas have increasingly been put into practice. The movement for corporate social responsibility, beginning from 1990s, is clear proof. International consumer organizations, environmental organizations, economic cooperation organizations, human rights organizations, labour unions and other non-governmental organizations, as well as the United Nations, have reached consensus on the importance of sustainability and have been promoting the concept of corporate responsibility. Much of the effort has been

Table 9: The Change of Concept.

Change	Traditional company	Future company
Position	Economic entity	Global citizenship
Measurement for achievements	Fiscal	Economic, social and environmental contribution
Structure	* Market share * Clear boundary (between departments and enterprises)	* Establish alliance * Vague boundary
Core technology	Hard-tech	* Hard-tech and soft-tech * Integration of the hard and soft-tech
Capital	Financial capital	Human capital/financial capital/social capital – money/talent/organization/custom
Management of asset	Tangible asset	Both of tangible & intangible asset
Value of staff	Cost/expense	Human capital, Stakeholders, Property and Wealth
Leaders responsibility	Make profits	* Sustainable development * Encourage innovation
Culture	* Benefit No. 1 * Maximum return to Shareholder * Independent/competitive	* Triple bottom line * Maximum to return to Stakeholders * Cooperative/innovative
Brand	Products/service	Credit standing/Products/Service
Competitiveness	Hard power	Hard power + Soft power

Source: Zhouying Jin

focused on implementation and practice, not just concepts. An increasing number of excellent enterprises have published annual reports on sustainable development or social responsibility. The Future 500 is an excellent example of an organization doing exemplary work in this domain (see the section of 'Future 500: The Successful Case of Assisting Enterprises to Upgrade their Soft Power' in Chapter 3).

A new round of economic competition has begun in which sustainability matters. Those enterprises whose leaders comprehend this trend early enough, and which accordingly develop new business models and new rules for the game, will be winners in future business competition.

Note

1. In China the phrase '*trade → manufacturing → technology*' is widely used to refer to the idea that the business activities in Zhongguancun are based first of all on trading any goods for profit; secondly on manufacturing; and only third on technology development and technology transfer – in other words, that, despite its name, the basis of business in the Zone is not really high technology.

Chapter 5

Soft Industries

A. Economic Softening and Soft Industry

1. *Economic Softening in the Twenty-first Century*

Economic softening could be defined as a 'soft' phenomenon of the entire economy, which arises when the added value created by 'soft' (or intangible) factors surpasses that created by 'hard' (or tangible) factors in economic activities.

As far back as the early seventeenth century, the experts had taken note of the softening trend of the economy. For example, in 1691, while reflecting on the situation then facing Britain, Sir William Petty pointed out that industry (i.e. the manufacturing and processing sectors) could typically turn out more profits than agriculture and commerce simultaneously. He therefore predicted that, as a consequence, the labour force would turn to industry from agriculture and then to commerce. Petty also stated that along with economic development, the centre of industry would gradually shift from tangible production to intangible service production. John Maurice Clark discovered the same economic rule after conducting research on the classification of primary, secondary and tertiary sectors in 1940: with economic development, the centre of employment structure would shift from primary industries to secondary industries and shift again to tertiary industries. This theory is now called Petty-Clark's Law (Nakayama et. al. 1971: 427).

Although the individual circumstances vary in each case, the above rule has been followed by the changes in economic structure during the last 200 years of the industrialization process in the countries that we now call 'developed'. Most countries, in their processes of industrialization, show that the proportion of their GDP accounted for by industry exceeds that for agriculture, and that the number of persons employed in industry also exceeds that of agriculture. When industrialization rises to a certain level, the development of the service industry that caters to the needs of production will accelerate. Both the proportion of GDP accounted for by the service sector and the proportion of employed persons accounted for by the service sector exceed those of secondary industry and agriculture. This becomes an important index of the progress of industrialization in a country. After World War II, the economic structure of all developed countries shared the above characteristics.

Table 10: The service industry as a proportion of GDP and a proportion of the number of persons employed (% added-value of service industry / % employment of service industry)

	1950	1960	1970	1980	1987	1997
Britain#	46.3/40.2	48.5/41	54.5/54	57.6/61.5	66.1/64.8	70.8/71.3
France	37/38.1	46.4/43.3	50.1/49.3	56/56.4	66.9/66.2	71.5/69.9
USA	55.1/50.2	60.2/56.3	64.8/62.3	66.3/67	68.3/69.9	71.4/73.4
Japan	37/24.1*	49/32.2**	47/47.3	53/54.6	56.8/57.9	60.2/61.6

#: For Britain, the year of employee is 1952, 1958, 1972, 1978, 1987 and 1997.
*: Year 1948/1947
**: Data for 1953.
Source: Data of 1987 and 1997 comes from OECD Proceedings: 'Innovation and Productivity in Services, OECD'. The rest come from Wang Shuying 'Tertiary Industry: History, Theory and Development', pp. 39; 44–45; 49; 52; 54.

Table 11: GDP proportions of the three industry sectors in various countries (primary/secondary/tertiary industries)

	1970	1980	1987	1990
USA	3/38/59	2.8/36/61	2.1/31.4/66.5	1.9/31.6/66.4*
Canada	4.4/36.5/59.1	4.2/36/59.6	3.0/33.6/63.4	2.7/31.8/65.5
France	5.4/40.1/54.5**	4.8/38.6/56.5	4.1/33.8/62.1	3.9/33.3/62.7
Germany	3.6/55.8/40.6	2.3/48.1/49.6	1.7/44.2/54.1	1.4/42.5/56.1***
Italy	8.5/45.2/46.3	6.1/41.4/52.5	4.4/36.5/59.1	3.5/36.5/60
Britain	2.7/44.2/53	2.2/42.8/55.0	1.9/37.1/60.9	1.86/35.8/62.9

*: 1992
**: 1977
***: 1991
Source: *International Statistical Yearbook* by Gong Feihong.

As shown in Tables 10 and 12, the contribution to GDP of tertiary industries surpassed that of primary and secondary industries in the United States as early as the 1950s, while other industrialized countries achieved this transition during the 1960s and 1970s.

Since the 1990s, the above changes have also accelerated at the global level including the industrial structure of some developing countries. According to the 2008 World Development Indicators from the World Bank, the service sector accounted on average for 69% of GDP for all countries, 52% for low-income countries, 54% for middle-income countries, 54% for middle-low income countries, and 72% for high-income countries (with the average figure for Euro-currency countries also being 72%). In those economically developed areas of China, such as Beijing & Shanghai, the proportion of GDP accounted for by the service sector has already reached 73.2% and 53% respectively.

It is obvious that the above characteristics are eventually expressed in the economic structure in most of countries of the world, so we are able to say that the world economy has been 'softening'.

In Japan, the number of people employed in tertiary industry exceeded 50% of all employed persons by the middle of the 1970s. Japanese scholars have been aware of the softening of the economy and have been studying it from as early as the 1980s. Ryozo Aoki thinks that 'economic softening' in general refers to the situation in which 'more importance is attached to the soft dimensions of business activity such as information and service activities than is attached to the hard dimensions of business activity such as materials, goods, and energy' (Aoki 1987). It manifests as follows: one way is whether or not the number of persons employed in tertiary industries exceeds 50% of all employed

Table 12: GDP proportions of the three industry sectors in the United States.

	Primary Industry	*Secondary Industry*	*Tertiary Industry*
1889–1899	25.8	37.7	36.5
1919–1929	11.2	41.3	47.5
1953	5.8	40.3	53.9
1955	5	38.3	54.7
1960	4.3	38.4	57.3
1965	3.7	38.2	58.1
1970	3.0	38.0	59.0
1975	3.6	35.9	60.4
1980	2.8	36.0	61.2
1985	2.3	33.1	64.6
1987	2.1	31.4	66.5
1990	2.0	28.1	69.9
1995	1.6	26.8	71.7
1997	1.7	26.2	72.0
2000	1.6	24.4	73.9
2001	1.6	23.0	75.3

Sources: 1) The statistics of 1970–2001 are from the world statistics of the World Bank.
2) The statistics of 1953–1965 are from the American Historical Statistics – calculated by the GDP structure (price of that year) from the colonial period to 1970 and edited by the American Census Bureau. Tertiary industries include all sectors except agriculture, mining, manufacturing, construction, transportation and electric power.
3) *The Abstract of American Statistics*, edited by the US Department of Commerce.

persons; the other indicator is the rate at which the input proportion of raw materials and energy to all industries decreases, while the informational and other non-material input increases. The Japanese Softening Economy Centre at that time designed and calculated two 'softening rate' indices by applying industrial input and output models (ibid.). The first softening rate (the rate of information input and other non-material input, divided by input of the endogenous sector) of seventeen industries out of the twenty-four industries chosen for the study were increasing during the 1970s to the 1980s. Thus, the softening rates for fine mechanics, food products and commerce reached 53.5%, 17.9% and 76.2% respectively. The second softening rate (the rate of information and other non-material inputs plus personnel expenses and the soft component of capital expenditure, divided by the sum of production), increased for all industries except the high energy-consuming industries in fields such as petroleum, coal, water, electricity and gas. The second softening rates for finance, insurance, commerce, education, research and the medical industry amounted to 66.8%, 61.6% and 73.1% respectively. Ryozo Aoki indicated that the softened economy is 'the economy of producing, circulating, and consuming information and service'.

As for economic softening in the twenty-first century, the concept and features of those so-called 'soft' parts have changed, so it is no longer sufficient to summarize them as only information input and other non-material input. The research on economic softening now involves more than just the fact of the service sector surpassing other sectors in the economy, and does not aim to observe the level of industrialization and service economy; moreover, during the twenty-first century, the industrialization process is no longer a useful indicator of social progress.

On the one hand, through analyzing the essence of economic softening, we could probe into what is affecting people's thinking mode and transformation in contemporary social values and thereby influencing changes in the economic structure; on the other hand, by studying the softening of various industries, including primary industries and secondary industries, we can make clear how the soft factor in industry creates value, how this process may be made conducive to sustainable development (controlling the direction of innovation and promoting industries favourable to sustainable development), how the development of new economy (new profit model) and green business model might be facilitated, and how innovation may be promoted in entire range of economic, social and environmental fields.

So what is it that makes the economy soften and what is its significance?

1. Highly efficient agriculture and industry

These increase the possibility of producing a sufficient supply of food and manufactured goods for an increasing population, using proportionally lower inputs of human labour and raw materials. This is the material basis for overall economic softening.

2. Changing values lies at the core of the transformation of the structure of value-addition in the economy

At present, the GDP per capita of many developed countries generally amounts to more than 20,000 US dollars and the GDP per capita of the entire world increased from

2113 US dollars in 1950 to 6995 US dollars (1990 International Geary-Khamis dollars) in 2005, and to 7275 US dollars in 2006. As their material requirements become relatively satisfied, people probably spend more money on things besides food and other material goods. For instance, the Engel's coefficient for Japan, which indicates the quotient of food consumption, reached 67% after World War II – the hardest time in Japan's history, and decreased in 1960 to 38%, and continued to drop to 23% by 1993. The Engel's coefficient of Chinese urban citizens decreased from 52.9% in 1992 to 39.2% in 2000, and continued to drop to 36.32% by 2007. The rapid decline of Engel's coefficient also indicates that, relatively speaking, human material needs are limited, but spiritual needs are infinite. The focus of consumers in the market has shifted from pursuing material goods necessary for sustaining life towards pursuing a high quality of material life and spiritual enjoyment; this in turn creates a demand for a more extensive range, and a higher quality, of services. The lifestyle and thinking modes of today are entirely different from those of the past. This marks the rapid growth of the market share of personal consumption, forming many different styles of personal consumables industries. Even in China it is common to go to a concert or watch a game at the cost of several hundred RMB; it is fashionable to usually live frugally in order to go on a trip with the whole family; and young people sometimes even spend money on experiences and entertainments designed to produce the thrill and excitement of exceeding normal human limits. Indeed, for most people, the realization of industrialization is not the end, and their ultimate concern is how much happiness, beauty and enjoyment they are able to create. Dennis Gobar, a British Nobel Laureate in physics, considers that this change of values is a feature of a mature society. Gobar believes that in a mature society, human beings pay more attention to the quality of life and the value of spirit than to the quantity of material possessions.

It is obvious that the underlying causes of economic softening is that, as the general improvement of material living standards, significant changes have taken place in people's values, making the market value of the service sector far beyond the manufacturing costs. While the change in values has brought new consumption patterns, thus promoting the rapid development of non-material production areas, it may ultimately result in shifting the focus of value-added from the material production sectors to the non-material production sectors, so that the whole industrial structure will change drastically.

3. The softening of manufacturing industry

The progress of globalization and the rapid development of information technology, in addition to improvements in soft-tech innovation ability, generate huge changes in the strategies and business models of enterprises. Outstanding companies recognize that the sources of innovation rest not only with products and technologies but permeate the entire manufacturing process – research, supply chain, production, marketing, logistics, after-sales service, financing, collaboration, as well as the business model itself – thus adding more value than may be created by increasing manufacturing productivity. IBM provides the best example of the softening of manufacturing industry. In its recently published

Soft Manufacturing, IBM provides substantial and persuasive examples that illustrate how the company has invested time and effort into the 'soft' factors upon which soft technology innovation depends (IBM Global Service 2008). These efforts have permeated not only the entire manufacturing process but also all operating processes, with the result that soft factors are responsible company-wide for more than 50 per cent of value-added.

4. The softening of agricultural sectors
(See details in the section of 'The Softening of Primary Industries and Agriculture Service Industry' in this chapter).

5. The trend of intellectualization of service industry
The quantitative and qualitative change of the service sector is what ought to make the greatest contribution to economic softening. The main elements of the increasing prominence of the service industry are the intellectual service-oriented soft industry and the intellectualized trend of traditional service industry. As mentioned in the second chapter, the boom of commercial technology in the latter half of the twentieth century not only brought on a new phase in the development of soft industry, but also promoted the world economy to enter the era of the intellectual service economy (see the section of 'Intellectual Service Industry' in this chapter for details).

Globalization and advanced information technology are the boosters of economic softening, and soft-tech development and innovation are its engine and steering wheel. Only these factors constitute the driving force of economic softening impelling the rapid development of soft industry. The challenge facing the soft-tech development is how to enable industrial softening to be advantageous to sustainable development, whilst promoting the new economy, especially economic activity based on the green business model.

2. *The Softening of Primary Industries and Agriculture Service Industry*

Along with the evolution of industrialization, rural 'depopulation' has become ubiquitous internationally, and serious problems such as the reduction of arable land, land pollution, and eco-environmental damage, etc., have emerged. Whether seen from the vantage point of the natural environment, the social environment or the economic environment, global agriculture is facing a deep crisis. China is also facing a series of challenges related to issues affecting rural areas, agriculture and farmers, including the reduction of farmer's income, rural 'depopulation', the reduction of arable land, food supplies, ecological damage, and unhealthy urbanization patterns, etc.

The major reason for the appearance of the above-mentioned issues appears to be that the economic value created by agriculture is inferior to that of secondary industry. However, it actually stems from a limited understanding of agriculture, with the result that the multiplicity of ways in which agriculture may generate value cannot be adequately maintained or created.

The following observations are therefore important. First of all, we should change our cognitive frame about agriculture to re-understand it – agriculture should not be considered simply as 'manufacturing' for enriching the dining table. Agriculture is the basis of civilization, and that agriculture has not only economic value (it creates economic value), but also environmental value (it can maintain a sound eco-environment) as well as social and life value (it may enrich human life style and social culture). Whether or not human civilization can be maintained and progress will depend on its development mode, in which sustainable agriculture is the key as a headstream of life and supply source of food. The Japanese report, 'Sustainable Agriculture Survey' (E-Square Inc. [2007] 2008), pointed out that 'farming' and 'agriculture' should be differentiated. The notion of so-called 'farming' is wider than 'agriculture', which refers not only to engaging in agriculture as a means of generating a livelihood but also connotes protection of the natural environment and the maintenance of vital functions and the food supply, as well as lifestyle, tradition and culture in rural areas.

Secondly, in seeking to affirm the three main functions of 'farming', it is necessary to seek ways to improve value-added of the economy, the society and the ecological environment, and finally to create a new concept of rural areas or the future countryside: it is necessary to maintain a sound ecological environment; to enable local agriculture to flourish while simultaneously protecting the sublime characteristics of rural of life style; foster adaptation of integrated agricultural enterprises to the local characteristics to attract more young people living and working in rural areas; and to ensure that rural areas are provided with complete infrastructure, including convenient access to information and transportation conditions not inferior to those of modern cities.

Thirdly, it needs to expand agricultural innovation space. The innovation in primary industry should be directed not only towards the aggregate system of the entire set of processes of various sectors of agriculture, forestry, animal husbandry and fishery, but also towards the various operational models focusing on above-mentioned sectors. For example, in the case of agriculture, all operating processes associated with land use, products, markets, services, recycling, financing, insurance, risk management and others, should be infused with innovation.

Fourthly, it needs to develop the social industries in rural areas gradually.

Fifthly, we must vigorously develop agricultural service industries.

It is obvious that the backwardness of agricultural service industries was the bottleneck for improving agricultural added-value. Traditional agricultural service industries refer to agricultural seeds services, agricultural technology services, agricultural machinery services, and agricultural business services, etc. However, such services focusing on the agricultural products have been insufficient to create the necessary economic, social and environmental values in agriculture and rural areas. Therefore, agricultural services should be expanded as follows:

1. Vertical services focusing on the service industry of agriculture, forestry, stockbreeding and fishery. Let's take agriculture as an example of where services may be developed following the vertical industrial chain of agriculture: to provide pre-production services such as land improvement, seeds, fertilizers, agricultural material, production tools and equipment; providing services in the field of production such as technical solutions, machine operation services, irrigation and drainage services, and large-scale equipment-based planting, tillage and harvest; to provide post-production services like processing, packaging, storage, refrigeration, transportation, marketing and other services (Q. Zhou 2009).
2. Horizontal services focusing on the agricultural industrial chain may be developed, for example, in the cultivation of agricultural talents; information transmission; diagnosis and prevention of animal and plant diseases and insect pests; disaster prevention and risk management systems; agricultural financial and insurance services; logistics systems; food security management systems; ecological security system; as well as information service system and the system of 'Internet of Things' integrating the various systems mentioned above.
3. Services focusing on collaboration between agriculture, forestry, stockbreeding and fishery services.
4. Comprehensive services of production, education, travel, experience, etc.
5. Innovation in the agricultural business model.
6. The design of the soft environment focusing on the issues concerning rural areas, agriculture and farmers, including the new rural planning and design, various systems of laws, regulations and policies, agricultural cooperative organizations, financing systems, farmers insurance, health care and pension systems, and so on.

Obviously, the above-mentioned areas of services have gone beyond the scope of agricultural services. Thus, it is not accurate to name this industry the 'agricultural service industry'. Industry involving 'the comprehensive services of production, education, travel, experience, etc.' (from the viewpoint of tourism) takes the content of 'farming' as only part of its tourism resources, and can be regarded as an industrial innovation related to agriculture. It is therefore mainly innovation as part of soft industry.

Let us now analyze some examples on the innovation in agricultural business model.

Grapevine in Fukuoka, Japan started in the hotel industry, and then became a cafeteria according to the idea of tourist farms (E-Square Inc. [2007] 2008). In order to build a welcome enterprise with strong cooperation with local agriculture, taking root in the region, they brought forward the idea of 'sixth industrialization'. Namely, agriculture and fishing together comprise primary industry; secondary industry consists of cooking and processing, using local agricultural by-products; and tertiary industry takes these products and related manufacturing processes, as well as the rich dietary environment, as the content of service. Grapevine also links closely to above three kinds of industries and makes them inter-cross to create the so-called 'sixth industrialization', so as to maximize the development and utilization of local resources, and promote the prosperity of the region.

The Grapevine cafeteria can play to the score in accordance with the types and supply of local food materials, and promote the idea of 'local production for local consumption', by which means it becomes possible for local agriculture and fishing to flourish. Grapevine set up a restaurant chain not only in Kyushu, but also even in the Kanto and Kansai regions. This kind of self-adapted undertaking can be called 'environment-adapted industry', which is advantageous in exploring how to continuously change corporate form according to changes in the region and the times.

The Village Moku Moku hand-made farm (ibid.) in Iga, Japan, was started by the investment of sixteen pig-raising farmers and started as a processing plant for ham and sausages. The project takes fixing the price by themselves and by constructing 'an agriculture young people yearn for' as the goal, which aims to promote sustainable development and maximizing the value of agriculture. Now, as the primary industry there is the cultivation of paddy, wheat, vegetables, strawberries and fruit tree, pasture, etc.; as the secondary industry, there are processing of pork (ham, sausage, etc.), local beer, local bread, tofu, dessert, pastries, milk products manufacturing; as the tertiary industry they provide direct marketing of products, correspondence sales, food and agro-learning programmes, catering, accommodation, and so on. As a result, more than 400,000 people come from their county and adjacent areas each year to congregate there to enjoy the 'relieved, safe, delicious food', and seek the ideal of 'agriculture – experience of manufacture'.

Guaranteed Chicken (Tonggui Liu 2009) provides an example for business model innovation in animal raising and breeding. Shangdong Liuhe Group, located in Qingdao, is the biggest livestock and poultry feed-processing company in China and also the biggest company involved in slaughtering poultry. The company set up a joint venture bonding company under the support of the local government, in order to provide a guarantee for rural farmers to access to bank loans, so as to promote the large-scale, standardized new breeding mode. It holds more than 200 factories, 40,000 employees and provides one-stop service for 80,000 culturists. The first successful experiment in this mode took place in Wudi County in 2007. All of the culturists who raised 'guaranteed chicken' became rich. For example, the farmers with only one standard henhouse have become 'hundred-thousand-Yuan households' after one year, while the farmers with large scale of henhouse have become 'million-Yuan households'.

Since 2008, along the same lines as opening chain stores, Shangdong Liuhe Group set up breeding bonding companies in various counties, and furthermore diversified from 'guaranteed chicken' to 'guaranteed duck', 'guaranteed pig', 'guaranteed cow' and 'guaranteed sheep'. Guaranteed Chicken is a breeding mode which integrates eight factors such as leading enterprises, farmers, bonding companies, banks, insurance companies, governments, relevant businesses and breeding cooperation. It represents a significant innovation in development mode of animal raising and breeding which takes guarantee as the key link to integrating the supply of production resources, technical services, production (breeding), physical distribution, financial support, insurance, as well as government credit. It affords lots of lessons and enlightenment for agriculture-related companies. From the point of view of the product (chicken), this kind of enterprise is an agricultural company; from the point of view of their leading enterprises, which are feed-processing or slaughtering enterprises, it's an industrial company; from the point of view of physical distribution, finance and insurance, it belongs to service sector; and it is a new type of enterprise which disrupts the traditional industrial division. Compared with the developed countries, the main problems of China are the divorce of processing, circulation, services, foreign trade and risk management of agricultural production, etc., and moreover, the influence of regional blockade, industry segmentation and sector monopoly still exist. Therefore, this example makes sense for China's current reform of the agricultural management system.

In order to get rid of the agricultural crisis and enable primary industry – especially agriculture – to create more value, we should place attention on the 'soft' factors and focus on the above approaches to creating a space of innovation, to carrying out integrated innovation and, moreover, to creating new agricultural business model. This business model would be conducive not only to maximizing the agricultural value, but also to avoid or reduce agriculture-related risks such as natural disasters and so on, eventually constructing a rural area for where more and more young people yearned.

Another significance of the agricultural softening lies in the fact that if the ideals for agriculture and rural areas mentioned above can be achieved, it will gradually change the classic 'Petty-Clark's Law', and increase the outputs and employment in the primary industry gradually (if agricultural service industry is also included in the primary industry, see Table 16), so that the direction and content of optimizing the industrial structure should also be transformed.

3. Soft Industry

The so-called soft industry is the industry in which soft technology is the core technology, in contrast with the industrial sectors that takes hard technology as the core technology. Up until now, in the jargon of industrial management, all non-material production sectors have been known collectively as 'service industries'. However, an overall industrial softening has not only produced variations of the service industry itself, but some new industries have gone beyond the service sector in the traditional sense and even beyond the category of 'new service industry'. At the same time, the added value created by industries concerned with culture, entertainment, software, information and communications are much higher than those in the manufacturing industry – and their scale and expansion increase at an alarming rate. These industries have increasingly become the driving force for social, economic and technological development, while – together with the high-tech manufacturing industry – they have already become an important pillar of the so-called 'knowledge economy'. Moreover, because of the extensive integration of soft and hard industry, it has become very difficult to distinguish industries based upon the orthodox categories of material or non-material production. For example, it is difficult to subdivide cultural industries and social industries, which have more and more categories and a growing economic contribution, and to classify them respectively according to China's current official classification system for the service industry – using labels such as 'Social Service Sector', 'Public Health', 'Sports and Social Welfare Industries', 'Education/Culture & Arts/Radio/Film', or 'Research/Integrated Technical Services'. So what is the meaning to define and research soft industry?

According to our current knowledge, we can incorporate traditional service industries, intelligent service industries (in the narrow sense), cultural industries, and social industries, etc., under the general umbrella of 'soft industries'. It should be noted, however, that along with the softening of primary and secondary industry, more and more soft industry becomes merged. For example, software industry and artificial intelligence industry in the manufacturing industry, and the soft industry related to 'farming' in agriculture, traditional Chinese medicine industry, life prolonging industry, beauty, health preservation and health industry in the medicine and health sectors (see Table 16). The characteristics of information industry have been recognized by the world and every country adopts special policies in this industry in order to enhance their competitiveness. Because of the features of its core technology, product boundaries, industry boundaries and national boundary become increasingly blurred and, moreover, the value created by 'soft' factors gradually surpassed that created by manufacturing. Then, what to do for other similar sectors?

Due to their entirely different core technologies, the innovation process in soft and hard industry – that is, their processes, ways and forms of value creation – are dissimilar. It is worth noting that even within soft industry, the processes, ways and rules of value creation are not the same for each different set of core soft technologies. Certainly, the performance evaluation criteria and methods, as well as required human talent, are also different in each type of soft industry. What should be also taken into account is that each kind of soft technology will be able (in an upward direction) to develop into a discipline in the future, and derive (in a downward direction) more soft industries (such as the present creative industries).

Therefore, it would be forward-looking to subdivide and normalize industries according to the characteristics of its core technology, while mastering its development rules in order to formulate respectively long-term strategies and industrial policies appropriate to its characteristics, and promote its innovations, including institutional innovation implemented by systematic management.

In order to clarify the relationships between industries which are increasingly complex and interpenetrating, let us analyze the basic process of industry formation (see Figure 22).

From the perspective of core technology, industry can be divided into hard industry and soft industry. But no matter whether we are concerned with soft or hard industry, or primary, secondary and tertiary industries distinguished by material production, they all originally came from a creative idea – yet the characteristics of their goals are different. Soft industry has aimed mainly at soft (or intangible) targets such as increasing profitability, promoting innovation, improving intelligence, avoiding risks, enriching and beautifying life and solving social problems; while hard industry has aimed mainly at hard (or tangible)

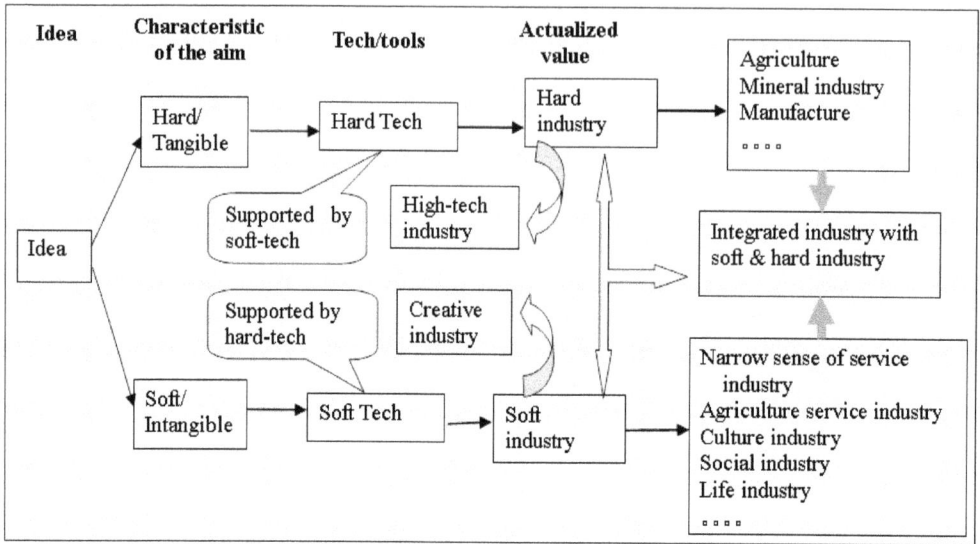

Figure 22: From creative idea to industry.

targets, for example, robots (the substitutes that are able to complete more accurate, longer time and dangerous work instead of humans); cars, trains, planes as well as rockets (the tools for arriving at a destination faster than is possible by human legs). The core means by which the latter achieves its goals mainly relies on hard technology, while that of the former relies on soft technology. I use the word 'core' here, because in order to make the process of achieving the goal more effective for achieving a higher goal, it must be integrated with 'sub-core' means. For example, for fabricating tangible goods, it is necessary to unceasingly improve management technologies during the phase of high-volume and high-quality production (such as the emergence of scientific management in the nineteenth century and the development of the Toyota management model). Service industry, furthermore, is increasingly supported by the unceasing progress of information technology and network technology.

Here, we pointed only to the tools or means – technology – which need to be used, depending on the different nature of targets. But in addition to the means (technology) of achieving its objectives, the actualization of its objectives starting from creative ideas or ideals will eventually need a place which enables its creative goals to be embodied or manifested (in other words, to be converted in to products in a broad sense) – industry. Industry is not only the result of 'technology transfer' but also the platform which enables different technologies to manifest their value through materialization. Thus, according to the core technology, hard industry is divided into different types of manufacturing industry, processing industry, mining and agriculture, etc., while soft industry is divided into service industry, cultural industry and social industry, etc.

On the one hand, with the evolution of hard technology – which has enabled products and technologies to be thinner, smaller, more accurate and faster, and which has provided an implementation platform – high-tech industry has emerged. It is a transitional industrial area formed by focusing on the existing tangible industry (in fact, there is also high-tech in soft technology). The so-called transition refers to the era character of high-tech as previously mentioned. As high-tech matures at different times, it becomes independent or penetrates into an industry, with the result that a new high-tech cluster appears.

On the other hand, along with the public recognition of soft industry, more soft technologies form an independent industry, and 'high-tech' industries focusing on soft technology – creative industries, which give rise to hot arguments – spring up like mushrooms. Thinking back to the time when the fire insurance industry was created as a result of the ideas (to make up for the losses by catastrophes and the majority share the loss of the minority) of a British dentist – what a great example of creativity! For hundreds of years, countless occasions of 'soft' targets have been transformed into actualized value through the manipulation of hard and soft technologies – with examples ranging from Hollywood movies, to the advertising industry and the consulting industry, thereby forming increasingly powerful soft industries. In fact, the current creative industries are the most active part of soft industry.

With the rise of the 'creative economy', the development of soft industry has taken the form of a high tide during the twenty-first century. Creative industries are the most dynamic sectors in world trade. According to UNCTAD's preliminary calculations, the products and services of the creative industries grew by 8.7 per cent annually from 2000 to 2005 (UNCTAD 2008). The exports of creative products increased from 227.5 billion US dollars in 1996 to 424.4 billion US dollars in 2005, while the exports of creative services during this period grew by an average of 8.8 per cent. This is good news, indicating that the function of another paradigm of technology – soft technology (soft technology can be used as the core technology to form an independent industry) – has been recognized by the public, and the theory of the 'innovation space' in a broad sense has been verified. This should encourage people to consciously research, invest and develop other means to solve problems (soft technology and the soft environment innovation), study the process of creating added value and attach importance to intellectual property rights associated with soft technology.

However, with the arrival of new high tide of soft technologies, we must remain sober-minded to ascertain the nature, characteristics and attributes of this wave. In recent years, we have experienced an upsurge in the establishment of high-tech development zones and incubator pilot areas. By 2003, the various economic and technological development zones reached 6866 with 38.6 thousand square kilometres of the planning areas, and after rectification there are still 1568 development zones by 2006 with planning areas reduced to 9949 square kilometres. Now people in China are falling over one another to establish creative industrial centres or parks all over the country (there are hundreds of development zone by 2009). And, moreover, in the field of information technology, since the Internet Technology Revolution around the year of 1995, each city is competing to set up various centres of 'Internet of Things' facing the current technological upsurge in 'Internet of Things'.

The so-called development zones, experimental areas or industrial parks are actually the pilot areas for the creation of a soft environment and, as such, their contribution is positive and significant at a certain stage. However, it is not necessary for each region to adopt the kind of government-guided approach for each new technology or new concept in order to avoid duplication of investment and to avoid unnecessary increases in social costs.

Our study on the soft industry has not only delineated clearly the trend and essence of the future industrial structure, but also opened up new areas which are familiar but neglected for a long time by entrepreneurial activities. And, moreover, it has expanded the entrepreneurial space and channels, helped the enterprises to change their strategic thinking and provided a theoretical basis for creating a new business model. The key is how to create an efficient environment for creation, innovation and entrepreneurship.

4. Soft Industries and So-called Creative Industries

At present, the term 'creative industries' frequently appears in the policy recommendations of the business community, academia and government; it has become the new economic growth point upon which various countries are focusing. Some representative examples of research on creative industries include the Creative Industries Mapping Document (CIMD) of the UK Government's Department for Culture, Media and Sport (United Kingdom DCMS 1998) in 1998, *The Creative Economy* by John Howkins (2006), and the 'Creative Economy Report 2008' by the United Nations Conference on Trade and Development (UNCTAD) (2008). China's creative industries are also emerging. The Government of China has introduced policies to encourage the creative industries, and governmental and social investment has begun to move in the direction of creative industries, while Creative Industry Bases have been building in Shanghai, Shenzhen, Chengdu, Beijing and other cities. At present, the products of Chinese enterprises lack brand vitality; the vast majority of products, which are fabricated in China and occupied the first place of the world sales, are foreign-designed foreign brands, while China can only get very little profits. In the path from 'Made in China' to 'Created in China', the rise of creative economy will be historic.

But what are creative industries? According to the definition of DCMS, the creative industries are 'those industries which have their origin in individual creativity, skill and talent and which have a potential for wealth and job creation through the generation and exploitation of intellectual property'. UNCTAD defines creative industries as 'the cycles of creation, production and distribution of goods and services that use creativity and intellectual capital as primary inputs. They comprise a set of knowledge-based activities that produce tangible goods and intangible intellectual or artistic services with creative content, economic value and market objectives'. Creative industries are at the cross-road among the artisan, services and industrial sectors. John Howkins believes that four industries, including trademark and design industries, as well as copyright and patent industries, constitute the 'creative industries' and 'creative economy' together.

Analyzing the classification of creative industries by different groups and organizations may be profitable for us in helping us to further understand the meaning and connotations of 'creative industries'.

The UNCTAD classification of creative industries is divided into four broad groups: heritage, arts, media and functional creations. These groups are in turn divided into nine subgroups, that is: traditional cultural expressions (arts and crafts, festivals and celebrations); cultural sites (archaeological sites, museums, libraries, exhibitions, etc.); visual arts (painting, sculpture, photography and antiques); performing arts (live music, theatre, dance, opera, circus, puppetry, etc.); publishing and printed media (books, press and other publications); audiovisuals (film, television, radio and other broadcasting); design (interior, graphic, fashion, jewellery, toys); new media (software, video games, and digitalized creative content); creative services (architectural, advertising, cultural and recreational, creative research and development [R&D]); digital; and other related creative services.

UNCTAD has also provided four different classification systems in their report and each model leads to a somewhat different basis for classification into 'core' and 'peripheral' industries within the creative economy. The four models are as follows:

1. The UK DCMS model classified the creative industries as advertising, architecture, art and antiques market, crafts, design, fashion, film and video, music, performing arts, publishing, software, television and radio, video and computer games.
2. The core cultural industries under the 'Symbolic Texts Model' are advertising, film, Internet, music, publishing, television and radio, video and computer games; peripheral cultural industries are creative arts; borderline cultural industries are consumer electronics, fashion, software and sport.
3. The core creative arts under the 'Concentric Circles Model' are literature, music, performing arts, visual arts; other core cultural industries are film, museums and libraries; wider cultural industries are heritage services, publishing, sound recording, television and radio, video and computer, games; related industries are advertising, architecture, design, fashion.
4. The core copyright industries under the WIPO 'Copyright Model' are advertising, collecting societies, film and video, music, performing arts, publishing, software, television and radio, visual and graphic art; interdependent copyright industries include blank recording material, consumer electronics, musical instruments, paper, photocopiers, photographic equipment; partial copyright industries are architecture, clothing, footwear, design, fashion, household goods and toys.

John Howkins enumerated fifteen industries which constitute the creative economy, including advertising, architecture, art, crafts, design, fashion, film, music, performing arts, publishing, R&D, software, toys and games, television and radio and video games. It is obvious that despite the fact that the creative industries may be defined from a variety of different angles, the overall classification of the industries by different authorities is more or less the same.

The above analysis reveals that the so-called creative industries are actually part of the soft industries, but also the most active part of cultural industry and intellectual service industry among soft industries. People have recently become very excited by the prospect of creative industries, because they have discovered that 'soft' things – including soft goals which are manifested by new ideas, designs and conceptions – can also create wealth. As a result people are willing to put more energy, time and money into research, exploitation and development of creative industries. This phenomenon indicates that with the development of contemporary soft technologies, the 'production' of non-material or intangible assets created bigger and bigger value and, increasingly, people are willing to pay for achievements of 'productive activity' or solutions in such non-material fields.

In fact, if we get to the root of the matter, a lot of hard industries, including high-tech industries, also come from the creative ideas.

Anyhow, the change in values and the overall economic softening that commenced during the last century enable new types of problem-solving, which is related to new consumption patterns, new business models, new lifestyles, new forms of leisure and entertainment, new work patterns and new concepts – and can lead to business opportunities. Thus, these changes have given birth to a larger volume of creative ideas, along with the development of hundreds of companies and even the trade that specifically provides new notions, concepts and ideas for others also formed into an industry: the so-called 'idea industry'. This is the background of the rise of dazzling creative industries.

As we mentioned previously (see the section of 'Revealing the Essence of Creating Value for Intellectual Capital' in Chapter 1), intellectual capital cannot create the value directly, and new ideas have only the potential to create value. In order to make ideas turn into value, it is necessary to carry on a series of technical operations – from fund-raising, organizing teams, 'interpreting' the creative idea to make it convince recipients of product or service (i.e. the solutions that are perceptible or visible by the 'consumers' of creative industries, so as to convince them to pay the bill): in short, to exploit the market. The operating process of moving from the idea to creating value is that of soft technology, and the innovation processes in view of different categories form the core technologies of different creative industries. John Howkins has said that 'intangible industry' is another appellation for 'creative industry', which has a wider coverage. I could say that the creative industries are part of the soft industries that have a broader coverage.

The current upsurge of creative industries is actually a soft industrial cluster emerged periodically along with a new round of business model innovation boom, and the industrial cluster which emerged in this cycle is mainly centralized in the field of culture and art. I believe that the creative industry for the future will also emerge in various areas, in particular in the cross-cutting areas of soft industries and hard industries. And the undertakings of sustainable development in many domains, including resources, energy, environment, society and other fields, have the opportunity and potential to develop creative industries.

This is the reasons why I did not solely list the creative industries in the future industrial structure, and as was the case with high-tech industry, creative industries have a time character. For hundreds of years, every Industrial Revolution has been driven by the creative industries of the day. When a new round of soft industrial cluster emerges, most of its entrepreneurial activities are irregular commercial arrangements; thus, it is called as creative industries. After these entrepreneurial activities obtain appropriate services and institutional arrangements (current entrepreneurial parks aim to speed up this course), thereby being mature, it will be incorporated into relatively independent or similar industries, and then it will meet the cycle of the next generation of soft industrial cluster (see Figure 9). In this sense, we should not regard simply developing and encouraging the creative industries as a temporary measure to increase employment in the course of getting rid of the financial crisis, and round after round of them will retain its vigour forever.

Now, we have analyzed the essence of creative industry from the angle of soft industry that's aim is to identify relevant core technologies, enhance its innovation efficiency, and speed up the process of moving from ideas to value creation.

5. *The Characteristics of the Soft Industry*

Compared with hard industry, soft industry has many distinctive features. These include the factors listed below.

1. Soft targets
Due to the 'soft' attribute (see the section of 'What is Soft Technology?' in Chapter 1) of soft technology, the target 'products' of soft industries are mostly intangible. 'Soft' targets could include increasing capacity, developing intelligence, advertisement, solving social problems, visual enjoyment, spiritual enjoyment, experience, learning, avoiding the risk, the sense of security, health, beauty, convincing others, and so on.

2. Core technology
The so-called 'core technology' refers to the core tool of value creation. A variety of soft technologies are the core technologies of the soft industries, and hard technology is a tool to improve the efficiency of soft technology. One of the differences among those industries is the targeted objectives, and another is the characteristic of the soft technology utilized in the process of actualizing the target; namely, the different process technologies required by value creation.

3. The transformation of industrial attributes
In some integrated industries of soft technology and hard technology, the transformation of industrial attributes often happens. For instance, in the software industry and the artificial intelligence industry, when the value created by their 'soft' factors exceeded that created by

'hard' factors, it can be regarded as soft industry. Namely, when the value is converted to a certain level, it may lead to the transformation of an industrial attribute, where the tangible products become means and tools to achieve the soft targets.

4. Price
Price is greatly influenced by values. Like the prices of famous brands, new recreational machines, special services and urgent solutions, etc. are based mainly on consideration of the benefits of the creators and new value created for the customers or the extent of meeting spiritual demands such as satisfaction (which can not be quantified). So, there is an obvious gap between price and production cost. This tendency is most noteworthy in the cultural industry; e.g. in the fashion industry it is difficult to explain the causes of the increase or decrease of gross sales by simply using orthodox economic concepts associated with traditional products such as price, quality and service.

5. Psychological factors
As mentioned above, one of the key variables of soft technology is the psychological factor. At present, the factors that give people excitement shift from novel product function, good product performance and low product prices to the dream of life, the meaning of life, interests and hobbies and self-fulfilment. Consequently, people's demands change from function type, essence type and rationality type to feeling type and emotion type. It is therefore important that 'products' of soft technology try to meet these demands. Accordingly, Kimindo Kusaka has even put forward such ideas as the exchange position of psychology and economics in the new age and the 'emotionalization' of service (Kusaka 1980). At the same time, when government departments contemplate institutional innovation, they should consider the psychological endurance and reaction of the mass of people; likewise, in industrial circles, successful enterprise managers need to first of all consider how to satisfy the psychological pursuits of their customers when developing new products so as to materialize the psychological needs of their customers in a timely manner.

6. Resources
Hard industries mainly develop material resources, whereas soft industries draw upon a variety of resources. In addition to material resources, soft industries develop richer nonmaterial resources, including part of natural resources (view, landscape, geographical environment, etc.), intelligence resources, cultural and artistic resources, social resources and human body resources, as well as industrial resources and environmental resources, etc., thus enabling development to take place across a broader array of fields (see Figure 15).

7. Integration and fusion
Soft industries involve the fusion and integration of different types of knowledge, technologies and art, etc. The boundaries between various soft-tech industries are therefore blurred and

they infiltrate each other or across a wide range of industries. Therefore, the classification of soft industries is very difficult.

8. Soft industry does not equate with intangible industry
Soft industries operate or manipulate either those intangible resources or tangible resources, or create value by dint of tangible facilities or conditions. The tangible part of the cultural industry includes theatre, library and other facilities, and that of social industry includes various public facilities.

9. The cradle of new conceptual enterprises and entrepreneurs
The changes in the understanding of technology have raised awareness that each kind of soft technology is the growth point for new industries and the inexhaustible 'mineral' resource. The integrated innovation of soft and hard technologies, in particular, is a seedbed from which immeasurable business opportunities may sprout. At the same time, the professionalization and industrialization of soft technologies will create a great diversity of entrepreneurs, e.g. cultural entrepreneurs, design entrepreneurs, educational entrepreneurs, social entrepreneurs, etc., even entrepreneurs of entrepreneurship.

The above features are only the common characteristics of soft industries but, because of their various resources, they may have identities distinct from each other. Thus, according to different operational resources, some of the soft industries will also exhibit different characteristics, which we need to further study.

Let us further understand soft industries and their impact on future industrial structure through the examples of intellectual service industry, cultural industry, social industry, agricultural service industry and so on.

The service industry has still occupied half of the soft industries. In order to further understand the essence of soft industries, let us first recognize service, update the concept of service and analyze how service innovation promotes the development of intellectual service industry.

B. Intellectual Service Industry

1. What is Service?

It seems that we do not need to define the word 'service' painstakingly. However, although service innovation gets most of the attention, the old notion of service and the misunderstanding of service innovation have become the principal barrier to the sound development of the service economy. Thus, renewing our understanding of service and carrying out the in-depth research on the essence and connotation of service and service innovation are necessary for developing service economy.

1. Service in the traditional sense

In the traditional sense, service is generally treated as labour service. There is the following description of service in China's thesaurus: 'The service is also called "labor [sic] service"; it is the activity to meet some kind of special needs of other people with human labor [sic] rather than material objects. However, along with the development of service economy, "service" has already grown beyond the scope of the labor [sic] service, and become an important object of research in its own right' (*CDCL* 1989).

When the Japanese scholar, Takao Kondo, mentioned the exploitation of value in 1994, he defined service as follows: 'The so-called service is the activity or function that can bring an individual or an organization certain conveniences and can be traded in the market. In other words, the service component in economic activities can, in themselves, be the object of transactions in the market for valuable productive activities' (Kondo 1997). Here the object of transaction is stressed for the reason that there are many service activities that cannot be the objects of transactions and, conversely, not all value productive activities ought to be thought of as service. For example, the domestic chores that belong to each member of a household cannot normally be traded on the market. However, if these activities are taken to a market to sell, they become service activities.

However, Hazel Henderson has put forward a query about the nature of domestic services; she believes that it is a prejudice of traditional economics to take the non-remunerative work in which women engage as 'non-economic' compulsory service, and to not include calculations for it in the national economic accounts (Henderson 2002).

The OECD defined service in 'The Service Economy' published in 2000 as follows: 'services are a diverse group of economic activities not directly associated with the manufacture of goods, mining or agriculture. They typically involve the provision of human value added in the form of labour, advice, managerial skill, entertainment, training, intermediation and the like' (OECD 2000).

The definition of services given by OECD in their report 'Innovation and Productivity in Services' in 2001 has been developed, and takes 'service' as: 'services deliver help, utility or care, and experience, information or other intellectual content – and the majority of the value is intangible rather than residing in any physical product' (OECD 2001).

Obviously, along with the development of time, the concept of service is also evolving. What kind of description or definition of 'service' in contemporary society is more reasonable?

2. The evolution of service concept

In western countries, the main service industries before the twentieth century were centred on the family servant. In the US, half of the country's service work was conducted by the family servant in 1900 (I believe that the case in China was similar, but there is no statistical data available); but the family servant accounted for only 25 per cent of the employees of service sector by 1968, and the main increase was automobile mechanic, worker of hotel and restaurant and other similar occupations (Bell 1984: 211).

With the progress of economic softening, the old idea of treating service as 'labour service' is no longer applicable; and, moreover, the elaboration of OECD in 2000, which reported that 'services had no direct relationship with the manufacture of goods (material objects), mining industry or agriculture', is also incomplete. This is because the respective service industries will be derived from the sectors related to material objects, such as manufacturing, mining industry and agriculture of all categories. Moreover, the description of OECD in 2001, which regards the service as '[to] deliver help, utility or care, and experience, information or other intellectual content' is also not comprehensive as the process of providing material products such as machine, equipment, etc. is also classified as service.

Therefore, I would like adopt a broad concept of service and to redefine the term 'service'. 'Service' refers to a 'process by which the demands of the one who accepts the service are satisfied through the necessary tools and means'. As shown in Figure 23, the service is a process, and in this process the service provider can use any necessary tools and means to meet the demands of the service accepter.

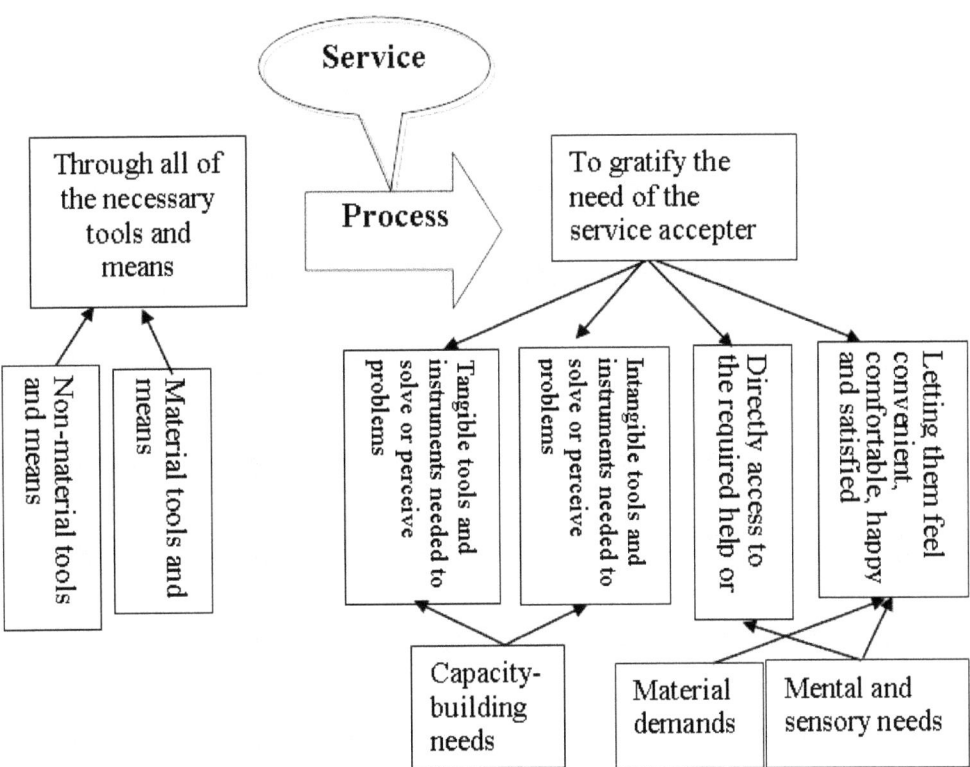

Figure 23: The signification of service.

Table 13: Differences between the traditional concept of service and that with a new concept.

	Traditional understanding of service	New concept or broad sense of service
Function of service	Mainly to meet the material demands	To meet the material demands, spiritual and psychological needs and provide problem-solving abilities
Object of service	Individuals, families and enterprises	Individuals, families, organizations (government, enterprises, groups) and society
Mode of resolving problems	Labour and intelligence	Any necessary tool and means
Provided 'products'	Non-material services	Material and non-material services
Core technology	Service industry is not considered to have a core technology	Soft technology
Relationship with manufacturing industry	To be independent with each other	To be interpenetrating, service-oriented manufacturing/ manufacturing of service industry

In this definition, the so-called necessary tools and means could provide service through either direct contact or indirect contact; could use either non-material tools and means, such as work or physical strength, wisdom, knowledge and soft technological means (for example consultation or advice, organization, management, mediation and arbitration, etc.), or hard technological tools and material means, such as goods, tangible tools, machine, equipment and all the necessary natural objects and artificial manufacture, but not including the manufacture of the latter. Certainly, these material means must be implemented via non-material tools and means to meet the demands of the one who accepts the service.

The so-called demands of the service recipient include: acquiring tangible tools and instruments (including goods, equipment, machine or others) needed to solve or perceive problems for them (organizations, enterprises, individuals, etc.); acquiring intangible tools and means (experience, intelligence, knowledge, information or other abilities) which they need to solve problems or make sense of puzzles; making them directly accessible to the required help or care; letting them feel convenient, comfortable, happy and satisfied. The former two concern enabling the service recipient to obtain problem-solving skills so that they can independently solve problems; the latter two are directed towards directly helping the one who accepts the service to solve the problems by the main body of service. It is noteworthy that the tangible tools needed when people solve problems are actually serving the intangible tools and serving the process of meeting other demands. In short, the demands of service accepted include material demands, mental and sensory needs, capacity-building needs, etc.

From the point of view of the service provider, the ones who accept the service are the objects of service behaviour or the consumers of service who are generally considered to be natural persons, legal persons, social organizations or government departments; in other words, individuals, families, organizations (enterprises, government, and groups), society and all other objects in need of services. Service providers also include these main bodies mentioned above. Table 13 shows the differences between traditional sense of service and the broad sense of service.

2. The Essence of Service and Service Innovation

For a long time, although people have enjoyed services everywhere, they have not realized what service really is. Although added-value created by global service industries accounts for 70 per cent of the GDP, the service sector is still regarded by many people as the 'assistant sector' or 'catalyst' of primary and secondary industry. The OECD report of 2000 on service innovation pointed out that 'services have not received due recognition in terms of their role in the innovation process in either academic or policy circles. Although this is gradually changing, innovation models still centre predominantly on manufacturing […]' (OECD 2001). This mainly comes from a lack of understanding about the essence of service, and a presumption that service has no technology and that there is no core technology in the service industries – unlike the metallurgical industry which takes metallurgical technology as its core technology; the electronics industry that takes electronic technology as its core, etc. It is therefore necessary to study whether or services have their 'technology'.

As mentioned in Chapter 1, soft technology is an intangible operable knowledge system for the solution of problems. Thus, the process of 'solving' the demands of service recipients through a variety of ways belongs to the category of soft technology. Any service, focusing on the demands of different service recipients, forms a solution through providing different means and ways of service; that is, the process of each service requires a specific solution – one kind of soft technology (which is part of process technology). Service has technology, hasn't it? The essence of service is the process of meeting human material demands, mental and sensory needs, and capacity-building needs for problem solving. So-called 'service technology' is the general term for these process soft technologies.

However, service is different with soft technology. Because the broad sense of technology itself is the tool to provide 'service' for social progress and economic development, in addition, all kinds of soft technologies are the core technologies of the service industries. That is why the new concepts of service industries and soft industries are intertwined with one another. But not all soft technologies are identical with the service technologies.

First of all, service technology is an operable knowledge system which exists in the supply-demand relationship between the specific service provider and service recipient;

namely, it is more limited and oriented specifically towards the various processes and methods in the supply-demand relationship, while soft technology is based upon another paradigm of solving problems. The level and orientation of problems addressed by each type of technology are different, and their research objectives, targets, scopes and function (the three main functions of soft technology) are also different. Hence, service technology can not cover the full scope of the operable knowledge system represented by soft technology. From those business technologies listed in the Chapter 2, we can make out that the majority of business technologies and social technologies may not be easily incorporated into the field of services.

It is important to make clear the attributes of service from the perspective of soft technology. Studying the methods and processes of providing services as soft technologies is not only conducive to in-depth study of the essence of services and service innovation, but also many new issues in the service economy can be theoretically explained – including the incentive system of service innovation, standards of service and intellectual property rights in the field of service. At the same time, service industry and the manufacturing sectors increasingly interpenetrate each other and are thus difficult to distinguish. Soft technology provides services innovation in hard-tech and hard industry, or integration of hard and soft technology during the process by which soft industries improve their intellectual content by use of hard-tech achievements, etc. Moreover, the concept of the 'service innovation system' can be established according to a broad innovation framework, so as to systematically study and solve the environmental problems of stimulating service innovation through policies, standards, laws, and intellectual property rights in the service sectors, to further promote service innovation.

As the world enters the post-industrial era, the professional division of labour becomes increasingly detailed, and demands of enterprises for external service providers are increasingly broad and specific. Especially since the latter half of the twentieth century, the service industries have become the fast-growing sectors and have become the focus for creating new employment positions. On the one hand, service has become the key element of changes and competition in all sectors of the economy; on the other hand, social, economic and technological development increasingly requires service innovation so as to promote the rapid growth of the service economy. We can say that the growth of the service economy is driven by service innovation.

It is obvious that service innovation is no longer just about innovation within the service sector, nor does it concern only the innovation associated with 'non-technological factors' or the packaging of technology; rather, it belongs as part of 'technological' innovation in the broad sense.

From an economic point of view, service innovation refers to the activity of embodying or increasing value-added through the different process of satisfying the demands of service consumers. Such activities are particularly prominent in the information industry. The rapid development of information technology (IT) enable the technologies and functions of products to attain an increasingly high level of homogeneity, and there is an increasingly narrow space for creating profit through improving product quality and reducing production costs. IT services in the narrow sense formerly referred to after-sales service for IT products (which in previous times were generally free of charge), while IT services in the broad sense emphasize the integrated innovation of soft technology and information technology; thus, the concept of IT services has undergone profound changes. Now, service becomes an important weapon for enterprises in market competition and the main source of add-value. IBM's global revenue reached 103.6 billion US dollars in 2008, of which software and services revenues accounted for two-thirds. IBM is no longer a hardware company. According to IBM's report which was submitted to the US Securities and Exchange Commission (SEC) in 2009, among the four major sources of profits of IBM in that year, pre-tax profits of hardware and financial practices amounted to approximately 3 billion US dollars, while software and services brought about 8 billion US dollars of pre-tax profits respectively, totalling 16 billion US dollars.

From a technical point of view, service innovation refers to soft-tech innovation activities for the purpose of meeting human material demands, mental and sensory needs, and capacity-building needs. The progress of soft technology in the dimension of the human 'mind' and in exploiting the 'relationship between mind and body' will provide many new opportunities for service innovation.

From a social point of view, service innovation refers to the social activities of improving human self-value and capabilities, and enhancing the quality of life.

From the perspective of methodology, service innovation is the activity to invent or create new service methods, delivery channels, service targets and service markets. Take the software service as an example. The approach to service innovation includes three aspects. Firstly, it varies according to the type of services in question, such as consultation, training, implementation, maintenance, secondary development, supporting products, environmental construction, version upgrades, etc. Secondly, it is based on the service process itself including pre-sales, sales and after-sales services, in which pre-sales services consist of consultation of software products, presentations of the solutions, etc.; sales services comprise training, implementation and so on; and after-sales services include technical supports and so on. Thirdly, it includes the 'services of commitment', namely, the commitment to providing services that many software companies promise to users. However, the above-mentioned methods are the software-centred services involving their own company. But, currently, the excellent software companies like IBM use the information sensor new technologies such as intelligent recognition technology to integrate themselves into the 'Internet of Things', thereby developing different solutions according to different objects – different conventions and agreements (soft technologies). Thus, the service recipient and service approach are extended to almost all areas, such as traffic management, urban management, resources management, etc. so as to make themselves incorporate veritably into soft industry.

In short: 1) service innovation is essentially soft tech innovation, that is, all kinds of process technology are the core from which service industries are formed; 2) the efficiency of service innovation depends on soft tech innovation, and on whether or not the integration of soft and hard technology is successful, as well as on the innovation in the soft environment, which involves 'keeping up with the times'. On the other hand, in the service innovation process, all the necessary natural matter and manufactured goods needed by the service targets are the tools of soft-tech innovation.

For a long time, because people did not think that services required special technology, the skills, methods, processes and rules of services were not taken as specialized technologies for study and development, and there has even been insufficient vocabulary to talk about service technology. It is generally known that there is an abundance of economic terms for explaining primary and secondary industry; moreover, creative technology developments in these industries has always been given special importance by economists and technologists, with no shortage of pertinent studies ensuing. In the case of service industry, however, there are few R&D institutions in the world devoted to carrying on systematic research related to services. In China, many intellectual services were uniformly referred to as 'planning' and this has had the effect of passing off much sham as genuine. In the United States, following the US Government's efforts in the 1980s to 'deregulate' the aviation, telecommunications and other industries, competition among enterprises has become more intense with the consequence that entrepreneurs in the above-mentioned industries have turned to academia to seek insights on how to win customers with high-quality service, and the service-related solutions for triumphing against competitors. This is an historic event, that is, that academia has initiated concentrated research about service sectors due to expressed demand from the market.

3. Service Economy and Intellectualization of Service Industry

An article entitled 'High-tech Industries and Coffee Shops Propel a Revolution' was published in the *Financial Times* in Britain on 2 March 1999. It pointed out that during the previous thirty years the economic output of the service industries of Britain had increased from accounting for one-half of the total economic output to accounting for two-thirds; the number of employees in the private services sector took up half the total number of employees; and if people working in government departments were added, the figure would be three-fourths. The growth in the proportion of service industries was not achieved primarily by the increase of fancy coffee shops, wedding ceremony photography or personal coaches but by providing professional services for enterprises and communities, which include companies specializing in management services, intelligence gathering from other companies, computer services, telecommunication services, financial services, accounting management services, etc. Furthermore, these services are becoming increasingly internationalized.

The great success of companies such as McDonald's and Coca Cola, etc., in the global market, and in addition, the increase in 'nabobism' within the service industries, has compelled people to develop new and more profound regard for service industries. In the year 2000, S. Robson ('Sam') Walton, Chairman of the Board of Wal-Mart Stores, Inc., became the richest man in the world, taking the place of Bill Gates who had previously been listed as the wealthiest man in the world for three consecutive years. When Walton's company was first founded in 1962, it was only a small store. Today Wal-Mart has 3500 chain stores in the United States and over 1000 stores abroad, with the total number of employees reaching almost 900,000. Since 1990, its annual revenue has increased by 20 per cent, replacing the oil giant Exxon Corporation as the biggest company in the world, remaining three years (2001, 2002 and 2003) and dominated the Fortune 500. GE, which ranked the first on the annual Fortune 500 list 36 times since 1954, has also been surpassed.

Service companies were included in the Fortune 500 in 1995 for the first time, thereby creating an important symbol for the service economy. Warren Batts, a professor of the University Of Chicago Graduate School Of Business, believes that this event signalled the big transformation of United States shifting towards the service economy. Of the top twenty companies worldwide listed in Fortune 500 in 2001, seven were service companies and of the fifteen richest people in the world, nine were from service industries.

In this context, the service innovation occupies increasing attention, and various countries scramble to increase the R&D expenditure of service (see the section of 'Strengthening the R&D for Soft Technology' in Chapter 4).

How does a company that 'produces nothing' generate so much revenue? It relies on its unique soft tech innovation, especially in business models – for example, Wal-Mart's supermarket operating methods, or McDonald's concept of self-service fast food. Unceasing innovation in soft technologies makes it possible to continuously create and integrate new modes of service delivery and high value-added services. Enterprises that have taken this pathway have actually applied for hundreds of patents (although not 'high-tech' patents) each year so as to facilitate the long-term sustainability of their business.

Table 14: The US GDP Composition (1982 dollars)

Item	1989/1980 ratio
GDP	30%
1. Agriculture, Hunting, Forestry, Fishery	31%
2. Manufacture	40%
Thereinto: Machinery (electrical equipment excluded)	106%
Electricity, Gas and Health Service	60%
3. Trade	43%
4. Finance, Insurance and Real Estate	30%
Thereinto: Mortgage/Commodity Agent and its services	255%
5. Service	47%
Thereinto: Management service	89%
Entertainment and Consumer service	67%
Other professional service	51%

The data in Table 14 further illustrate that the most outstanding change in the economic structure of the United States during the last several decades has been the rapid development of the intellectual service industry. The table summarizes the GDP growth of the US during the ten years from 1980 to 1989, and compares the growth proportions of several industrial sectors. The growth of GDP as a whole during the period averaged 30%, while the mortgage/commodity agent and its services in finance, insurance and real estate industries increased by 225%. Service industries as a whole grew by 47%, in which management services grew by 89% and entertainment and consumer services grew by 67%.

Table 15 documents that the added-value of tertiary industries in China during 1991 to 2001 increased by 359% and the top four fastest developing sectors in the service industry were postal services and telecommunications (1501%), social service industries (762%), scientific research and comprehensive technological services (619%) and education, culture, art, broadcasting, movie and TV industries (507%). These industries are, in the broad sense, intellectual service industries of soft-tech industry and they have the greatest future potentiality in China, illuminating that intellectual service industry is the main driving force of promoting the development of whole service industry. However, the growth rate of all tertiary industries combined is still lower than that of the secondary industries. This lagging growth of China's tertiary sector will seriously limit the sustainable growth of its economy.

On 19 March 2007, the State Council issued 'The State Council on Several Policies and Measures on Accelerating the Development of Service Industry' as well as enforcement opinions of the General Office of the State Council, which clearly indicated that accelerating the development and upgrading of service industry is a national strategy related to China's modernization. Therefore, each province and municipality of China has carried out a series of arrangements, increasing funds and policy inputs in the service sector.

Yuexiu District in the centre of Guangzhou City has been planned as a congregation zone for modern service industries; there are plans for construction of six concentrated areas of modern service industries in Guangzhou/Yuexiu, including international purchasing, international commerce, creative economy and network economy, health industry, logistics operation and cultural tourism. At present, there are many well-known international companies gathered there, including not only Samsung Electronics of South Korea, Seiko of Japan, Alcatel of France, Swire Group of UK and the South China regional headquarters of UPS, etc., but also more than 3000 modern service enterprises related to advertising design, legal services, software dissemination, investment consultants, consultation assessment, financial insurance and so on.

In the 1990s, the development of information technology in India has brought along with it the intellectual service industries, such as communications, software, finance and banking, which grew by 8 per cent annually and has become one of the fastest developing sectors. It is called the new economy of India. Network TV, music and entertainment are important sources of wealth in India and they are not only the engines of Indian economic development but are also sources of wealth for India's civilians. The Indian middle class

Table 15: The value-added in Chinese tertiary industry (+/-: Higher or lower than the total GDP growth).

Industry	2001/1991	+/-	Rank
National GDP	350%		
GDP of Primary Industries	191%	–	
GDP of Secondary Industries	436%	+	
Total Value added of Tertiary Industries	359%	–	
Sub-categories:			
Services in Agriculture, Forestry, Stock raising & Fishing	394%	+	6
Management of Geographical prospecting and water conservancy	346%	–	8
Communications and Transportation/Storage/Postal and Telecommunications	323%	–	11
Communications and Transportation and Storage	184%	–	14
Postal and Telecommunications	1501%	+++	1
Wholesale, Retail, Trade, Food and Beverages	279%	–	13
Finance and Insurance	333%	–	10
Real Estate	411%	+	5
Social Services	762%	++	2
Health, Sport and Social Welfare	358%	+	7
Education/Culture and Art/Broadcasting/Movies and TVR&D/Comprehensive technological service	507%	+	4
Government offices/party/Social groups	619%	+	3
Other service industry	283%	–	9
	333%	–	9

Source: Chinese National Statistics Almanac (2000, 2003)

includes a hundred million members by 2000 and 50 per cent of the income of Indians is derived from the service industry (L. Tang 2001).

Now, the rapid development of intellectual services changes the position of service industry thoroughly. The new trends of service industry manifest as follows (Jin & Ren 2004):

1. Service industry is the sector with the strongest vitality, the most active innovation, the fastest absorption of technical staff, and the greatest increase in employees.
2. The boundary between service industry and manufacturing industry has become increasingly blurred.
3. Continuous advances in the intellectual content of service industries have prompted a large number of the high-quality labour force, including highly talented intellectuals to turn-to service industries.

4. Service industry is no longer a low-wage employment centre.
5. Intellectual property issues in service industries are now on the agenda.
6. The majority of service industries are comparable with the manufacturing sector vis-à-vis their economic scale and profit levels.
7. New business model innovation has become the main driving force for the supererogation of traditional service sectors.
8. The service globalization and the services trade gives full play to people's abilities.
9. The relationship between service providers and service consumers is being transformed.
10. Service brands are becoming increasingly well-known.
11. Various countries have strengthened R&D related to service industries.

In the above context, intellectual services fully penetrate every walk of life, so any demand can stimulate the formation of a service market, and each industry can generate its own service sector. That is why the service innovation has become increasingly active, and why many service industries have become 'high-tech' industries – intelligence-intensive service industries – so that the era of product diversification of the twentieth century has been replaced by the era of service diversification of the twenty-first century.

4. The Intellectual Service Industry in the Narrow Sense

According to the attribute of technology as a tool (see Chapter 1), there is a 'service attribute' in technology itself. Hence, if we think about this from the perspective of service, broadly defined it is possible to incorporate all of the soft industries and hard industries into the broad category of service industry. This is also true of the emergent service sectors of various industries, including the agricultural service industry, the industrial service industry (such as service-oriented manufacturing), the commercial service industry, the cultural services industry, the health service industry, education service industry and the sports service industry. However, analyzing the differences (see the section of 'The Essence of Service and Service Innovation' in this chapter) between soft technology and service technology, it is clear that in studying soft industry it was never our intention to incorporate all of the 'non-material production' fields into the large 'box' of intellectual services.

What we need is to define intelligent service industry in the narrow sense, in a way that exceeds the limits of industry and according to the approaches and means that are consistent with providing intelligence. For example, despite the fact that the various industrial service industries each have their own unique characteristics, they have a common core in the soft technologies (common need or common way to provide intelligence) that in turn can form the basis of some intelligent service industries, in the narrow sense and in their own right. Let us now look briefly at several commonly recognized intelligent service industries in the narrow sense.

1. Consulting Industry

The consulting industry is an industry that has consulting technology as its core technology. It creates value by providing users with knowledge, wisdom, sense, insight, experience and solutions of problem solving. Consulting industry is the most mature of the intelligent service industries not only because its history commenced more than a century ago, but also because it exhibits a high division of labour on the basis of specialization. It has expanded to a variety of fields, including strategic planning, enterprise management, trade negotiation, policy consultation, legal consultation, technological advice, psychological consulting and health consulting, etc. The development of hard technologies has provided newer and better tools for the consulting industry, and the development of soft technologies will further expand the service field and the influence of this industry.

2. Intermediary Service Industry

Strictly speaking, the intermediary service industry is one type of consulting industry. It is listed here as an independent category because of its special role in modern economic and social development and in international cooperation.

3. Public Relations Industry

Along with the globalization of economy, technology and services, the public relations industry has expanded to cover all fields of social and economic activity, such as media relations, government relations, image design, product and service markets, large-scale events, community relations and crisis management.

4. Various Specialized Management Industries

Various specialized management industries provide professional services for enterprises, communities and individuals, including companies devoted to asset management, accounting management services, specialized production, design services, talent services (including head-hunting companies), tax administration services, logistics services, legal services, financial and insurance services, information services, investigation services, translation services and computer services, as well as even matchmaking and wedding services, etc.

5. Information Service Industry

The information service industry's role is to provide valuable information for clients or facilitate multi-way information exchange. For example, media technology is a set of tools for changing the views and concepts of the public and guiding public behaviour by using all kinds of information-conveying means, such as news, broadcasting and television, to transmit and explain the content of publicity. Modern information service industry generally make full use of modern exchange means, such as publishing books, periodicals, newspapers, movies, broadcasting, television, telephone, telegraph and the postal service. The higher the technological content of these means, the more effectively hard means are integrated with soft ideas, and the more advanced the information service and exchange technology becomes, the greater the benefits will be.

6. Internet Industry and 'Internet of Things' Industry

Along with the development of network technology, modern Internet technologies and 'Internet of Things' technologies are rapidly coming into being a large industry. The Internet technology which enables people to communicate with each other by means of the network has shaped a large industrial scale all over the world. For example, Google has become the multi-national corporation involving the Internet. The 'Internet of Things' is based on the Internet and realizes the connection between human beings, objects and things through the integration of the next-generation of information technology and soft technology (solutions); plus the appropriate rules of the game (the conventional protocol), has already begun to apply in urban management, resources management, transportation, medical treatment, security and other fields, forming a large industry. The scale of Internet industry will be incomparable with that of 'Internet of Things' industry.

7. E-commerce Industry

The e-commerce industry specializes in providing new technologies and new services for e-commerce. Like information networks, e-commerce can only create wealth through the specific content of transactions (the products or services around which transactions occur) or business. The objects of e-commerce services can be divided as tangible commodities, media, information commodities and online services.

8. Venture Capital Industry

See the section of 'Venture Capital Technology' in Chapter 2.

9. Modern Finance and Insurance Industries

They are intelligence-intensive traditional service industries. Taking financial industry as an example, there are many technologies of asset accumulation, property accumulation and management for supporting this industry. For instance, various kinds of financial instruments, financial technology, share technology, venture capital technology, transaction technology, stock technology, as well as omnifarious financial derivatives, support the development of financial industry.

As early as 1987, a book called *A General Guide for Financial Products* was published in Japan, which summarized 107 kinds of financial products, in which 26 were used in banks, ten in trust banks, twenty-three for stock exchanges, eight for banks and stock exchanges, eleven for post offices, nine for life insurance companies, six for insurance companies that deal with loss and damage, two for agriculture associations, five for other organizations and seven for loan products (Nihon 1987).

10. Design Industry

The design industry is an industry that takes design technology as its core technology. Design technologies describe the pathway by which goals may be achieved and by which new concepts, ideals and goals may be converted into operable plans or projects. They can be categorized according to their different goals, objects and standards (economic, social, artistic, aesthetic, national and international, etc.), and can be separated into two classes: 'hard design', which includes industrial design, construction design, product design, interior design, city design and transportation design, etc.; and 'soft design', which includes business division design, image design, enterprise design, enterprise image design, advertising design, fashion design, environment design and plot design, etc.

Since ancient times, any successful design including construction design – i.e., architecture – has been seen as the quintessential embodiment of culture, art and technology, as well as strongly reflecting the cultural and economic background of that time. Therefore, a successful design should exceed the range of traditional technology to integrate and syncretise the above factors.

11. Industry of Intellectual Property

Intellectual property includes patents, copyrights, trademarks and so on. The intellectual property industry refers to the sector that provides identification, generation, protection of intellectual assets and other services oriented towards addressing the technological innovation activities of an enterprise or organization. As mentioned in Chapter 1, it is noteworthy that intellectual assets contain both hard and soft aspects. Since Ideal Final Result Consultants Ltd. was founded six years ago in the UK by Dr. Darrell Mann, it has built a worldwide reputation with more than 100 success stories, through providing services for systematic innovation methods, intellectual property generation and licensing.

12. Re-engineering Industry

The re-engineering industry is the industry that specializes in creating new games through re-designing, re-engineering or re-organizing existing games. It creates value by the means, such as the technology of technologies (e.g. the design technology of technological institutions and technological standards), the design of designs, the engineering of projects, the business of businesses (designing new business modes), the marketing of markets (e.g. a market broker in the stock market) and the system of systems, etc.

13. R&D Industry

The R&D industry is focused on providing new technological resources through systematically organized research activities and development activities. The R&D industry provides research results that have social or market value, including new products, new methods, new services, new tools and new conceptions. It also includes new systems, institutions, organization and management methods for research and development itself. Besides generating hard-tech resources, it may also generate soft-tech resources for further commercialization and industrialization, thereby promoting the development of R&D service industries.

C. Social Enterprise and Social Industry

The social industry is formed with social technology as its core technology (Jin & Ren 2004). It creates and embodies economic and social value in the process of solving social problems and handling social affairs by developing and applying social resources. From another perspective, the social industry or social economy refers to a third sector in economy between the government (public sector) and market in the traditional sense.

The background of social technology development, the connotation of social resources as well as the classification and value of social technology, have been introduced in the section entitled 'A Retrospective of Social Technology Development' in Chapter 2. This section will explore the significance and characteristics of the social market and social industry so as to facilitate understanding of social enterprises and social industries.

1. More Attention is Being Given to Various Types of Social Industries

Although the development of various non-profit organizations has a long history, our understanding of social relations and social activities as a kind of social resource (in other words, developing social activities as one special field of economic life – social industry) began only a few decades ago.

The WIR system, which was founded in 1934 by sixteen members in Zurich, Switzerland, is a community that survives on mutual economic help by exchanging materials and services within the community or getting credit loans from related organizations. After sixty years of development in this most conservative country with the highest living standard in the world, WIR's residents and SME members are continually growing and the system has developed to a respectable size. It had 80,000 members in 1994 and had a trade value the equivalent of two billion US dollars during that year.

The LETS system, which was developed in the Canadian province of British Colombia during that country's period of high unemployment in 1982, has expanded to 25 to 30 regions in Canada after ten years of development, becoming the most widely applied system in the world (Lietaer 1999). Similar community economies are developing rapidly in the US, New Zealand, Australia, Japan, the UK, France, Germany and other European countries.

According to Charles Leadbeater, the no-profit sector of Britain is vast and amorphous (Leadbeater [1997] 2006). In its broadest definition, the sector is made up of nearly 400,000 organizations, employing perhaps 950,000 people – about 4 per cent of employment in the entire economy. This sector has an income of perhaps £29.5 billion a year, mainly in culture, recreation, education, health and medical services, as well as social utility.

An article by Roger Sue entitled 'The Main Source for Social Economy' was published in the French newspaper *Le Monde* in 1999, which proposed that social activities are an important source of economic resources that form a basis for the 'social economy' (Sue 1999). Roger points out that social economy has always been the basis for all kinds of associations whose role is to solve problems which cannot be solved by the government and the market, such as medical and health treatment, training, social relations and ethics and moral issues. Roger says that, in addressing such needs, social industries play an important role in the creation of wealth. In France it is estimated that the social economy accounts for 4 per cent of the GDP, comprises one million employees and involves 80 per cent of all the French people in its constituent social activities.

In recent years, Neal Kocurek of David's Health System of Austin, Texas, has engaged in a special incubator career. Working with the government to provide services for citizens, he and his colleagues help pertinent departments of local governments solve problems in strategic planning, traffic and the labour force; work together with universities to solve problems concerning education and tentative planning for a new type of university; cooperate with enterprises and the government to solve problems of e-commerce; coordinate with hospitals to solve problems related to healthcare; and help leaders in different fields to expand relationship networks, etc. Through the regularization of these activities, which are then managed by specialized individuals and organizations, a new type of enterprise is forming – a community system company, a traffic system company, etc. People who establish these kinds of enterprises are called 'civil entrepreneurs'. These activities fall under the rubric of 'entrepreneurial incubation'.

Now, social enterprise has already not been limited to civil enterprise and various types of social enterprises spread all over the world. In Asia, including China, Japan, Singapore, India, the Philippines, Indonesia, Thailand and other countries, the scale of social enterprise is growing larger and larger, and the range of type of social enterprises affected by each type of social resource and the number of social entrepreneurs keeps increasing. It has even been proposed in the United States that a third generation of social entrepreneurs is rising.

In the book of *How to Change the World: Social Entrepreneurs and the Power of New Ideas* David Bernstein enumerated a lot of fascinating stories emanating from the United States, Brazil, Britain, Hungary and South Africa, in which the most famous case was the story of Mr. Muhammad Yunus, the 2006 Nobel Peace Prize Laureate (Bernstein 2004).

Professor of Economics, Muhammad Yunus, who started by making his first loan of 27 US dollars to a group of 42 penniless village women, set up a 'bank for the poor' in Bangladesh. The Grameen Bank (as it is known and which he founded) has shown how 'micro-credit' can be the basis for a successful social enterprise. At the end of February 2006, the Grameen Bank, which has the elimination of poverty as its goal, had cumulatively disbursed more than 5.3 billion US dollars to 5.77 million of the poor in more than 100 countries worldwide through its successful business operation.

2. The Significance of Social Industry

People used to believe that as long as market is running well and governmental management is appropriate, the problems of human society can be solved. However, with the advanced development of the market economy and with the entrance of economic globalization and the information age, the areas where both government and market are not effective have expanded. Such broad 'in-between' domains outside of the market and the government are commonly known as the 'civil society' or the 'third sector' where, because of their 'relationship with the citizens, their flexibility, and their ability to stimulate individuals to support public goals on their own initiative' (Salamon & Anheier 1992: 5), all kinds of social organizations not only have become a thriving economic power, but have played an increasingly important strategic role in political and social life. Their 'in-between approach' goes beyond placing trust only in the market and in the nation. Lester M. Salamon has also pointed out that 'a veritable "global associational revolution" appears to be underway, the social and political significance of this revolution emerged in the end of twentieth century can compare favourably with the rise of nation-state in the nineteenth century'.

At present, various social organizations are flourishing worldwide. The development of the 'third sector' is the foundation of the so-called 'social economy'. Firstly, it solves 'third sector' problems and plays an important political and social role through filling the areas where both government and the market are ineffective; secondly, along with a deepening understanding of social resources and social capital, the 'economic' activities of the non-profit organizations have become standardized and heightened to the level of an industrial activity thereby enabling many social organizations beyond the non-profit organization to become social enterprises, whilst promoting the development of social technological innovation and social industry.

From the viewpoint of the needs of third sector development on the one hand, along with the rise of civil society, a large number of non-governmental organizations and non-profit organizations have played an active role in public welfare undertaking, but the range of services and functions of these organizations has been comparatively limited and insufficient to fully undertake the mission of the third sector. On the other hand, in order to enable the third sector to develop sustainably and to become economically independent, it is not enough to rely only on the 'aid' and 'charity' of the traditional economy. Only by actively developing social enterprises and by integrating various resources according to corporate business models may the third sector shift gradually from mainly being supported by the government, charity and donations, to taking the operation of social industries and social responsibility investment as the major source of income, and to becoming a sound and independent economic sector which is more conducive to completing the mission of the social industry.

From the point of view of social demands, the social industry has emerged as a requirement of the 'social market' (see the section of 'Social Market' below). There are now growing needs of the social market not only in North America and Europe, but also in Asian countries such as Japan and China, etc. The conditions are favourable for the development of various social enterprises in the domain of education, health care, social welfare, women and children, community activities, disaster prevention and environmental protection, etc.

Analyzed from the vantage point of public affairs management, developing social industry is a kind of attempt to engage in institutional innovation. The challenge of separating the appropriate functions of government from those of the enterprises exists not only in China, which is undergoing a transition from a planned economy to a market economy, but also in developed countries. This challenge is not well resolved concerning the issues on public administration, because it involves a fundamental transformation of government functions. With the intensification of the contradictions of sustainable development, various factors of production are increasingly in active state and a variety of social issues have also emerged; thus the conventional approach of expecting all social public affairs to be managed directly by government faces a great challenge. For example, in China, over time, this resulted in the burden of government growing increasingly heavy, leading inevitably to an overload in government officers; on the other hand, enterprises and even research organizations in China have had to be responsible for society. These features of Chinese society run the risk of weakening the country's social capability for self-management.

Actually, along with the maturing development of the market economy, government should out-source the administration and give part of the right to managing public affairs back to society at large, establishing a concept of 'government in society' in the relationship between government and the rest of society (Z. Gan 2001). The administration of part of the public affairs by social organizations improves the capability of social self-management and self-discipline, and gradually set up the autonomous mechanism for economic development – the 'wild horse'. This may illustrate the reason for the development of social industry. The All America Brokers Association is a stellar example of an organization that promotes financial self-discipline in the US.

The function of government has changed from directly managing public affairs to entrusting and authorizing the right of management to enterprises, through contracts to enterprises engaged in social industry. The government is granted with supervisory rights through laws and policies and it can also intervene in social enterprises by means of purchasing or subsidizing when necessary. The introduction of competitive mechanisms into public affairs (as in the above examples) may not only benefit the public, but may help avoid problems of unclear property rights in public agencies, lack of competitiveness and low efficiency in government administration.

Take urban management as an example. More than half the world's population now lives in cities; and of those urban dwellers, more than 75 per cent of the population lives in cities in developed countries. Whether in developed or developing countries, cities are facing similar problems, such as employment, pollution, traffic congestion, public security and housing shortages. In this case, it is problematic that city affairs have been managed by local government. Various civil entrepreneurs who manage city and town affairs now treat the urban management and civil issues as their business. This is a good attempt on separating conventional government function from enterprises management, creating new cities with new functions, establishing urban civilization as well as strengthening city management.

Other resources for social industries include urban infrastructure, training and education, medical treatment, providing services for the aged, housing and employment services, etc.

These kinds of managerial activities need not be restricted to urban areas only but can also be the basis for creating series of new industries in small towns or villages.

In conclusion, the social industry not only means founding companies by taking the social DNA as the basis for business opportunities (and to thereby have far-reaching consequences for the consolidation and strengthening of sustainable development of the third sector), but it also means that the social industry will play an increasingly important role in the reorganization of social structure for the future, building a healthy world order, national governance and global governance. Some people even believe that social enterprises will change the world.

3. The Social Market

The so-called social market is different from the traditional market, which is the summation of the demands related to the society (Jin & Ren 2004). The issues that generate this demand include:

1. The demand to resolve social problems
With the maturing of the industrialization society, many complex social problems such as environment, energy, population ageing, urban and traffic issues, family disintegration, teenage crime and violence, ethics, the gap between the poor and rich, etc. are increasing.

2. The fields in which both the government and the market are inefficient
As we enter the age of economic globalization and information, the issues in which both the government and the market are inefficient or ineffective are gradually increasing, which the third sector needs to intervene.

3. The demand for improvement in the social function of self-discipline
With social progress and the improvement of people's educational levels, common awareness of the necessity and possibility for social self-discipline is rising.

4. The demand of solving problems concomitant with science and technology development
The rapid development of hard technology has amplified the negative effects of high technology as embodied not only in issues related to freedom of scientific inquiry, deep conflicts between the pure scientific development and ethics, morality, world view and culture, but also in issues related to economic and social sustainable development and public safety.

5. The existent and developmental demand of external social resources
Working hours are generally decreasing worldwide, spiritual demands are increasing, lifestyles are becoming diversified and people are living longer, thus accelerating the rise

of an ageing population. People are paying more attention to, and are involved in, more activities associated with informal organizations, such as community organizations, social groups, associations, learned societies and reunions. Social resources are therefore becoming more abundant and increasing their economic value.

6. The demand of developing social industry
In order to facilitate the sustainable development of the 'third sector', as well as taking social industry as an important and independent power in the national economy to create value, it is necessary to encourage social technological innovation – and we need more social entrepreneurs to form strong social industries.

7. The demand of developing regional economy and community economy
With economic globalization, we need to develop ways to protect and develop regional economies and to protect the diversity of culture, minority resources, ecology resources and history resources, thus requiring a different type of economy.

8. The demand of changing the government functions and the challenges of rapidly increased social affairs
(See the relevant part in the previous section of 'The Significance of Social Industry').

9. Increasing cross-cutting decision-making issues
Along with the challenges of sustainable development and globalization, there are increasing comprehensive fateful decision-making issues that are cross-sectoral and interdisciplinary, as well as involved synchronously in social, economic, environmental and natural resource aspects.

10. The demand of developing the 'new economy'
Social enterprises must be good companies which are able to fulfil the social and environmental responsibilities so as to serve as an important economic pillar of the new economy.

4. The Characteristics of Social Capital and Social Industry

To discuss the characteristics of social industry, we must first make clear what the features of social capital are. The results of a rigorous study along these lines have been published in the report of PIU (Performance and Innovation Unit) (Aldridge, Halpern & Fitzpatrick 2002).

- Social capital is not the exclusive property of any one individual. Social capital is shared by a group or by groups of individuals. To the extent that all members of society or a community have access, it may constitute it a *public good*. But to the extent that groups of individuals can control access by other individuals, it may correspond more to a *club*

good. This distinction has important consequences for whether and when the impacts of social capital are likely to be economically and socially beneficial, and for the role of government in promoting and shaping social capital.
- The use of the term 'capital' in this context has its implication that there is a stock of social capital assets on which returns are earned. This is analogous to the term 'human capital'. Analyses of capital as financial, physical and other tangible assets neglect the value that lies in social networks and in the shared values that facilitate cooperation between actors.
- The term capital also helps to highlight the potential 'fundability' between financial capital, physical capital, human capital and social capital.
- In general, financial capital is embodied in the bank accounts, human capital exists in human minds, and social capital is endogenous within the structure of human relationships. Therefore, the majority of social capital is a kind of intangible capital, but with many tangible carriers (external resources).

In fact, measuring social capital requires developing correct quantification techniques for social capital stock and returns. This goes beyond the concept of capital stock, as traditionally understood in economics, and is one of the major issues for research on social industry in the future.

Based on the feature of social capital as described above, and upon the definition of social industry in this book, the characteristics of social industry can be described as follows:

1. **The aim:** to exploit furthest the positive and active value of social resources so as to make profits for most of people, solve social problems and promote social progress.
2. **Business philosophy:** to create the social, environmental and economic value in the process of exploiting and applying social resources to solve social problems and deal with social affairs, in which the premise of economic value creation is to complete its social mission, to accelerate social advancement in the field of education, environment protection, rural development, decreasing poverty, human rights, health care, services for the disabled, etc.; namely, to fulfil the social and environmental responsibility. Therefore, the leadership of social industrial organizations, including the social entrepreneurs, must be outstanding talents themselves and must aspire to promote social progress and embrace the ideal of protecting our planet and creating a better world.
3. **Resource:** the main resources of social industry can be divided in to *club good* resources, and the *public good* resources, according to their attributes; and can be further classified into external social resources and internal social resources, according to the manner in which they are manifested.
4. **Product:** the objects of exchange in the social market are not commodities in the traditional sense. Rather, they mainly consist of various kinds of services; moreover, ethics and morality are inseparable aspects of social industries.
5. **Distribution:** social industry is about managing social capital and, therefore, in accordance with its social, cultural, ecological and other objectives, the beneficiaries

of corporate business income or profits must be correlative communities, groups or targets, rather than a minority of operators or a particular administrative layer of the organization. This determines the non-profit feature of the social industry.

6. **Scale and quality:** the scale, channel of income and quality of social industry is strongly influenced by the history, cultural tradition, religion, educational level, values and social structure as well as the economic development level of a country or a region. As the level and form of social capital (as well as its social and economic impact) will depend on a series of factors of internal social resources, such as the political system, various institutions, beliefs, values, ethics, customs and social aspirations of the people, then (compared to other industries) the development of social industry needs to be attuned to the distinctive national conditions of the society in which it takes place.

7. **The diversity of the income source:** it is an un-shirkable responsibility of government, enterprises, communities and citizens to settle social problems by ensuring that there is diversity of income source for social industries. These must include government investment, charities, donations from enterprises or individuals, and socially responsible investments, as well as operating income of the social enterprises themselves.

8. **Exchange tool:** for traditional industries which are involved in a globally competitive economy, it is common to take financial currency as the primary means of exchange and as the primary unit of measurement, because of its higher efficiency in facilitating competition.

    There are two types of exchange tools or measurement units in social industry: traditional financial currency and complementary currency (Lietaer 1999). These two are mutually complementary, and with the development of social industries, the second set of tools will become more prominent. The latter has many advantages. For example, complementary currency disappears in the process of being used, thereby avoiding many potential problems in issuing formal currency of the conventional financial kind; it is resistant to the influence of the bubble economy and inflation; it typically does not interest; it is beneficial to community and regional economic development; it may be more convenient for financing SMEs and for the operation of non-profit organizations; and it can be conducive to the cultivation of harmonious interpersonal relations. Once the complementary currency system is running in a benign circle, it is possible to form a self-reliant system for solving social problems without social subsidies or payments from taxpayers. It is most appropriate for the 'Yin' economy with the complementary currency (the 'Yin' type of currency) to be developed and coordinated in a complementary and cooperative manner with the 'Yang' economy and the financial currency (the 'Yang' type of currency). Bernard Lietaer predicts that complementary currency systems will top 10,000 worldwide by 2008; complementary currencies will represent 20 per cent of total domestic trade in the most industrially advanced countries by 2020 (ibid.).

9. **The core technology and innovation means:** although social technology is the core technology of the social industry, all soft and hard technologies, including business

technologies, could be widely used as an innovation tool in the process of value creation.
10. **Outputs**: according to the aim and business philosophy of social industry, there is a multiplicity of types of 'outputs' across the variety of social enterprises. Some of them may be quantified, such as the creation of economic value or even the efficiency of increasing employment and reducing crime, but the majority of the 'outputs' reside in social value that is impossible to quantify, such as the degree of satisfaction of social market demands.

5. The Types of Social Industry

The social industry in the narrow sense refers to the group of social enterprises, while the broader social industry includes various non-governmental and non-profit social organizations and groups, etc. According to our present understanding of social industry, social industry in a broad sense can be classified according to its carriers of external social resources (supporting its main body), its social market and the service level which social industry can provide.

1. **Social industry can be classified according to the carriers of external social resources**
    a. Various non-governmental and non-profit social organizations and groups includes industrial organizations (association, chambers of commerce, labour union), research institutes (science, technology, culture, art), academic groups (institute, association, academy,), social communities (related to social welfare, woman, ecology and environment, charity, deformity, human right, voluntary, associations of countrymen or schoolmate, foundation, benefit association, culture and art communities, international organizations, religious organizations), etc.

    In the foreseeable future, social organizations which belong to above categories will arise increasingly, and their activities will expand into broader areas. In China, for example, the Chinese Science and Technology Association consists of more than 7,000 organizational members and over 9.9 million individual members.
    b. All kinds of public sector and organizations, such as various schools and universities, hospitals and other related education and training sectors, as well as the organizations and institutes of healthcare and so on.
    c. Social enterprise includes those known as 'civil enterprise'. It is based on the type of commercial enterprise that develops and operates social resources. Pursuing maximum profit is not the purpose of social industries, while accessing economic benefit is a means to create greater social value rather than an objective. This is the biggest element that differentiates social industries from industries in the traditional economy.
    d. Communities, regional networks, etc.
    e. Other non-profit organizations, including political party, parliaments and courts.

2. **Social industries can also be classified in terms of the social market**
   a. Resolving social issues, handling social affairs, coordinating relations between government, enterprises and the public. These are major activity areas for the carrier of exterior social resources, such as the association.
   b. To participate in the operation and management of the public utility of urban, town and rural areas.
   c. Activities surrounding the socio-economic development of a region or community. These are major activity areas for social grass-roots organizations, such as communities, consortia, neighbourhood committees and so on.
   d. Concentrated information exchange activities, including seminars, exhibitions, commodity fairs and various types of exchanges, etc.
   e. Educational undertaking refers to an area of social activity that aims at intellectual development and the cultivation and training of people, but it does not include the factories run by schools, etc.
   f. Medical treatment and healthcare fields.
   g. The undertaking of protection for the rights and interests of woman and children.
   h. Social welfare work.
   i. Disaster prevention and environmental protection fields.
   j. Planning, strategy, decision-making and forecasting in all dimensions.
   k. The field of risk prevention and the maintenance of social order; security services for individuals, families, corporations and countries, including disaster prevention, prevention of crime and violence, legal services, and so on.
   l. The field of religion.

According to the research of Lester M. Salamon and others on the employment structure of non-profit organizations (averaged over sixteen countries) employees engaged with education and research accounted for 26.2%, 20.9% were related to health care, 19.2% to social services, 14.1% to culture and entertainment, 5.6% to religion, 4.5% to associations of vocational and professional, 3.9% to development, 3.3% to the environment and environment-related initiatives and 2.2% to other topics. Despite the fact that the proportion of NPO employment and income dispersed these sectors varies between countries (and despite the fact that the areas mentioned above are not classified by social industry or social market), the data mentioned above nevertheless illuminate the development potential of social industry for the future.

After discussing social industry, as well as the functions of the government related to social industry, it should be realized that while social capital and social resources are the base of capacity building for a society and the source for enhancing social functions, they nevertheless need to be developed and applied by social technologies so that their potential value (benefits or wealth) may be realized.

6. Education Industry

The core technology of education industry is education technology that was institutionalized earliest as a means of developing intelligence, producing human capital and training talents. It was also the first soft technology to be recognized and employed by society in the form of a social organization – school. The first known school in the world was founded in 3500 BC in Mali and the first known university was the Moroccan Islamic University-Qaawiyin University was founded in 859 A.D (Wang & Huang 1990).

From the perspective of 'taking human resources as the operational resources', education technology is a kind of intelligence developing technology (see the section of 'The Classifications of Soft Technology' in Chapter 1), which is characterized by providing and implementing educational methods and contents. Its aim is to enhance and develop human intelligence and abilities. However, from the viewpoint of the carrier of external social resources, because the main carriers for implementing education (school) are a part of the public sector within the social resources (see the section of 'The Types of Social Industry' in Chapter 5), education technology can be taken as a social technology. Moreover, from the perspective of the characteristics of social industry, especially from the angle of its non-profit nature as well as the mission undertaken by education, it is more appropriate to incorporate education industry into the category of social industry.

From the viewpoint of the education mode, education should include multi-levels, such as family education, preschool education, school education, military education, adult education, vocational education, distance education, all-round social education (e.g. social ethos, government action and enterprises behaviour, etc. are the indirect mode of civil education); from the perspective of the educational contents, education can be divided into knowledge education, intellectual education, moral education and physical education, etc. This is why the education system has faced with an eternal subject from its emergence, that is, along with the development of the times, we must carry on the reforms unceasingly in the different aspects, such as the education mood, educational content, etc., thereby resolving the relationship between education and practice, and cultivating the talented person with virtue and intelligence.

As a platform for embodying the value of education, educational direction and educational outcomes, education industry is the crucial industry for any country, which is related to the rise and fall of a nation and its development, and it is also an industry with the most extensive innovation space.

D. Cultural Industries

The improvement of material living conditions has increased the public demand for psycho-social products worldwide, making cultures, arts, sports, and entertainment technologies rapidly transform into hundreds of thousands of goods and services and forming a gigantic cultural industry. Furthermore, due to the unique role of the cultural industry in social progress and its contribution to enhancing national soft power, more and more importance has been attached to it by countries around the world.

Since the 1990s, the culture industry has become one of the world's fastest-growing industries. The United States, as the one with the most developed cultural industries in the world, is exporting American culture in the form of cultural products such as movies, TV programmes and fast food to the world.

1. Culture and Cultural Values

The definition of culture has many meanings; it can be explained differently, at different levels, and from different perspectives.

It was Sir Edward Burnett Taylor, a British anthropologist, who first gave the term 'culture' a clear definition. He stated in his book *Primitive Culture* (1871) that culture or civilization as a whole is complex because everything has to be taken into consideration, including knowledge, beliefs, arts, morals, laws, customs and any other abilities that a community member acquires through learning (*Encyclopaedia China-Sociology* 1991: 409–418).

British social anthropologist Bronislaw K. Malinowski further developed Taylor's definition of culture in his *The Scientific Theory of Culture* published in 1944, who viewed culture as the expression of the totality of individual and collective achievement, where 'every custom, material object, idea, and belief fulfils some vital function' (Malinowski [1944] 1946). He also further divided it into the material culture and spiritual culture, namely, so-called two main components including 'transformed environment and changed human organism' (ibid.).

Boyd and Richerson considered that culture is knowledge, valued judgment and other factors influencing behaviour that can be transmitted from the previous generation to the next generation through education and imitation (Z. Zhou 1999).

The 1989 edition of *The Comprehensive Dictionary of Chinese Language* defines culture as: 'in the broad sense, the integration of material and spiritual properties created in the process of historical practice of human society, and in the narrow sense, establishing not only social ideology but also institutions and organizations accordingly' (*CDCL* 1989: 1731).

In the sociology chapter of the 1991 edition of the *Chinese Encyclopaedia*, culture is defined as: 'in its broad sense, [it] refers to the summation of all the material and spiritual products and, in its narrow sense, refers only to the spiritual products, including all ideology, such as languages, literature, and arts, etc.' (*Encyclopaedia China – Sociology* 1991: 409–18). In addition, further definitions from the perspectives of cultural sociology, cultural ecology and cultural psychology are explained.

Some sociologists and anthropologists define culture in the following way: 'Culture is the shared fruits of human groups and society' (Popenoe [1995] 1999). These shared fruits include not only the non-material elements of culture such as values, languages, knowledge and ways of handling things, but also the material elements of culture such as tools, money, clothes and artwork.

Generally, 'culture is the overall way of life of a society' and is learned, taught, imitated and accumulated through social intercourse and down from generation to generation, as well as accumulated and altered through innovation by means of coexistence and integration of different lifestyles. Culture is usually a product of inner values, concepts, attitudes, regulations and customs. Each society must understand its own unique culture so that culture may play a vital role for every country in deciding what it should do, how it should do it and what it should do first during the development of its politics, economy and the society. Culture does not dominate these important issues but rather functions as 'an invisible hand' in the decision-making process of a state, collective or individual.

Recently, many people's behaviour as consumers has been augmented by a new emphasis on spiritual consumption. In order to meet this requirement, a huge array of cultural knowledge and cultural resources has been developed into cultural products, as well as services and markets. As a result, the development of a non-material economy has advanced. Culture has an increasingly important effect upon the development of technology and economy and sometimes its effects surpass the functions of marketing and government. Since the works of culture and the arts have been marketed, culture will no longer function as 'an invisible hand' nor will cultural knowledge and cultural value be considered as non-technological factors or as non-economic factors. Culture has either economic values or social values.

The economic value or market value of culture refers to the economic benefits that come into play when the works or creations of culture and arts are produced as commodities.

The social benefits of culture refer to the question of the degree to which cultural products, commodities or services contribute to achieving the goal of social and economic development. The goals of social and economic development are multi-dimensional: the amount of income should be raised while decreasing 'classism' by narrowing income gaps between classes; increasing the demands of material life while satisfying the spiritual life of residents; and improving the social and natural environments of life (Yining Li 1990). Like poems, paintings and statues, works of art and culture that were once created, appreciated and well-known will be expected to function in the real world of social and economic development goals even though they may not be marketed and converted into commodities. These are the social benefits of works of art and culture and the key point is that the effect can be positive or negative; it can either assist in the healthy development of human morals and minds or it can produce harmful effects in human minds.

It is worth recognizing that because of the duality of cultural value, culture creation and cultural innovations differ from hard-tech innovations. The Chinese economist Yining Li has extensively studied this subject. He stated that the use-value of a product does not possess ethical quality in the field of material production, since opium, morphine and cyanide are commodities. The problem lies with those who use them and with how they are used, not in the use-value itself. Therefore, the production and sale of these products should be strictly controlled to decrease their improper use.

However, works of art and culture are different from material products. Because they are products of spiritual endeavour and contain normative implications and features, their use-value varies according to social evaluations. For example, although books, magazines and audio-video products that contain pornography and scenes of murder and violence or other socially offensive material will sell once they are marketed, they can harm some customers and may harm society instead of helping to enrich it. Therefore, with regard to these products, the issues are not who uses them, how to use them, or how to 'use them properly'. Under no conditions should the production and sale of these harmful products be permitted. The key point is that the so-called 'strict control over production and sales' over these works of art and culture cannot even exist (ibid.).

Thus, it is evident that the relationship between cultural technology and marketing is different from the relationship between traditional technology and marketing. Why is this? Because simply from the perspective of the market, cultural products and services are produced to help pursue the pleasures of life and to meet different lifestyle demands, and once they are produced they are almost certainly bound to sell. But if social impacts are negative, they should not be developed, produced or sold, even if customers exist for the products. For instance, the 'Law of Unhealthy Publication' is one institution that assists in eliminating this kind of cultural trash before it is even developed. In the past thirty years in China, the non-commodity feature of cultural products has been overemphasized. Today, in contrast, in the market economy we need to take a stand against the tendency to put 'benefits first' or 'commercializing all things' primary in cultural industries.

2. Understanding of Cultural Industry

Cultural industries are closely related to the level of economic development and the civil quality of a nation. Adam Smith remarked as early as the eighteenth century: 'Whether a man is wealthy or poor can be judged by seeing to what extent he can enjoy necessities, conveniences and recreations' (Smith [1880] 1997: 26). Western developed countries made important moves in the development of cultural industries as early as the 1930s and 1940s. Today, cultural industries occupy a pivotal position in the national economies of advanced industrial countries. Furthermore, cultural industries have long become an important export industry. Since 1996, American cultural exports have surpassed the aerospace industry, thereby becoming the first largest export-oriented industry. Since the beginning of this century, Japan's animation industry has become their third largest industry. America's Hollywood is the most remarkable example. Some US politicians have said that 'America's largest export now is no longer crops in the farmland, and also no longer the products of factory, but the mass-produced American culture'. For instance, Hollywood's movie production products cover global markets, producing massive profits, such as in 1998, when *Titanic* generated about 1.845 billion US dollars in box office grosses, and as of March 2010, the box office grosses of *Avatar* reached 2.6 billion US dollars. India is the world's largest film producing country at present, possessing a huge industry with nearly 2 million staff and an annual output of nearly 1000 movies. While the box office revenues of China's domestic movies in 1999 were only 850 million Yuan (just over 100 million US dollars), by 2009 those of the domestic movies market (excluding the rural market) had reached 6206 million Yuan (about 900 million US dollars), in which that of Chinese-made films accounted for 56.6 per cent. The exports of China-made cartoons in 2009 reached 30.566 million US dollars, an increase of 150 per cent compared with that of 2008.[1]

Kimindo Kusaka, a representative Japanese scholar in the field of cultural industries, has proposed that there will be an age of cultural development after the economic development age, social development age and qualified personnel development age. In this age, members of society will devote a great amount of income, time and ability to self-fulfilment and the pursuit of happiness. In his 1978 book, *New Culture Industry*, Kusaka calls for a policy of 'nation-building on the basis of culture' to be implemented in Japan, arguing that culture can produce high profits and can be sold at high prices; and, in addition, that cultural symbols can enhance the income of commodity businesses. Kusaka advocates the exportation of Japanese culture, art, movies, TV programmes, music, books, tea ceremonies, IKEBANA, language schools, entertainment, martial arts, I-go, etc. He also proposes that the conditions for creating cultural industries are as follows: 1) the general improvement of material life (which can turn cultural products into popular spiritual products, rather than the luxury of a few); 2) the general improvement of the national cultural quality of people; 3) rich cultural resources; 4) in institutions, providing lots of opportunities for introspection (introspection on the weaknesses and deep-rooted habits of a nation, as well as those 'illnesses' that are caused by the cultural factors in social and economic life and the cultural reasons); and 5) a high level of manufacturing capability suitable to support the production of cultural products (Kusaka 1978). Obviously, the development of cultural industries and cultural innovation shows the maturity of a nation, while it is also an important indicator that symbolizes a certain stage of its socio-economic development.

The splendid cultural heritage of China is a great treasure in which immense wisdom can be found, and it can also be the source for creating countless spiritual products and developing future soft technologies and soft industries in thousands of forms. However, along with China's full integration into the global community, it is obvious that Chinese cultural industries and even cultural traditions are going to experience profound and serious shocks. On the one hand, the connotation of cultural industries is increasingly enriched, the information industry will comprehensively promote the digitized competition of cultural industries, and the internationally powerful cultural enterprises will exploit and compete vigorously for China's cultural market with their rich experiences and strong capital. On the other hand, a more important task is how to maintain the noble traditional culture in the face of foreign cultures that are entering China on a large scale and, at the same time, to identify how Chinese culture may be exported positively with its unique cultural predominance and establish its crucial position in the world cultural market.

Understandings of the meaning of cultural industry vary according to different starting points of the research. Examples are below:

Cultural industry is the complex whole of all industries that provide services for the demand of spiritual life and entertainment. (Wang & Zhen 1994)

Cultural industry is an industry engaged in the production and service of spiritual and cultural products. It takes cultural products and activities as the main objects, dealing with production and management to exploit and build up administrative and service sectors. It mainly includes culture and arts, education, sports, science and technology, tourism, religion, etc. (Siyan Cheng 1999: 61)

Cultural industry is the special industry dealing with the production of cultural products and providing cultural service. Its essential feature is to transform all human activities of knowledge, intelligence, spirit, art and information, and their successes into cultural products that can be consumed, enjoyed, exchanged and transacted by using certain materials as their carriers. The biggest difference between the cultural industry and other industries is that its cultural content and function is much higher than its material content and function. Its peculiarity lies in taking original spiritual activities as its foundation. (Le 2000)

Cultural industries are the industries born for creating culture, which entails creating a certain culture, selling this culture and its symbols. (Kusaka 1978)

Cultural industries are a range of activities that produce, reproduce, circulate, store and distribute cultural products and services in accordance with industrial standards. (Deng 2003)

From the perspective of soft technology, the cultural industry is an industry that takes cultural technology as its core technology. Thus, it utilizes cultural technology to exploit cultural values and cultural resources (see the section of 'Cultural Technology' in Chapter 1) so that they are able to become products or services 'consumed' or enjoyed by the customer – social values and economic values. Cultural resources include education, science, literature and art, morality, laws, customs, beliefs, natural environment and historical heritage, etc.

In view of the dualism of cultural values, an analysis of the features of cultural products is an important first step.

1. Cultural content or cultural values are the main value components of the products and services of the cultural industry.
2. Less energy consumption and high value-added products are the features of cultural industry.

3. Cultural and artistic products require more individuality and creativity than is required for the production of industrial products.
4. A large disparity may often exist between the price and cost of cultural products.
5. The exportation of cultural products not only creates economic benefits for the exporting country but also plays the role of disseminating its culture. American movies, McDonalds, Japanese cartoons, Japanese video games, Japanese sushi, Chinese martial arts, Chinese cuisine, Korean kim chi, Korean Tae Kwan Do, etc., are creating quite large markets worldwide, thereby actually helping to disseminate the cultures of these countries.
6. The development of culture commodities should adhere to the principle of healthy culture and 'uphold the good and repress the bad'.
7. Owing to the non-commercial nature of spiritual production, some cultural activities or products with good social benefits may have little or no economic benefits in the medium or short run. For example, cultural heritage development, elementary education and research on basic science may cost more than they generate. The development of cultural industries, therefore, needs conscious coordination of social and economic benefits. Not all spiritual products or activities can be commercialized and we should not seek to commercialize the works of culture and art as well as cultural activities simply for economic reasons.

3. *The Classification of Cultural Industries*

There are a wide variety of ways in which cultural industries may be classified, depending upon the frame of reference of those doing the classification. We will now look at some important contributions in this area.

The classification of cultural industries offered by Kimindo Kusaka is as follows (Kusaka 1978):

1. Industries that make active use of spare time, including the life-long education industry, the interest industry, sports industry, long-distance education industry, music industry, publishing industry, self-fulfilment industry, etc.
2. Industries centred around conducting household choices and enjoying family activities, including the household agent industry, the catering industry, the processing industry, the family reunion industry and new forms of agriculture and horticulture, etc.
3. Industries that nourish life, including telecom selling, vending machines, sole-agent stores, advertising industry, large-scale retail activities, etc.
4. Industries that expand the circle of friends and acquaintances, including the dating industry, urban entertainment industries, the lovers industry, tourism, the beauty industry and other cultural industries, etc.

Professor Yining Li of Peking University has divided cultural and artistic articles into products in material form (such as music and audio-video products, artistic works, books and magazines) and spiritual service products (services provided to the society by cultural and artistic sectors and organizations like artistic performances), from the vantage point of economics (Yining Li 1990).

According to the project *Research on Developing Cultural Industries in Shanghai,* which was organized by the Shanghai Municipal Government in 1998, cultural industry is the industry of producing, selling and utilizing cultural and artistic commodities (Project Team of Shanghai Municipal Government 1998: 82). The Shanghai project classified cultural industries as listed below.

1. Cultural product manufacturing: the book, newspaper and magazine publishing industry; the duplication of recorded media; the manufacture of musical instruments and other cultural entertainment products; and the manufacture of handicraft articles, etc.
2. Cultural product retail trade: the retailing of books, newspapers and magazines and retailing of stationery commodities, handicraft articles, etc.
3. Cultural service industry: entertainment service industry, artistic industry, publishing, relic industry, books, archives establishments, mass and popular culture, news, cultural and artistic dealers and agents, broadcasting, movies, television and other cultural service industries.

Cultural industry can be classified according to different categories of cultural resources, which include education, science, literature and art, morality, laws, customs, beliefs, natural environment and historical heritage. Moreover, it can also be classified into 'paper-based culture' and 'non-paper-based culture', in which 'non-paper-based culture' comprises folklore, customs, historical physical remains, myth, as well as traditional opera, drama and stories handed down by oral instruction. The task that carries on the collation, transmission and research about 'non-paper-based culture' in China is burdensome.

However, the classification of cultural industries is becoming more difficult as all kinds of cultural products and services tend to intersect and merge with each other. Take the entertainment industry as an example. The rapid development of digital technology promotes the 'high technologification' of entertainment technologies and alters our concepts of entertainment technology, the entertainment mode and even entertainment itself. It also blurs the boundaries between television, PCs, recreational machines and mobile phones and enables the convergence of all entertainment-providing media such as music, movies, television, electronic games and the Internet, thereby giving birth to digital entertainment. In the twenty-first century, digital entertainment will surely change the mode of life and business of people dramatically. With the popularization of high performance broadband Internet access, all the single direction media such as broadcasting, television, movies, magazines, newspapers and books will be challenged ('The Future of Entertainment Technology' 2001).

Based on the understanding that cultural industry may be seen as the aggregation of professions engaging in cultural production and services, we may classify the typical cultural industries as follows:

1. Entertainment industry

The entertainment industry creates entertainment, produces and disseminates entertainment and manages entertainment. Entertainment factors are important for increasing the added value of products. The entertainment industry includes the music industry, the movie and television industries, the performing arts industries and various game industries, etc.

2. Artistic industry

The artistic industry is the aesthetic cultural industry or the appreciation industry of the cultural industries. According to different means and methods of its performance, art can be divided into the performing arts (music, dance), plastic arts (painting, sculpture, and architecture), linguistic art (literature) and integrated arts (drama, movie and video). According to the nature of space-time of performance, it can also be divided into time art (music), space arts (painting, sculpture, architecture), and the space-time coordinated arts (literature, drama, movie and video) (*CDCL* 1989: 627). Therefore, the arts industry is the one that has the most abundant categories, so that its occupational categories are also richest.

3. Sports industry

The sports industry refers to an industry that creates value by sports-related economic and social activities. From the viewpoint of activity mode, sports industry can be divided into three categories: engaging in sports (professional and mass sports); appreciating sports (the fan groups are loyal customers of the sports market); and operating sports (sports-related organizations, facilities, game and performance market, intermediary, dissemination as well as sports service trade, etc.). From the angle of the 'products' which it delivers, the sports industry can be divided into sports-related behavioural products (sports games, sports performance, fitness, sports tourism, etc.), material products involving sports services (sports wear, sports equipment, sports facilities, sports food and beverage, sports lottery, etc.) and information products involving sports services (advertisement, telecast, press, sports lottery, sports information transmission, etc.).

From the viewpoint of the created value, the sports industry, with a variety of value as well as its extensive influence in international relations and even its contributions to world peace, is unparalleled in any industry. The facts that the sustained growth of the proportion of the GDP in the developed countries accounted for by their sports industry, the profits created by the sports clubs, for instance, the enormous profits created by NBA through industrialized sports events with the convergence of game, performance and entertainment, as well as the development of Olympic economy in recent years, sufficiently illustrate the economic value of the sports industry; social value is manifested in the advancement of the national health level and moral standards; the rich content of human life, bringing entertainment for the common people; as for the cultural value, the same as music, sports has become a common language of the people of all the countries and a medium promoting interpersonal and international friendship and exchanges, and has even become a diplomatic tool.

4. Tourism industry

The tourism industry is a larger cosmopolitan industry. The tourism industry in a broad sense refers to the sector providing broader services for activities outside, while in contemporary society, people go out not only for sightseeing, holiday, entertainment, leisure or visiting friends, but also for other purposes including business, conferences, etc. For example, 10 million tourists visit Hong Kong each year, of which 30 per cent do so to participate in trade exhibitions or meetings. Each year more than 20,000 enterprises go to Hong Kong for participating in conferences and exhibitions. In 2000, conference revenues of Hong Kong generated 7.5 billion Hong Kong dollars in revenue, which provides 9000 permanent jobs, and increases the hotel occupancy rate by 10 per cent. Therefore, the tourism industry has for a long time exceeded the scope of the traditional cultural industries. On the other hand, tourism can also be classified into social industry. For instance, tourism no longer simply meant visiting scenes at different spots. From a viewpoint of tourist destination tourism is a platform of the regional economic and social development, thus particular effort has been made to expand eco-tourism, healthcare tourism, and cultural tourism, etc. – from the perspective of a tourist, it provides a learning and exchange platform for diverse culture and lifestyle. Studying Asia's tourism industry, Dutch scholar Marien van den Boom put forward several issues, such as how to exploit and develop the intellectual capital and intellectual assets of tourism industry (van den Boom 2009).

5. Leisure industry

It is the industry with the aim of helping people to have a more colourful life, providing more interesting ways of spending their spare time, providing tranquillity and peace of mind, coordinating work and rest. The music, art, painting and calligraphy, fine arts, poetry, repose, travelling, communication and dating, etc. are examples of the leisure industry.

6. Hobby industry

The hobby industry is closely related to the cultural and educational background of consumers. The rapid development of the hobby industry is mainly due to increases in income, free time and the improvement of the average educational levels of a population. There are also 'spare time development centres' in Japan. People's hobbies, like handicraft making or collecting artefacts, are the basis for the development of the hobby industry.

7. Feeling and experience industry

Now, consumers are increasingly willing to pay the bill for satisfying their spiritual and sensory needs (see 'Soft Industry and So-called Creative Industries' in this chapter). Namely, feeling and experiencing comfort, pleasure, satisfaction, stimulation, novelty, surprise and other desires, which were previously rarely experienced, have now become a kind of demand. Feeling and experience industry just meet such demands. Therefore, designing various activities in a non-normal environment, the extreme activities in a normal environment and simulation of virtual environments, etc., enables people to experience personally a special feeling or experience so as to create business opportunities. The feeling and experience industry is about creating value by meeting these kinds of demands. Disneyland and extreme sports are relatively typical examples. In *The Experience Economy*, published as early as in 1999, B. Joseph Pine II and James H. Gilmore described the principles of how to design experiences, and the basic characteristics of the experience economy, for the first time (Pine & Gilmore [1999] 2008). Since then, the feeling and experience industry has been widely evolved into entertainment, health care, education and other fields.

8. Beautification industry

This is an industry in which the core technology is beautification technology, and it aims to beautify living conditions and life. The fashion industry, hairdressing industry, decoration industry, image-building industry, etc. are the most popular beautification industries today, and they also belong to the larger category of the artistic industry in the broad sense. The beautification technology involves pursuing beauty, presenting and embodying personality, peculiarity and self-confidence for the satisfaction of spiritual demands. A successful beautification technology must combine aesthetic standards, culture quality, morality and habits (towards beauty) with the psychological pursuit of beauty in human beings; and to make full use of all the instruments and artifices that result from the modern hard technological achievement.

9. Media industry

10. Language industry

The language industry is the one developing and applying language resources, including the research and development of language knowledge and technology, language education and training, language dissemination, language translation, and language services for specific industries, etc. The language industry has a history of decades in the United States, and there is also a Language Industry Association in Europe. The Chinese Language Resource Development and Application Center was established in 2008.

11. Self-fulfilment industry

The core technology supporting the self-fulfilment industry can be called 'technology of personal independent creative space'. This technology provides space for a wide range of personal independent creativity and provides opportunities for people to create freelance careers. The traditional 'self-fulfilment' professions provided creative space for I-go and chess players, judo players (with its level system), cooks, professional athletes, painters, engravers, lawyers, architects, photographers, accountants, tax accountants, auditors, doctors, professional information services providers, comprehensive information services providers, freelancers and all kinds of agents (technological projects, culture, sports, etc.). More recently, musical producers and various types of designers – including image designers, programme producers and programme hosts or hostesses, independent filmmakers, network designers, simultaneous interpreters, etc. – have also joined this group, making this space larger and larger.

Nowadays, many young people prefer to set up their own companies rather than stay in big companies awaiting promotion. It is reported that since 1983 the number of new companies set up yearly in the US has never been lower than 600,000. During ten years since 1986 the number of self-employed workers has increased by nearly one million. It is estimated there will be 12 million self-employed workers by 2010 (*Reference News* 2001). With the development of the market economy, this trend is also clearly apparent in China. Although there is difference in the statistical scope, the self-employed individuals in cities and towns increased by 6.14 million in 1990 to more than 36 million in 2008 (*China Statistical Yearbook* 1991; *China Statistical Yearbook* 2009).

4. Culture Service Industry

The cultural service industry provides services focusing on various cultural industries such as the copyright industry and the cultural and artistic management industry, etc. The cultural and artistic management industry includes cultural and artistic brokers and agents; firms that provide organizational and management services for entertainment; the organization and management of the sports industry; the auction industry and programme hosting services, etc. Along with the development of cultural industry, the added value of cultural service industry occupies higher proportion in the cultural industry. It's similar with the trend of 'servicization' of manufacturing industry. Similarly for China, the cultural service industry has gradually become the main body of the cultural industry. In Shanghai, where the cultural industry is relatively well developed, the total output of the cultural industry was 271.9 billion RMB in 2007, an increase of 15.7% over the previous year, accounting for 5.6% share of GDP (for the whole of China, the figure is 2.3%; the figure for Hunan province is 6.5%, which the highest in China); the value-added of cultural industry is 68.3 billion RMB, among which the proportion of cultural service industry has been 64%, an increase of 18.5% over the previous year.

The copyright service industry may also serve as an example. The core copyright industries are those involved in the creation of works eligible for copyright or the production and/or distribution of copyright-protected materials as their primary product. The industry also includes the re-creation, duplication, production and dissemination of copyrighted works such as publishing industry for newspapers and books, radio and TV broadcasting industry, publishing industry of recording programming and video tapes, film production, drama creation and performance, advertising industry, as well as the services like computer software development and data processing. Since the 1990s, electronic publication, digitization, network transmission and other new technologies have been widely used in the cultural field, enabling copyright industries to become a new economic growth point of developed countries. For instance, the copyright industries in the US have been regarded as an independent industry in the national economy. The American core copyright industries had created 348.4 billion US dollars of the output value, accounting for 4.3% of US GDP in 1997, and 889.1 billion US dollars in 2007, accounting for 6.44% of GDP; the scale of the copyright industries was 1525.1 billion US dollars in 2007, occupying 11.05% of GDP; from 2006 to 2007, the average annual growth rate of the copyright industries and the core copyright industries was 7.91% and 7.26% respectively, faster than that of US GDP (2.03%) during the corresponding period (Siwek 2009). The exports of core copyright industries increased from 116 billion US dollars by 2006 to 126 billion US dollars by 2007, an annual increase of 8%, far exceeding exports of other US industries (Xiao 2009).

5. A Reflection on the Commercialization of Culture and Arts

As discussed above, we have expanded the concept of innovation to include all value-creating processes. Nonetheless, many negative results have been developed from the innovation activities aiming at 'gainful economy'. Terms such as 'knowledge economy', 'knowledge commercialization', 'digital economy', 'entertainment economy', 'holiday economy' and 'exchange economy', etc., have become quite fashionable. It seems that everything has a connection with the economy and that everything is closely related to commercialization. Upstarts have become icons of the Internet age and have consequently become the focus of worldwide media attention for their strategies, competitive market positions, property, and roads to financial success, and even for their lawsuits. Their success has become the dream of millions of young people around the world.

It is a common phenomenon throughout the world that the more the economy develops, the less warm-hearted relationships become, the harder it becomes to find family warmth, juvenile delinquency rates grow higher and divorce rates rise. Although material life may become rich, love becomes poorer and 'the milk of human kindness' becomes harder to find. Are those outcomes the reason we aim to pursue economic development? No wonder many middle-aged and elderly Chinese are now yearning for the days of the 1950s and 1960s. During that time, although they were comparatively poor with regard to material life, they were rich in human relationships and their social ethos was captured by sayings such as 'when a difficulty should arise, support from all directions will gather'. Is it possible for us to keep a healthy balance between material civilization and spiritual civilization, and work collaboratively for the development of society in which all of us enjoy our human lives? Is embracing the market-oriented economic system, stressing competition and efficiency really the best thing?

Economic 'knowledge-ification' is an inevitable trend for which it is worth striving. However, we should be careful about overemphasizing the 'economic-ification' of knowledge and culture. A 2000 article from France entitled 'A Survey for New Master of Event', analyzed today's Internet stars: the culture is eaten up by the Internet which is swallowed by its simplicity. Culture is one of the human activities that may most easily be destroyed by computers. Although the Internet will play a more important role in education in the future, it can never take the place of teachers who teach by making themselves an example, especially an example of how to be a decent human being, and it can never take the place of family education. Online communication can never replace education through love, caring, feeling and face-to-face emotional exchange (although some people are already researching how the Internet may become a medium for smells!); online movies and music can never take the place of the spiritual happiness that occurs when hundreds of people enjoy seeing art in a theatre or a concert hall. In the wake of these trends, and influenced by the pressures of the market, the proportion of people who would like to dedicate themselves to the development of 'science itself' has decreased.

We must recognize that developing cultural technology and culture industry should not mean the commercialization of culture and arts. Industrialization has damaged our natural environment and ecological resources. We should learn from our past mistakes and never allow the commercialization of culture and arts to further erode and pollute humanistic and social environments. Our challenge is to correctly grasp the innovation direction of culture and arts, to encourage cultural and artistic products to go to market, but we should oppose taking money as the purpose of cultural and artistic creation, resulting in flooding people's lives with 'products' by manufacture in a rough way; to keep in mind the purpose of culture and arts that provides spiritual enjoyment for the majority of the public, produces more products people love to see and hear, but also encourage enterprises and individuals who engage in cultural and artistic 'production' to comply with the triple bottom line – the value created by taking into account the economic value, social value (including cultural value) and the value of ecological environment. Thus, it is essential that institutional innovation keeps pace with the times and that the cultural quality of the whole nation is enhanced.

E. Think about the Future Industrial Structure

In the traditional sense, human productive activities can be roughly divided into two categories – material production activities and non-material production activities. Since the Industrial Revolution, material production activities have been classified more clearly into agricultural production activities focusing on farmland and industrial production activities, which focus on processing natural substances and fabricating artificial materials and other products, based on equipment; non-material production activities have gradually developed to become the service industry. While the former two are typically treated as primary and secondary industry, service industries tend to be referred to collectively as tertiary industry. Material production activities are intended primarily to provide tangible products, and non-material ones are intended mainly to provide opportunities for applying labour and intelligence.

However, along with the third Industrial Revolution driven by globalization, information technology, and other modern soft technologies, there has been an acceleration of changes in industrial structure. Economists try continuously to track such changes and to re-classify industries from different angles or to offer new definitions and explanations for newly emerged industries.

In 1962, Fritz Machlup, an American scholar, introduced the concept of the knowledge industry for the first time (Machlup [1962] 1969). First, he divided knowledge into five categories: practical knowledge (professional, practice, operational, political, military knowledge and others), intellectual knowledge (including satisfying curiosity or the awareness of cultural values), entertainment knowledge, religious knowledge, redundant knowledge and knowledge obtained by chance. Machlup separated the knowledge profession into six levels: transporting knowledge, transmitting information, modifying knowledge (changing the form of knowledge, e.g. a shorthand typist), dealing with or processing knowledge (changing the content of knowledge and information, e.g. accounting), interpreting knowledge (lecturing, translating, analyzing) and creating knowledge (original innovation of new knowledge). He believed that the knowledge industry could be understood from two angles: the aggregation of industries producing knowledge; and the aggregation of professions (no matter in which industry) producing knowledge. Machlup later added education, R&D, media (printing, publishing, magazines, photography, stage and theatre, movie and broadcasting), information machinery, information service, etc. to the knowledge industry.

In 1977, Marc Uri Porat and his partners in the US proposed the idea of the fourth industry (quaternary industries), in which they divided national economic activities into agriculture, industry, service industry and information industry; they elaborated the concepts of primary, secondary, tertiary and quaternary industries in terms of industrial divisions and the speed of development (Porat & Rubin [1977] 1987).

Roger Sue in France proposed that the primary industry is actually the agricultural subsistence economy as the mainstream; the second industry is the industrial-equipment economy; and the third industry is the paid-service economy (Sue 1999). Now the world is ushering in the fourth industry sector, which is a human-centred economy and it is an economic stage that concerns individual 'production', e.g. individual educational status, personal ability, health, personal social relationships and even genetic heredity.

During the 1998–1999 upsurge of studying the 'knowledge economy' in China, many scholars discussed changes in the modern industrial structure. Most of the points of view expressed at that time were derivations of Porat's ideas, pointing to the information industry as the fourth industry sector. Others, quoting Machlup, proposed to classify the so-called knowledge industry as the fifth industry following the information industry. Jianjun Yan, in his book *The Rise of the Fourth Industry*, defines the fourth industry as the 'industry creating social wealth using knowledge and the "brain industry" to communicate with and coordinate between the primary, secondary and tertiary industries through information. It is searching for the key points for productivity by developing and making use of information [...] intellectual industry' (Yan 1993). In addition, Yan also believed that the fourth (quaternary) industry could be divided into hardware and software categories.

In fact, it is not appropriate to treat a sector with a certain technology as an independent industry and then treat it as belonging to the same spectrum as the primary, secondary and tertiary industry, thus treating the information industry as the fourth industry sector. Since there will be many new technologies in the future, the width and depth of their infiltration into primary, secondary and tertiary industries will be pervasive and will be no less diffused than that of the information technology.

Take biotechnology as an example. The Biotechnology Revolution is regarded as a significant event that occurred at the end of twentieth century and is another milestone after the application of the steam engine, electrical energy and information technology in the modern history of technology. The great economic powers in the world all regard biotechnology to be a critical technology in twenty-first century and are fiercely competing with one another in the field of biotechnology. The Blue Books on biotechnology in the twenty-first century, published by the US National Science and Technology Committee each year since 1992, indicate that biotechnology is commencing its second tidal wave after the first surge (in the medical and health care area) ('S863 Soft Science Strategy Report' 1999). They emphasize development of the following: 1) agricultural biotechnology; 2) environmental biotechnology; 3) biomanufacturing and bioprocess technology, as well as energy research; and 4) oceanic biotechnology.

Modern biotechnology has provided means for human beings to thoroughly understand nature and to potentially overcome a series of critical problems such as the population explosion, food shortages, environment pollution, harmful diseases, severe shortages of energy and resources, destruction of the ecological equilibrium, as well as the extinction of species. At present, 30 per cent of the 10,000 patents issued each year worldwide come from biology technology (ibid.). Research, development and industrialization in biotechnology has brought about great changes to the world's industry structure and these new industrial factors will also help in forming new high-tech industrial clusters in the various fields such as agriculture, medicine, food, chemistry, energy and environmental protection.

In addition, nanotechnology has brought about another wave of technology development (Gloss & Port 1998). Nanotechnology has emerged from across the fields of chemistry, physics, electronic engineering and biology. The application and diffusion of 'micromation' devices and molecular-electronics ('molectronics') devices into traditional industries such as the electronics, machinery and chemical industries will cause revolutionary changes in healthcare, aviation, space, medical care and possibly every field of human life. Scientists predict that the Nanotechnology Revolution will bring more striking large-scale industrial changes than did micro-electronic devices in the twentieth century. The annual output value of this technology in the world has reached 50 billion US dollars in 2006 and there are 323 nanotechnology enterprises in China up to May of 2001 (2001 *International Nanometre High-level Forum and Technology Application Seminar* 2001). Scientists working in the areas mentioned above believe that the next twenty years will be the era for biology and materials. They believe that humans have conquered space in the industrial age, conquered time in the information age, and will conquer substance itself in the era of biology and materials (D. Wang 2001).

Thus, if we call the information industry the fourth industry, then in the future whether will biotech industry, robotics industry, nanotechnology industry and genetic industry, etc. be necessary to refer to the fifth, sixth, seventh and eighth industry, and so on? It is clear that it is not feasible to continue to list new industries in this way whenever new technologies sprout and develop.

I would suggest that we take all those sectors whose root technology comes from natural-scientific knowledge, such as the information technology, software, biotechnology and nanotechnology, and refer to it as the 'contemporary high-tech industry'.

1. *About High Technology and High-Tech Industry*

Here we stress high technology coming from natural scientific knowledge, owing to the existence of high technology in soft technology. A fresh understanding of high-tech industry is essential at this point. High technology is relative and temporary. From the perspective of technological history, the leading group of hard technologies which has caused the outbreak of every technological revolution, thereby resulting in the Industrial Revolution, such as the steam power technology, spinning technology, electric power technology and automation technology, etc., were all high technologies within the hard technologies of that time, while the corresponding currency technology, financial technology, stock technology, enterprise merger technology, venture capital technology and so on were high technologies within the soft technologies of that era. Along with the periodic updates of these high technologies, the high-tech industries of that age emerged.

As they were applied they became mature, they were filtrated, absorbed, fused and diffused into the agriculture, industry, traditional service, intellectual service and life industries, just as the widespread application of information technology has happened in different industries, such as agriculture, manufacturing, finance and biotechnology industries today. As the result, on one hand, the technological intensity and efficiency of existing industries has been improved; on the other hand, mature high-tech industries contract out those universal or common parts which come into being specialized manufacturing sectors or service sectors. Dale W. Jorgenson, an American economist, proposes that the manufacturing of computer and telecommunications facilities as well as software should be treated as the output of information technology at the gross level (GDP) and be included in the information products manufacturing industry, which has already been adopted by the US National Income and Production Audit (NIPA) (Jorgenson 2001).

E-commerce is another example. E-commerce is a contemporary soft technology with high-tech content. As network technology matures, e-commerce now penetrates into all sectors, such as finance, manufacturing, circulation and media industries, thereby changing their modes of conducting transactions and improving their competitiveness. It is even a vehicle for the development of 'E-Government'. E-commerce has become a universal tool for increasing added-value for relevant industries, and those professions that support e-commerce technology form an intellectual service industry in the narrow sense. A profession that specializes in offering e-commerce services has already been developed in Shanghai centred on the activities of e-commerce consultants, whose function may be defined as the profession that involves sharing non-structured and structured business information by using e-tools (EDI, WEB, etc.), and managing and accomplishing various kinds of transactions in business, management and consumption activities by using those tools (R. Li 2001). This profession has set up five grades. It has been calculated that 25,000 enterprises in Shanghai will need at least 150,000 e-commerce professionals within their businesses.

It is evident that high-tech industries are continuously developing, separating and flowing. With the maturity of these industries, industry structures will be continuously upgrading and optimizing, and then the high technology and high-tech industries of the new generation will emerge. Therefore, high-tech industries of different ages have different content. Thus, the industries associated with information, software, spacecraft, bioengineering and nanotechnology, etc. should be called 'the contemporary high-tech industry'. The content of high-tech industries of the next generation should be different from earlier ones. The representative example of the unceasing separation of high-tech industries is electronic information industry, which has been subdivided into e-commerce industry, mobile phone industry, Internet industry and 'Internet of Things' industry, etc. The high-tech industries can also be divided into hard industries and soft industries (see Table 16). The software industry, the artificial intelligence industry, the Internet industry and the 'Internet of Things' industry belong together as part of the integrated industry of soft-tech and hard-tech, but in which the value created by the innovation of soft factors will exceed the value created by hard factors.

Moreover, the 'life industry' is also part of the contemporary high-tech industries, but it is separated solely because of its particularity.

2. Subdividing Service Industry

It is obvious that the most profound change of industrial structure exists in the service industry. Therefore, experts have conducted a good deal of research on the classification of service industry.

In 1978, Kimindo Kusaka, a Japanese scholar, proposed that tertiary industry could be separated into three parts: the service industry that is physical-power oriented; the service industry that is intelligence oriented (so-called knowledge and information type); and the service industry that is psychology oriented. The third part refers to the emotional satisfaction type or informational presentation type and is proposed as the fifth ('quinary') industry, while the second part is proposed as the fourth (quaternary) industry (Kusaka 1978).

In 1983, Noboru Makino argued for classifying the new type of the service industry into intermediary service sector (consultants, counsellors, tax accountants, accountants, designers and recruitment agents), computer software service sector, security and insurance service sector, social service sector (education, health, traffic) and personal service sector (express delivery, medical treatment, lawyers, tutors, etc.) (Makino 1983).

In 1986, Kazuhisa Maeda suggested that the service industry should be divided into the third industry (physical-power service industry), the fourth (equipment industry), the fifth (intelligence service), the sixth (emotional service) and the seventh (religious service) (Maeda [1986] 1987). According to him, the 'Third Industry' consists of the provision of hands-on services, e.g. baggage carriers, tailors, taxi drivers, etc.; the 'Fourth Industry' consists of combined services that developed by using intelligence but are protected by physical powers, e.g. railways, real estate, hotels and banks, etc.; the 'Fifth Industry' consists of providing intelligent services by the brainwork, including industries centred around consulting, think tanks, writing, designing, teaching, broadcasting and providing news; the 'Sixth Industry' provides emotional services to please people, for instance, industries such as the entertainment, movie, art, theatrical and music industries; and the 'Seventh Industry' provides religious services to set people free from the uneasiness regarding death and to help them gain psychological calmness.

Service industries have become so intermingled, which, on the one hand, put services at different levels (industrial service sector and services focusing on individuals) side by side and, on the other hand, the industry that is far beyond the scope of mere service sector is incorporated into the so-called tertiary industry in the name of service, such as cultural industry and social industry. Along with the economic softening, the proportion of service industry has been increasing. Without classifying, distinguishing and adjusting, it will have a negative effect on both development for service industries themselves and development for other soft industries.

3. About Life Industry

In view of the importance of and future prospects of the 'life industry', it needs to separate the industries related to the human body, human life and health from the general service industry and manufacturing industry, and that it be named the 'life industry'. As a special industry, its isolation may be justified on several grounds. First, it is centred on the human body; second, the orientation of its development is more closely related to culture, ethics and morality than is the case with the intellectual service industry; third, since ancient times, there have been different understandings and approaches in the East and West about the life and health, thus it is necessary to carry out a integrated development for East and West science and culture in this area; fourth, no other industry needs such a wide range of disciplinary integrated technologies. As life expectancy rises and the geriatric society looms over us, there is an urgent need to provide for high-quality life for people. Apart from what may be offered by the general medical and healthcare industries there are new technologies and industries, although ones such as those concerned with disease prevention, prevention of ageing and anti-senescence are emerging. In addition, special manufacturing industries that produce artificial organs, blood vessels, bones and skin, etc., will also be separated out into their own industry.

As the special relationship between soft technology and life, and moreover mentality, life and human body are important resources of soft technology operations, so that large numbers of soft-life technologies are brought into being. At present, the relatively matured soft-life technologies include: 1) the technologies of coordinating and harmonizing of human physiological and psychological factor – technologies such as longevity extension technology, health care technology and health preservation technology, etc. have expanded and applied the concept of the 'human body' in the sense of soft technology, and have persisted in the principle of harmonizing physical health and spiritual health, and in this sense they are quite different from hard-life technology, which includes gene technology or regenerative medical technology; 2) telepathy technology and technology of DTBOAIPC (Diagnosis and Treatment Based On an Overall Analysis of the Illness and the Patient's Condition) in traditional Oriental medicines such as the diagnostic technology that applies the skills of 'looking, hearing, asking and feeling'; therapeutic technology in traditional Chinese medicines; and the medicine of ethnic minorities, including Tibetan medicine.

Soft-life technologies are taking shape as the most dynamic soft industries within life industry. At the same time, the above unique soft-life technologies have been infiltrating deeply into various hard-life technologies and hard industries, increasing their added-value and making them soften.

Therefore, the life industry will consist of two parts: one part includes hard industries that are mainly composed of medicine, traditional healthcare, medical, human genetics and human organ manufacturing industries; another part includes soft industries, such as diagnostic and therapeutic industry in Chinese medicine, Chinese types of life prolonging industries, health preservation industries and new concept of health and wellness industries.

As the special theme of the industrial age, the division of industry into manufacturing and non-manufacturing is losing its practical significance and it is no longer feasible to strictly classify industries into the orthodox meta categories of primary, secondary and tertiary industry. Under these circumstances, some experts are still committed to studying and developing new criteria for the classification of industries, accounting system and statistics institution (Jorgenson 2001).

I would propose to classify the economic activity into agriculture, industry, service industry, cultural industries, social industry and life industry and, moreover, in accordance with the differentiated history of industry, it could be labelled in proper sequence as primary industry, secondary industry, tertiary industry, fourth industry, fifth industry, sixth industry respectively (see details in Table 16). Understanding of the future industrial structure will be conducive to studying industrial policy and adjusting industrial structure, as well as formulating the right socio-economic strategy.

Table 16: Soft-tech Development and the Change of Industrial Structure.

Industry Category	Name	Nature	Characteristic	Example
Primary industry	Agriculture industry	Agriculture production	Industry taking the natural resources as the object	* Hard industry: farming industry, forestry, stock raising, fishery
		Agriculture service	Comprehensive exertion of three main functions of 'farming' to improve its add value	* Soft industry: Services for farming industry, forestry, stockbreeding and fishery * Services the agricultural industrial chain * Comprehensive services * New business model * Design & planning new rural * Soft environment design for agriculture/farmer/rural
Secondary industry	The industry sector	General industry sector	Equipment-based material producing industry	* Mining and traditional industry * Manufacture and artifactitious industry
		High-tech industry sector	Industries with modern high tech as the core technology	* Soft industry: software, artificial intelligence * Hard industry: micro-electronics, photoelectron, spacecraft, biotech, nanometre, etc.
Tertiary industry	Service industry	Equipment & material-based service industry	Fixed assets based immaterial producing industries	Transportation, shops with tangible form, hotels, restaurants, real estate, etc.
		Intellectual service industry (in the narrow sense)	Immaterial producing industries that providing intellectual service	Consulting industry, intermediary service industry, design industry, a variety of professional management industry, information service sector, Internet industry, 'Internet of Things' industry, intellectual property industry, e-commerce sector, modern banking sector, insurance, recycling industry, and R&D industry.

Fourth industry	Cultural industry	Production and service of cultural products	Industries that developing and applied of cultural value and cultural resources	* Industry of Arts, Entertainment, Sports, Tourism, Leisure * Cultural service industry
Fifth industry	Social industry	Social economy	Industries which value is created by development and application of social resources	* Social enterprises * All kinds of public sectors and institutions * Communities and regional networks * Various non-profit organizations
Sixth industry	Life industry	Economy takes life as the centre	Industries which value is created surrounding human body, human life and health	* Hard industry: medicine, health care, medical, human gene and human organ manufacturing industries * Soft industry: traditional Chinese medicine, Chinese type of life prolonging industries, health preservation industries and new concept of health & wellness industries

Source: Zhouying Jin, January 2009

- Primary industry is certainly agriculture, which can be divided into the agriculture production sector and agriculture service sector. The former refers to the sector taking natural resources such as land, water, etc. as the object and being engaged in material production, including farming industry, forestry, stock raising, fishery; the latter aims to exert comprehensively three main functions of 'farming', improving its add value.
- Secondary industry is the industry sector, which can be classified as general industry and high-tech industry. The former refers to equipment-based material producing industry such as mining industry, metallurgical industry, manufacturing industry and processing industry, etc.; the latter refers to sector taking contemporary high technology as the core technology, including soft industry and hard industry.
- The tertiary industry is service industry, which can be classified as equipment and material-based service industry and intellectual service industry. Cultural industry, which has developed rapidly and formed a considerable scale, and social industry, which has an unusual significance for social development, as well as agricultural service industry, are separated from the service industry. Equipment and material-based service industry refers to those non-material 'production' sectors based on equipment (fixed assets) such as transportation industry, physical store industry, hotel industry, catering and real estate service industry, etc. Intellectual service industry (in its narrow sense) refers to those non-material 'production' industry providing intellectual services, including consulting industry, intermediary service industry, design industry, a variety of professional management industry, information service sector, Internet industry and 'Internet of Things' industry, intellectual property industry, e-commerce sector, modern banking sector, insurance, recycling industry, and R&D industry, etc.

Here, I did not use the term of 'new-type service industry'. First, the new and old are always two relative concepts and these concepts are not accurate enough to be used to define an industry. Moreover, the 'traditional' factors in the equipment and material-based service industry will gradually decrease over time while applied high-tech factors will increase.

- Fourth industry is cultural industry, which can be divided into production sector of cultural products and cultural service sector.
- Fifth industry is social industry, referring to the sector developing and applying the value of social resources.
- Sixth industry is life industry, which creates value surrounding human body, human life and health.

In the above discussion, I consciously avoided using terms such as 'knowledge industry' or 'knowledge-based industry' for the following reasons. First, most of the contemporary industries are knowledge-intensive, especially manufacturing, which have already become an equipment-based, knowledge-intensive material production sector, and even the traditional mining industry and metallurgical industry must increase its knowledge and technological contents in order to enhance the efficiency and sustainable survival. Therefore, I think it is inappropriate to take the knowledge-intensive degree as the industry classification standard. Second, modern agriculture is gradually becoming a knowledge-intensive industry. An American farm owner said that eight high technologies, including satellite communication and computer technology, are in use on his farm. This enables America to provide enough food for the entire country with only 2.7 per cent of the total workforce employed in agriculture, whilst still remaining one of the world's main exporters of grain.

Chapter 6

Soft Technology and the Fourth Generation of Technology Foresight

Salient trends in world development include a shift from the internationalization of production towards economic globalization, and a shift from the internationalization of R&D towards the globalization of science and technology. Industrial community increasingly cooperate each other through alliances in science and technology, thereby further strengthening the trend towards the globalization of the economy and of science and technology.

Under these circumstances, technology forecasting surveys and long-term technology foresight studies have become important foundations for formulating science and technology policy and industrial policy; reallocating human, material and financial resources; mapping long-range strategies; and adjusting economic and industrial structures. Thus, issues associated with the rationality, effectiveness and operability of technology foresight have again been added to the agenda, thereby stimulating new theoretical developments in technology foresight for the new age. The third generation of technology foresight theory which combines society, the economy and the environment, and fourth generation theory which places equal stress on soft technology and hard technology, began to be recognized and developed.

A. The Evolution and Development of Technology Foresight

Although technology-forecasting techniques have probably existed throughout history – and nascent versions of modern techniques such as trend extrapolation, brainstorming and scenario development, etc. emerged during the last several centuries – fully developed normative technology forecasting did not appear until the end of nineteenth century. During the past century, technology foresight has experienced three climaxes; its contents have undergone three stages of development and is now entering its fourth stage, while the concept of technology forecasting is changing into what has come to be known as 'technology foresight'.

The term 'technology foresight', as discussed in this book, refers to comprehensive foresight at the national level. Specifically, technology foresight is the process of identifying technologies (including soft technologies, hard technologies and their supporting fields) that are likely to emerge in the future, by systematically identifying the long-term development trends of science, technology, the economy, the environment and society, with the aim of working out strategic plans and policies and making related decisions. Research, development and application in the fields addressed by this type of technology foresight may be of strategic significance or may bring huge economic, environmental and social benefits.

1. The Three Climaxes of Technology Foresight

Technology foresight has experienced three climaxes in the development of its theories and methods:

The first climax: During the 1920s and 1930s, following the end of World War I, the European and American countries began to shift their attention to domestic economic development issues. 'Scientific' technology forecasting became the prerequisite for establishing a science and technology strategy and articulating associated policies. Technology forecasting during this period consisted mainly of the personal anticipation by technical experts with the maximum potential, and probability of the development of various isolated technologies that were still at a stage of the epic ideal or that were clouded with unrealistic expectations.

The second climax: During the 1960s, technology forecasting became accepted and developed further with a kind of formalized theoretical framework. In addition, it began to be widely applied in military departments and industrial circles in the developed countries of Europe and America, and it played an important role in the formulation of national plans in countries like France, the United States, Britain and Switzerland. This led to an upsurge in the field of futurology. Many forecasting methods emerged during this period, including the famous Delphi Method that was developed through the sponsorship of RAND. New scholarly and professional journals in the field, such as *Technological Forecasting and Social Change*, also began to appear.

Some historical significant forecasting events included the international conference on the methodology of long-term forecasting, held in March 1966, in Paris; a seminar conducted by the US Air Force team, discussing long-term forecasting and planning, in August 1966; and the industry-oriented technology forecasting conference held in May 1967 (Jantsch [1967] 1968: 290).

Japan studied and adopted American forecasting methods during its rapid growth period towards the end of the 1960s. Japan has subsequently conducted national level long-term forecasting exercises every five years over the three decades since 1971, thereby accumulating valuable experience and making valuable progress in the theory and practice of technology foresight.

The third climax: During the 1990s, for reasons mentioned at the beginning of this chapter, many nations, including developing countries, conducted national level comprehensive technology foresight exercises. The defining features of this phase were the following:

first, technology forecasting was used worldwide to assist decision-making related to strategic planning and policies at the national level; and, second, the concept changed from technology forecasting to technology foresight. National level comprehensive forecasting exercises in Japan, Germany, the US, the UK and Sweden, etc., have advanced to the stage where technology foresight covers technology, the economy, society and the environment together, rather than concentrating simply on technology alone, as tended to be the case under the rubric of technology forecasting.

Against the above backdrop, APEC founded a technology foresight centre in 1998. It was the first regional technology forecasting research institute established in the world. The International Conference on Technology Foresight, held in Tokyo in 2000, attracted participants from fourteen countries and two international organizations. The conference proposed to carry out international foresight research and technology foresight exercises that aimed at satisfying social and economic needs beyond the limitations of individual countries, based on organizations like APEC, EU and etc.

2. *From Technology Forecasting to Technology Foresight*

Because of its failure to predict the 1973 'oil-shock', scepticism about the validity and utility of forecasting became widespread. Many firms disbanded their long-range corporate planning groups, and the boom in futurology, which had begun in the mid-1960s, ended quickly. Criticisms of long-term technology forecasting became even more trenchant and widespread by the beginning of the 1980s (Martin & Irvine 1989: 109). In this context, the scientific and professional communities engaged in technology forecasting found it difficult to withstand the pressure placed on them for their technology forecasting exercises to deliver up reliable and rational management tools for decision-makers. They also had trouble living up to the pressure for technology forecasting to accurately anticipate how choices made today would shape or create the future. Technology forecasting activities were therefore gradually superseded by a variety of other activities covered by labels such as 'outlook', 'foresight', 'issues management', 'strategic thinking', etc. as a way of escaping what had come to be considered as impractical and conceptually vague 'prophecy making' and forecasting. The first two of these labels were used mostly within government organizations, while the latter two were popular mostly within industrial community.

Martin and Irvine were early advocates of the change from an emphasis on technology forecasting to an emphasis on technology foresight (ibid.). They recommended adoption of a definition of technology foresight proposed by Coates in 1985:

A process by which one comes to a fuller understanding of the force influencing the long-term future which should be taken into account in policy, planning and decision-making […] Foresight includes qualitative and quantitative means for monitoring clues and indicators of evolving trends and developments and is best and most useful when directly linked to the analysis of policy implications. Foresight prepares use to meet the needs and opportunities of the future. (Cetron [1969] 1970: 4–5; 12)

This definition indicated the first 'is the emphasis placed on foresight as a process rather than a set of techniques, and foresight also encompasses the consultative procedures required to ensure feedback to and from relevant actors, including policy-makers, the scientific community, and research users in industry and elsewhere'. Second, 'forecasting techniques can be – and indeed often are – treated as a "black box" for translating input assumptions into outputs taking the form of predictions about the future'. Third, 'the notion of forecasting and foresight involve very different ontological assumption about the future'. In conventional forecasting, the aim is to arrive at predictions which can be justified 'scientifically' – i.e. demonstrating how, when starting from certain premises and using given methods and data input, the predictions was made. In contrast, the goal in foresight is to survey as systematically as possible 'what chances for development and what option for action are open at present, and then follow up analytically to determine to what alternative future outcomes the developments would lead', and more important 'is the deployment of monitoring mechanisms to provide an early warning of emerging trends and opportunities'. Martin and Irvine further classified technology foresight into four levels according to the level and characteristics of decision-making: the overall level, the macro-level, the meso-level and the micro-level.

The so-called 'scientific' approach to technology forecasting involved the use of relativity principles, probability principles, continuity principles and principles of cause and effect. However, because the economic, social and environmental factors of technological driving forces are so complex and because contemporary technological and human factors are so intermingled, it is extremely difficult to accurately forecast the future 'scientifically' based on the above principles.

It is therefore reasonable to substitute the technology foresight theory for traditional technology forecasting theory. The OECD, during a Paris conference in 1996, produced the following definition of technology foresight: 'the process involved in systematically attempting to look into the long-term future of science, technology, the economy and society with the aim of identifying the strategic research areas and the emerging future technologies likely to yield the greatest economic and social benefits' (Kuwahara 2001). Greg Tegard of the APEC centre for technology foresight considers technology foresight to be a tool of strategic planning, thus it must face the multiple challenges of economic outlook and moreover the results of technology foresight should be reasonable, efficient to produce and feasible to implement. The prerequisite for technology foresight is therefore the involvement of as many experts as possible from different fields, different disciplines and sectors, and the creation of a permanent network that contains the results of each round of exercises in technology foresight (Tegard 2000).

3. The Theory of the Four Stages of Technology Foresight

During the Tokyo conference in 2000, Professor Luke Georghiou, of England's Manchester University, elaborated upon what he saw as the three phases of the development of technology foresight and identified the different characteristics of the three phases. According to Professor Georghiou, Britain experienced its first generation of technology foresight during the 1980s, its second from 1993 to 1998, and its third generation from 1999 to 2003 (Georghiou 2000).

From a global perspective, as the level of comprehensive technology foresight is different in the national level, the time to enter the various stages is also variable for each country. Therefore, each country can judge the level of sophistication of technology foresight in their country by reviewing the degree to which the characteristics of each generation are present. The characteristics of each generation will now be adumbrated.

1. The first generation of technology foresight

The first generation of technology foresight is the 'pure' technology forecasting stage and is concerned primarily with the expansion of scientific fields as its main content, and with the activity of natural scientists in forecasting the probability of potential developments in science and technology. In this stage, technology forecasting is the sacred field of scientific and technological experts, who make forecasts on the direction of technological development and on the types of technologies that will possibly emerge in the future and that will need to be developed, from the point of view of pure science and technology. Hence, technology foresight belongs to the future research activities of scientists and engineers.

2. The second generation of technology foresight

The second generation of technology foresight involves the combination of technology and markets. It is the stage when academia and industrial community join together to study future developments in science and technology, as well as economy.

Many countries believe that their technology forecasting has been integrated with market demands from the very beginning of their engagement in forecasting activities. However, the reality is often quite different. A review and assessment of foresight and critical technology in the US by Bruce Don, Director of the Science and Technology Policy Institute of RAND, is very persuasive in judging the actual degree to which the integration of markets into technology foresight exercises has happened (Don 2000). A closer look at Don's assessment is in order.

In 1990, the US Congress formed a national critical technology panel to produce a 'National Critical Technologies Review'. This exercise played an important role in the orientation of national R&D policy and the adjustment of the allocation of national key resources, and addressed issues in the comprehensive competitiveness of the United States.

In the critical technology foresight of the US, technologists paid special attention to market needs. They 'focused exclusively on industry, depending on several sources: first, firm-level interviews; second, review of industry sector technology roadmaps; third, selected conferences on technology in the industry'. They managed, therefore, to reach agreement about critical technologies in the industrial context. They came to a consensus that software, microelectronics, telecommunication, advanced manufacturing technology, materials, sensor technology, image technology and others were critical technologies for the United States. When it came to the reasons why these technologies were critical technologies, however, there were obvious differences between opinions of industry managers, technologists and public policy-makers. Industry people stressed the key role that these technologies played in the economic performance of the US and focused on the issues associated with technology commercialization. Technologists tended to view these technologies more as tools to express expected discrete technological functions. Industry leaders tended to describe a system rather than just individual key technologies. In addition, the government sector is more concerned about the long-term issues like energy, environment and living systems, etc.

Bruce Don's analysis pointed to the narrowness of the concept and method embodied in the US government's 'critical technologies' exercise in 1990s. He recommended that the US should step beyond the limited scope of the current approach to analysis of key technologies (emphasizing the circle of technology-product-application) and should instead establish an approach that looked at the whole system in which technologies were embedded, in the broader sense. In other words, an approach was needed that involved a broader range of participants that paid more attention to the processes of conducting successful projects, rather than just to the eventual 'products' of the projects in the broad sense of an innovation system. Accordingly, the National Science and Technology Council sponsored a National Summit on Innovation at the end of 1999 to discuss such broader concepts.

3. The third generation of technology foresight

The third generation of technology foresight is oriented towards hard technology. It is also concerned with the dimensions of the market, society, the economy and the environment. In addition, a wide variety of stakeholders are included in the technology foresight exercise, which, in turn, incorporates a problem-solving approach that addresses an array of social factors, rather than just technical considerations.

Japanese technology foresight experts believe that Japan is experiencing a transition from the second generation to the third generation of technology foresight. The fifth technology forecasting exercise of Japan covered sixteen areas, including: agriculture, forestry and aquatics; information and electronics; materials and manufactures processing; life sciences; space, the oceans and the Earth; minerals; water resources; energy; the environment; production; cities, construction and civil engineering; telecommunications; transportation; healthcare and medical treatment; and social life ('The 5th Technology Forecasting Survey' 1992). The sixth technology forecasting exercise covered fifteen areas, including: materials and materials processing; electronics; information; life sciences; space, the oceans and the Earth; resources and energy; the environment; agriculture, forestry and aquatics; production and machinery;

cities, construction and civil engineering; communications; transportation; healthcare and medical treatment; and welfare ('The 6th Technology Forecasting Survey' 1997).

Japan has always stressed that technology forecasting should include not only natural science technology, but also technologies from a wider variety of fields, including production, healthcare, environment, security, city construction and social security, etc. Forecasting experts in Japanese technology forecasting exercises, furthermore, are constituted not only from the natural science fields, but also from the social science fields. For example, in the sixth forecasting exercise, 37% of experts were drawn from enterprises, 36% came from universities, 15% came from national research institutes and 12% originated from other types of organizations.

From 1994 to 1999, Britain's technology foresight activities covered sixteen areas: agriculture, horticulture and forestry; chemistry; construction; defence and aerospace; energy; financial services; food and drink; health and life sciences; information technology, electronics and communication; leisure and learning; manufacturing, production and business process technology; marine; materials; natural resources and environments; retail and distribution; and transportation (Wood 2000).

In Sweden, eight areas were chosen as the subject of the national level technology foresight exercise: health and medicine and care; biological natural resources, including agriculture, forestry, water usage, food, timber products, raw materials for bio-energy; community infrastructure; production systems; information and communications systems; materials and material flows in the community; service industries; and education and learning (Deiaco 2000).

The technology foresight activities of these two countries share one thing in common – they do not follow the traditional classification of disciplines in science and technology. They were organized around interdisciplinary themes rather than pure scientific fields. They clarified that the aim of technology development is to provide services for social progress and economic development in the future. Therefore, the priorities of technological fields should be determined from different vantage points such as science, technology, society, economy and the environment.

The mere inclusion of non-traditional technological fields like society, economy and environment within the purview of technology foresight does not mean that the third generation of technology foresight has necessarily been reached. According to Bruce Don's analysis and evaluation, it is important that such inquiries should not be limited to the technological highlights of the above fields to seek the high technologies or key technologies they need in the future. The technology foresight should be based on the development needs of these fields, not only address related key technologies, peripheral technologies and applicable technologies pertinent to those fields, but should also combine them from the perspective of commercialization so as to form a system in favour of solving problems (such as the need to increase competitiveness).

Table 17: Comprehensive Technology Foresight

Development stages	Content and Characteristics
First generation	Forecasting of hard technologies
Second generation	Combining hard technology and the market
Third generation	Focus on hard technology foresight and integrating social, economic and environmental dimensions
Fourth generation	Foresight on both hard & soft technologies, hard & soft environment, hard & soft industries, and on integrating the multiple dimensions of society, the economy, and the environment

4. The fourth generation of technology foresight

I believe that the fourth generation of technology foresight is now upon us. In fact, the technologies we mentioned in the above stages refer to hard technologies and even the third generation of technology foresight also focuses on hard technology under a broad systems framework, including society, economy, environment, etc.

Owing to our fresh understanding of the significance of technology foresight and soft technology, the fourth generation of technology foresight should be carried out according to the needs of social progress and sustainable development within the framework of technological innovation system (as broadly understood), and incorporating the multiple-dimensions of the market, the society, the economy and the environment, etc. It should also carry out foresight on a variety of soft technology and the soft environment they need.

B. The Fourth Generation of Technology Foresight and Soft Technology

1. The Goals of Technology Foresight

Let's start with studying the necessity of the fourth generation of technology foresight from the angle of goals of technology foresight.

Analyzing, at the national level, in view of the accelerating pace of technological progress, as well as the complexity of domestic and international environment, the range of possible choice that decision-makers face is growing quite wide. As a result, decision-makers are increasingly in need of assistance from technology foresight to set priorities for adjusting resource distribution, especially arranging the structure of budgets so as to adapt to international competition. From the perspective of enterprises, in order to constantly innovate in a new competitive environment, each enterprise needs to manage the 'interface' between their firms and customers, suppliers, partners, rule-makers and policy-makers with new thinking, new attitude and new methods. As a tool to help create a shared strategic plan, technology foresight can help reduce the uncertainty faced by managers and policy-makers.

Martin and Irvine have identified aims, objectives and functions of foresight as follow: 1) direction-setting – this is related to the use of foresight in determining broad guidelines for science policy (generally focuses on the holistic and macro-level of analysis); 2) determining priorities – this is arguably the most important objective of foresight and for this reason constitutes the central theme of US case studies; 3) anticipatory intelligence – the primary aim is to contribute background information on emerging trends in science and technology, in particular concerning development with major implications for future policy-making; 4) consensus-generation – this refers to the use of systematic analytical and consultative procedures to promote greater agreement among scientists, funding agencies and research users on identified R&D needs or opportunities; 5) advocacy – the deployment of foresight to promote policy decisions in line with the preferences of specific stakeholders; 6) communication and education – three principal functions of this type of foresight can be distinguished: a) promoting internal communication within the research community, b) external communication with industrial and other potential users of research, c) wider education of general public, politicians, government officials and others with a less direct, but ultimately no less important, stake in shaping a desirable future for science and technology (Martin & Irvine 1989: 22–24).

John Wood believes that technology foresight is, firstly, the process that enables 'to bring together business, the science base, the voluntary sector, and government' to plan for the future; secondly, to create a culture of 'forward thinking'; and, thirdly, 'to inform decision-makers' and to assist them to pay attention, and learn to better apprehend and analyse facts while formulating policies (Wood 2000). Wood considers that foresight is a 'think and do' tank. It not only gathers 'knowledge and ideas about future possibilities, needs and requirements' but also analyses the position and situation of the nation. He considers it to be a vehicle for matching a country's vision of possible futures to its circumstances, strengths and potential capabilities, and for measuring (and preparing to face) possible opportunities and threats it may be facing during the forthcoming five to ten years. It is obvious that depending on technological foresight about hard technology alone will not be sufficient to enable a country to deal with the above issues.

Technology foresight as a tool for strategic planning is itself a form of soft technology:

1. Formulating strategic planning must be combined with the framework of innovation system in the broad sense, thereby systematically planning the future of science, technology, the economy, the environment and society.
2. So-called 'strategic technologies' may realize great economic and social benefits for the community only through combination with soft technologies, otherwise it is difficult to determine and, moreover, it also can not be implemented.
3. Soft technologies also produce huge economic and social benefits in their own right.
4. Institutional foresights actually act as measures and guarantees of the realization of future strategies.

From the foregoing discussion we can see that technology foresight is an activity that may not be conducted successfully through the independent efforts of natural scientific community. In addition, it also can not merely rely on 'predicting' hard technology.

2. *Multiple Driving Forces of Technology and Technology Foresight*

I have carried out the research on the structure of technological driving force, which can be classified on three levels (Jin 2001).

The first level deals with the relationship between science and technology and economic development needs. The human motivation to exploit and challenge previously unknown fields, and the vast opportunities of the market, propel the development of science. On the other hand, new discoveries in science and the continuous renewal of knowledge not only deepen human understanding of nature and society, thereby enriching our library of knowledge, but they also provide new sources for technology and open up new directions for technological change. New technologies that profoundly influenced society and the economy in the twentieth century, such as integrated circuits, atomic energy technology and biological engineering were all closely related to great breakthroughs in science.

Conversely, however, the development of science (especially the verification of new discoveries and new theories) actually depends more often than it used to on new technological means and methods. In short, the development and wide application of technology is actually a key factor in the promotion and speed of development of science. The pull of market demands in technology development is related primarily to the motivation of enterprises to survive and develop in the fierce competition, as well as the requirement for new technologies and new products, while technology development is the engine for economic development (See the details in the section of 'Soft Technology and Thrice Industrial Revolutions' in Chapter 2). The direct driving forces of technological development are, first, the economic motive of profit-making and, second, human dedication, human desire to exploit new knowledge and the motive for self-fulfilment. In short, technological change drives both the scientific process of knowledge-creation and the process of economic needs, and these two processes in turn drive the process of technological change. For convenience, we call the virtuous cycle of causes the 'knowledge-economy circle'.

The macro-environment, which we call the 'environmental field', and which is composed of both the 'social field' and the 'natural field' (or soft environment and hard environment), constitutes the second level of technological driving forces. The social field includes human resources, institutional, cultural and social organizational factors, while the natural field is composed of natural resources, environmental and ecological factors. The reason why we call it a 'field' is because it is the backdrop and conditions of 'knowledge-economy circle', the same as magnetic field or electromagnetic field, which is affecting frequently the first level of factors.

During the industrial economic age, it was commonly believed that the push of science and the pull of market were the direct driving forces of technological change, while the environment was an indirect force. With the rapid development of hard technology, however, and the concomitant social problems, values transformation and the critical need for sustainable development, many informed observers began to place more emphasis on the 'environmental field' of technology as a driver of technological change. Environmental field may, in fact, be primary causes of technological change, which plays a revolutionary role rather than indirect force (for example, the influences of Italian Renaissance and the French Revolution on the revolutions of new technologies). Sometimes, however, if the 'environmental field' is unsuitable for the rules of economic development and technological progress, then it will become bottlenecks which slow down the pace of technological progress of that country or even retrogress (as instanced by China's Cultural Revolution).

The third level is centred on interaction between the above-mentioned driving forces. Only the coordination and harmony of all the factors from the knowledge-economy circle and environmental fields, as well as various elements in and out of the 'field', can accelerate the development of a country or a region.

As a process of observing systematically the long-term trends of scientific, technological, economic and social development, technology foresight must take those important driving forces which can promote technological development as the important factors for building a framework of technology foresight system.

3. Soft Technology, Soft Environment and Technological Foresight

The understanding of technology in a broad sense provides a more systematic theoretical basis for the fourth generation of technology foresight. The characteristics of the fourth generation of technology foresight are as follows: firstly, it takes into account the sustainable development in multiple dimensions of society, economy, environment, and resources, etc.; secondly, it carries out the foresight on both hard technologies and soft technologies; and thirdly, it carries through the foresight on an environment forming the competitiveness, including hard and soft environment.

In the other words, a comprehensive technology foresight of a nation should be implemented under the broad sense of innovation system framework, for instance under the framework of '6+1' innovation system. The opinions like 'paying more attention to the process leading to a success than the products' stated by Bruce Don, or 'it's a process rather than as a set of technique, […] and it is a qualitative and quantitative means for monitoring clues and indicators of evolving trends and development' pointed out by Martin and Irvine, have emphasized the role of soft technology; while the statements like 'raising the awareness of the potential development and its feasibility to develop a monitoring mechanism so as to provide an early warning of emerging trends and opportunities' have stressed the importance of the soft environment.

Successful technology foresight must be integrated with the soft-tech and hard-tech foresight, the foresight of soft and hard environment (the factors of technical feasibility) and that of soft and hard industrial foresight (the future demands and opportunities) in the multiple dimensions of society, the economy and the environment. Accordingly, the fourth generation of technology foresight is not only the joint responsibility of the communities of natural scientists and technologists, social scientists and industrial leaders, but should also actively involve as participants the related social communities and governmental agencies involved in making institutions, policies and regulations.

The fourth generation of technology foresight has begun to emerge in some countries in an embryonic form.

1. The framework for technology foresight in the United Kingdom included the fields related to soft technologies such as financial services, leisure and learning, manufacturing, production and business processes, materials, retailing and distribution, etc.
2. Sweden's approach to technology foresight has included the following fields related to soft-tech – along with orthodox S&T fields: community infrastructure, including housing, city planning, transportation systems, logistics, distribution and regional development planning; production systems; materials and material flows within the community; service industries, including media, leisure, trade, insurance and finance; and education and learning, etc.
3. Japan's seventh technology forecasting exercise, in 2001, comprised sixteen fields in six major systems ('The 7th Technology Forecasting Survey' 2001). Among them, technology of manufacturing and management systems and social infrastructure systems were juxtaposed with the technology of information systems, biological systems, environment systems and material systems. From the perspective of soft technology, Japan's technology forecasting process has some striking features: firstly, social infrastructure fields, including city planning, construction, civil engineering, transportation and services, are included as important fields. Most remarkable was the fact that service technology was treated as a subject amenable to forecasting, based on the demands of the development of the intellectual service economy, e-commerce and the knowledge-intensive society. Secondly, in order to follow the trend of economic softening and 'informatization', the manufacture/management system was selected. The system consists of three fields; namely, manufacturing, circulation and management. They belong to, or are closely related to, soft technology. In the field of management, furthermore, a subject that is traditionally seen as non-technological – that of institutions – is also taken into account.

While these embryonic examples are encouraging, no cases of technology foresight in which hard technology has been integrated with soft technology and in the framework of innovation system in the broad sense have yet been discovered.

4. The Causal Analysis of the Failure of Technology Forecasting

In the past 40 years, Japan has always adhered to the comprehensive technology foresight in the national level, making a significant contribution for Japanese technological development. Nevertheless, the Japanese academia has never forgotten an adage that 'failure is the mother of success' and summed up experiences and lessons learned in good time.

Yosaku Hasegawa, Director of the Institute for Future Technology in Japan, has conducted a thought-provoking study of 'the failure science of technology forecasting' (Hasegawa 2001). Hasegawa analyzed several forecasting mistakes in the information field in Japan and drew the following three conclusions:

1. Non-technological factors such as the economy, society, institutions and culture were neglected in technological forecasting. For example, it was forecasted that the 'newspaper delivery and residents information system under the influence of informatization' would be completed by the year 1985, but it had still not been completed by the year 2000. This was due to the negligence of such non-technological factors as institutions and national feelings that revolve around industries. As for the forecasting of information and communication technology, the forecasting is reasonable from various individual technological perspectives. Nevertheless, it was not expected that the Internet and mobile phones would become so ubiquitous today. The failure to forecast the trends of the Internet and mobile telephones was due mainly to excessive focus on technology itself and to insufficient knowledge about the synergistic effects of the sharp decline in the cost of information and communications equipment and social acceptance. The relationship between cost and acceptance has always been a problem. One key factor that played a decisive role has been the popularity of the worldwide 'mobile phone culture' with young people. Who could have predicted that such a phenomenon could be so influential? Thus, the problem of dealing with the social acceptance and the high cost of new technology is no longer a technological problem in the traditional sense.
2. The widely accepted Delphi forecasting technique has limitations. The others, like Trend Extrapolation, furthermore, place too heavy an emphasis on mathematical analysis and tend to neglect the emergence and influences of new technology, as well as new applications and new complementary development surrounding the technology in question. Even the 'expert interview' method is often ineffective, sometimes because of the limitations in the awareness of most experts but also because 'leading edge' opinions (which, by definition, are minority opinions) tends to be neglected in favour of majority opinions, and are often simply rejected because they are at odds with dominant social and political preferences (for instance, the global environmental problems).
3. Experts must be invited who pay attention to the non-technological subject matter of the social sciences and humanities when they are considering technology development; and experts in non-technological fields, who may nevertheless be able to understand the problems and difficulties of technology itself, along with techno-economic experts, need to be invited to participate in future technological forecasting exercises.

5. Technology Foresight in Developing Countries

Developing countries have exhibited almost no presence in traditional technology forecasting activities. This is due mainly to the fact that, in the straggling situation of overall technological level, most scientists in developing countries have no chance to work at the forefront of contemporary science and technology like their counterparts in developed countries, meaning they lack practical experience in the application of the most advanced technologies. Most information available to developing countries about frontiers in technology stems from the technology that developed countries have commercialized or issued publicly. Much of the leading-edge technical knowledge, of course, is not made publicly available. The vast majority of experts on future trends in science and technology who make presentations at annual meetings of the World Futures Society, for example, come from developed countries. This poignant fact raises the question as to what path developing countries should choose to catch up with, and hopefully surpass developed countries, in the domain of technology.

For most developing countries, there is certainly a big gap between developing countries and developed countries in R&D capability and in cutting-edge high technologies. However, the critically important point to recognize at this juncture is that it is the macro-environment, especially the imperfect soft environment and underdeveloped capability in soft technology, which leads to the failure of developing countries to transfer advanced technology or to absorb advanced technology by converting it into enterprise technology. This gives rise to unequal opportunities between rich countries and poor countries vis-à-vis new technology and it erects a strong barrier to improving technological competitiveness and industrial competitiveness in poor countries (see details in the section of 'The Essence of the Gap between Developed and Developing Countries' in Chapter 3). Thus, foresight in soft technologies, as well as the soft environment they needed, is an indispensable component of technology foresight. Regretfully, however, technology foresight in most developing countries has neglected soft-tech foresight.

First, if we defer to foresight objectives focus only on the field of natural science and technology, and place excessive attention on hot cutting-edge hard technology in the aspect of future technological development, it is very difficult for developing countries to be a new force suddenly rising in their own advantage areas and urgently needed areas, while eventually they will never be able to cultivate their distinctive strengths. Notwithstanding the popular icon of the 'leap-frog' strategy, it is actually very difficult for developing countries to catch up with developed countries from the perspective of technology qua technology. This is because technological invention and technological innovation are not determined solely by hard technology.

Second, even developing countries have been carrying on various studies on such topics as industrial and technology policy, while they have actualized layout and technology foresight on the environment, transportation, water and national life, but it is rarely conducted under the same system framework as natural science and technology foresight. As for the foresight for soft-tech areas like intellectual service technology, social technology and cultural technology, as well as the foresight for institutions in favour of targeted technology – in other words, topics that fall into soft environment categories – almost nobody in developing countries places stress on them as part of the content of technology foresight. Accordingly, it will be very easy for developing countries to commit the same error (described above) that Americans made in the area of critical technology assessment. Technology foresight professionals in the United States are currently engaged in a process of self-reflection about this issue. It would be a pity if leaders in developing counties were not able to learn from the previous self-acknowledged errors in technology foresight in America.

Third, a long history and cultural heritage, abundant human resources and unique social resources are important sources of technology (both of hard and soft) innovation and industry innovation. For this area, developing countries are almost at the same starting line with the developed countries.

Fourthly, it could benefit by establishing a network for mutual sharing of information about markets, economies, societies, the environment and technology, and they could benefit by sharing this within an international coordination network. It would also help fulfil an essential condition for the successful conduct of technology foresight in developing countries.

In summary, only under the broad sense of technological innovation framework, and according to their own specific resources and the soft and hard environment which may be provided (industrial structure and international division of labour, technological advantages and disadvantages, social life, institutions and culture, etc.), will it be possible to identify those target technology or areas appropriate for producing tremendous economic and social benefits, and ultimately to achieve the goal of promoting national comprehensive competitiveness and realize leapfrog development in more fields. Therefore, it has more important significance for developing countries to step into the fourth generation of technology foresight as soon as possible.

Note

1. Source: *XinHua net, www.news.cn*. Accessed 12 March 2010.

Postscript

The Principles for Development in the Twenty-first Century – Harmony, Balance and Coexistence

As we move into the twenty-first century there is a good deal of exciting news about new discoveries and new inventions all around the world, which will undoubtedly benefit human beings and develop a higher level of material civilization. However, the problems we are facing, such as the appropriate application of technology, social problems, the problem of human civilization and ethics, are becoming more profound – and various international conflicts and crises are increasing. As early as the 1960s and 1970s people became aware of natural resource shortages caused by high-level industrial development and of global maladjustments in the natural ecological system. This awareness led to painful predictions regarding the necessity of ending the 'resource economy' and re-establishing a human development mode. What is more alarming is that human beings are losing a great deal when they pursue a high degree of material civilization. Along with material advancement has come a decline in spiritual civilization, indicated by such things as the pre-eminence of money, the growth of crime, moral degeneration, indifference, and the destruction of the Earth and nature upon which we depend for our existence.

Human beings are now facing even more severe challenges. These include: conflict between economic growth and the environment; the conflicts between nations and between religions; the great gap between the rich and the poor; and the conflict between the development of science and technology, on the one hand, with social development on the other (e.g. the relationships between research freedom, research orientation and the development of human society, especially the ethical and moral problems caused by research on human genes and the breakthrough of mammal cloning technology, the unreasonable use of nuclear technology, computer viruses, etc.); from increasing family violence to internationally organized crime and violence, etc. Human beings are now destroying the civilization which they themselves created in an even more tragic way, with the help of high technology which they also developed. There is a very wide spectrum of problems, ranging from individuals and social interests, from national to global strategy, from domestic social relationships to international relationships, and from the economy and politics to national defence. Obviously, human beings are facing problems that now threaten their own mode of existence and development, as well problems related to sustainable development in resources, environment and ecology – which are not amenable to solution through the application of high technology (e.g. the Biosphere II Experiment proved that Natural Science & Technology Circle can not replace Biosphere), any management, alone or by the actions of politicians and governments alone.

In this context it is important for us to consider again the journey of sustainable development from the macro perspective (see Figure 24).

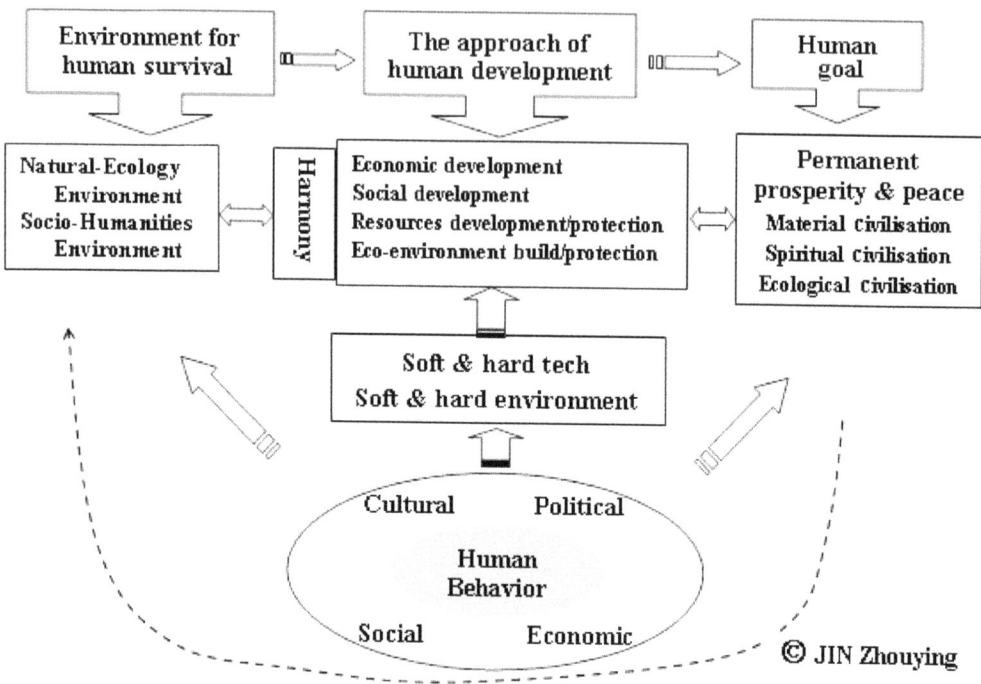

Figure 24: The Concept of Sustainable Development.

1. The ultimate goal of sustainable development is the progress of the human race and the creation of a world of prosperity and peace, owing to the development of material, spiritual and ecological civilization.
2. The social productive activities of human beings, including developing the economy and technology, promoting social progress, carrying out social and even political reforms, are not the goal. Rather they are but ways and means for achieving the goal of sustainable development. Among them, developing and innovating technology are the means and method for the development of those aspects mentioned above. This requires that we do not reverse the relationship between the means and ends, and that we do not treat means and methods (technology, economy and politics) as ends in their own right when we design a new strategy and new modes of development.
3. Good ecological environments and humanistic social environments are the result of sustainable development, as well as the essential conditions and prerequisites of sustainable development.

According to the above perspective of sustainable development, if we want to achieve sustainable development in our society, we must succeed in balancing the development of three systems: the natural ecological and resource environment, the humanistic social environment and human social productive activities. And, furthermore, we must succeed in coordinating economic development, social development, the construction and protection of ecological environment, as well as the development and utilization of resources. The process of implementing sustainable development actually is the process of coordinating human behaviours, which centred on nature, society and human activities. The core of coordination is the coordination of behaviours between humans and nature, humans and society, as well as human and human. And the means and method of coordinating and controlling human behaviours (external behaviour and internal psychological activities) is the heart of what soft technology and soft environment are all about. The challenges of sustainable development create profound significance for the concept of soft technology. They also create profound reasons for studying soft technology.

More importantly, important insights can be obtained by researching the new paradigm of technology – soft technology – to achieve sustainable development, to achieve transformation from the society where material civilization is paramount in the past thousand years, to the society where the material, spiritual and ecological civilization will be compatible, it needs firstly a thorough change in thinking mode in order to facilitate self-reflection and adjustment of the old development model. Therefore, those who are well educated and the elite of every country, including the national leaders, the leaders of companies as well as global citizens, need to adjust their values and change their world view. In other words, they need to work on the basis of common principles – harmony, balance and coexistence. These principles are not only suitable for dealing with problems related to the global governance mode, economic development modes, pursuing new lifestyles and business models, but are also suitable for dealing with problems related to the relations between humans and nature, between the West and the East, different nations and races, multi-culture, the development direction of science and technology, etc.

For example, the progress of human society should shift from emphasizing to conquer and control nature towards the direction of harmony between human beings and nature, human and human, and in the direction of mutual respect of all countries, nations and races, as well as cultures, thereby shifting from emphasizing competition towards emphasizing cooperation, peace and common development. Economic development also should shift from the old economy which emphasizes massive production (ergo massive consumption and mass abandonment) towards the new profit pattern emphasizing saving (moderate consumption and the recycling of resources). In the practice of science and technology development, a fundamental shift is required. Namely, it is necessary to shift from emphasizing hard technology development aimed at improving efficiency and conquering nature towards emphasizing balanced development with human society, creating an environment which is reassuring and suitable for securing substantial and healthy living (see Figure 25).

And, moreover, the long-term national and regional development strategies should adhere to making its strategy for economic development, social progress strategy, eco-environmental strategy and natural resources strategy suit the principle of mutually coordinated, harmonious and sustainable development (Quadruple Bottom Line); enterprise is necessary to shift its business model from the pursuit of maximizing economic interests to a profit pattern which takes account of economic, social and environmental responsibilities at one and the same time (Triple Bottom Line).

The essence of Chinese traditional culture speaks directly to the challenges now facing us. For example, the ideas of 'peace is precious' and of 'unifying humans and heaven' may be of significant theoretical and practical value through their contribution to the cultivation of sustainable development as a new concept and as guiding principles for establishing the future world order.

The principle of 'Harmony, Balance and Coexistence' is in line with ancient Chinese culture and philosophy of Yin-Yang, which is just like the words of the first sentence in *Plain Questions: On the Changes of Yin and Yang* of the *Yellow Emperor's Inner Canon*, 'the so-called "Yin" and "Yang" are the laws of heaven and Earth, the discipline of the universe, the origin of immortality and death, the parents of change, and the mansion of gods and spirits' (Shide Cheng 1982: 68).

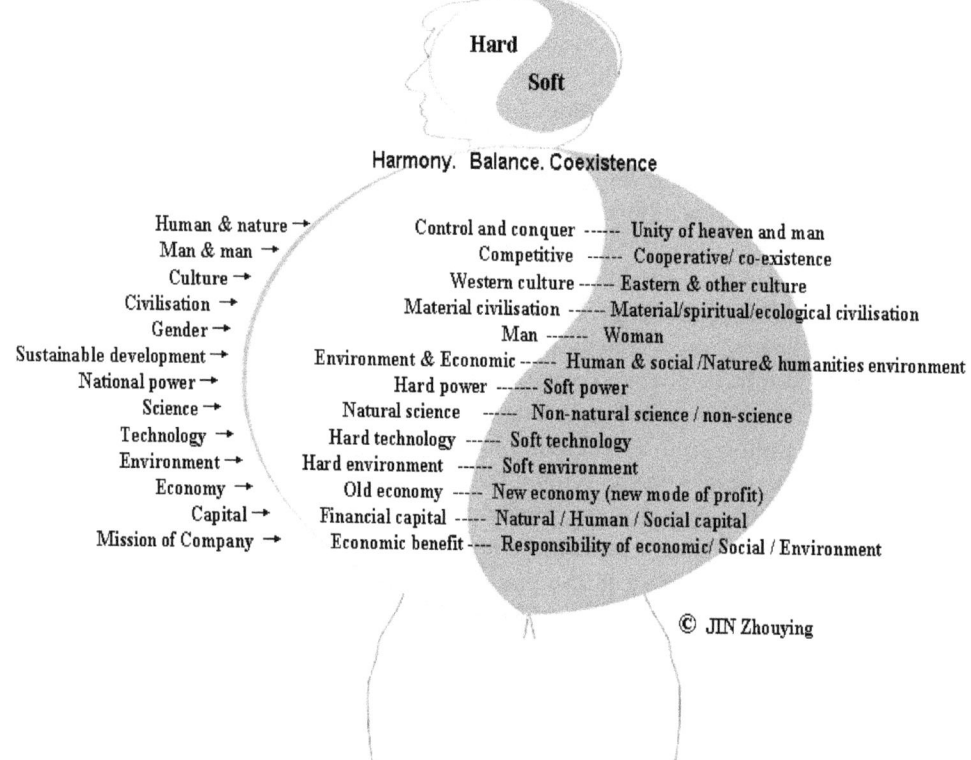

Figure 25: The Principles for Development in the Twenty-First Century – Harmony, Balance and Coexistence

About the Author

Zhouying JIN is a senior researcher and professor at Institute of Quantitative-Economics and Technical-Economics of the Chinese Academy of Social Sciences); director of the Centre for Technology Innovation and Strategy Studies (CTISS) of the Chinese Academy of Social Sciences. She is president and founder for the Beijing Academy of Soft Technology. She is also Chairman and founder of The Future 500 (China).

She graduated from Chinese University of Science and Technology in 1965; was a researcher, Institute of Mechanics, Chinese Academy of Sciences (1960s); Deputy Chief Engineer of Changchun Electric Industrial Administration (1970s); studied Management and Consultant in Japan Productivity Center (1980 and 1984); and vice-secretary-general for China enterprise Directors (Managers) Association of State Economic Committee of China (1980s). Between 1996 and 2000 she was the Team Leader of Strategy Research Experts for the High-Tech Research & Development Plan of China (Nation's S863 Plan, 2001–2010). She was visiting professor of Case Western Reserve University and University of Nebraska-Lincoln, U.S.A (1993–1994); Senior Research Fellow, Institute for the Future (U.S. CA, Menlo Park 1996); special researcher at the Institute of Science and Technology policy of Japan (2001–2002); and visiting professor of University of Aix-Marseille III of France (2003–10).

She is the author of thirteen books, six translated books, fifty research reports and hundreds of papers. Among her publications are: *Management and Enterprises Diagnosis*; *Soft Technology – The Essence and Space of Innovation*; *Service Innovation and Social Resource*; *Virtual Institute and Organizational Innovation*; *Technology Driving Force for Sustainable Development – Principle of Harmony and Balance*; *Soft Technology*; *Technological Institution and Soft Technology*; *Global Technological Change – From Hard Technology to Soft Technology*; *Long-term strategy Integration and Sustainable Development*; *Soft Technology, Soft Infrastructure and Knowledge Economy*; *From Nation's Soft Power to Soft Power of Enterprises* and *Global Technological Change and Business Mode Innovation – The Fourth Industrial Revolution*..

She led more than thirty important projects asked by central government, ministries, the Chinese Academy of Social Sciences, or international organizations including 'The Strategy Research for the Nation's High-tech Plan'; 'Long-term strategy integration and sustainable development'; "National strategy for Petroleum and Natural Gas'; 'Evaluation of Nation's High-Tech Research & Development Plan'; 'Assessment for The Breakthrough Project of China'; 'Service innovation and NGO organizations in China'; 'Strategic Management and institutional innovation on Coal bed Methane industry'; 'Green Car Development Guideline and Environmental policy in China'; 'Environmental strategy for transnational Auto companies'; 'The water problem in China'; 'The policy of robot development in China' and 'China's GPI—Genuine Progress Indicator'.

Bibliography

100 Questions and 100 Answers about Portfolio (1988), Tokyo: Nikkei Publishing.
'1999: New Record for Global Enterprise Merger' (1999), *EFE*, New York, 23 December.
2001 International Nanometre High-level Forum and Technology Application Seminar (2001), Beijing, July.
'The 5th Technology Forecasting Survey' (1992), *National Institute of Science and Technology Policy*, Japan Science and Technology Agency.
'The 6th Technology Forecasting Survey' (1997), *National Institute of Science and Technology Policy*, Japan Science and Technology Agency.
'The 7th Technology Forecasting Survey' (2001), *National Institute of Science and Technology Policy*, Japan Science and Technology Agency.
Adler, Paul and Kwon, Seok-Woo (2000), 'The Good, the Bad and the Ugly', *Knowledge and Social Capital*, Boston: Butterworth Heinemann.
'Advantage' (2001), *The Economist* (UK), 24 February.
Aikawa, Haruki (1941), *Introduction to Technology*, Tokyo: Mikasashobo, p. 131.
'The aim of social technologies – to make people feel at ease and substantial' (2001), *Asahi Shinbun*, 2 February.
Aldridge, S., Halpern, D. and Fitzpatrick, S. (eds.) (2002), 'Social Capital', *Performance and Innovation Unit*, 26 March.
American Accountants Association (1996), *Essential Accounting Theory Bulletin*, p. 1.
Amidon, Debra M. (1997), *The Ken Awakening*, Boston: Butterworth Heinemann.
Ansoff, Igor (1965), *Corporate Strategy*.
Anthony, Robert N., Reece, James S. and Hertenstein, Julie H. ([1995] trans. 2000), *Accounting: Text and Cases*, Ninth Edition, Boston: McGraw-Hill College, (trans. Xun Luo et. al.), Beijing: Peking University Press, pp. 7–8.
Aoki, Ryozo (1987), *New Industrial Theory*, Tokyo: Nikkei Publishing.
Barnard, Chester I. ([1938] trans. 2007), *The Functions of the Executive*, Cambridge, MA: Harvard University Press, (trans. Yonggui Wang), Beijing: China Machine Press.
Beijing Science and Technological Enterprise Incubator Annual Report (1999).
Beijing Youth Daily (2001), 22 November.
Bell, Daniel (1984), *The Coming of Post-Industry Society*, (trans. Kuo Gao et. al.), Beijing: The Commercial Press, p. 211.
Bernal, John D. ([1965] trans. 1970), *Science in History*, Third Edition, London: Watts & Co., (trans. Yasio Sizume), Tokyo: Misuzu Shobo, pp. 338–339; 767–768.
Bernstein, David (2004), *How to Change the World: Social Entrepreneurs and the Power of New Ideas*, New York: Oxford University Press.

Bourdieu, Pierre (1980), 'Le capital social', *Actes de la Recherche en Sciences Sociales*, 31.
Cetron, Marvin J. ([1969] trans. 1970), *Technological Forecasting*, New York: Gordon and Breach Science Publishers, (trans. Yukimazsu Takeda), Tokyo: Sangyo Noritsu Tanki Daigaku, pp. 4–5; 12.
Champy, James and Hammer, Michael (1994), *Reengineering the Corporation*, New York: Harper Business.
Chandler, Alfred D. (1962), *Strategy and Structure*.
Cheng, Deyu (comp. and ed.) (1997), *A Course of Insurance*, Beijing: China Communications Press, p. 15.
Cheng, Shide, Wang, Hongtu and Lu, Zhaolin (eds.) (1982), *Collection of Explanatory for 'Plain Questions' of the Yellow Emperor's Inner Canon*, 1, January, Beijing: People's Hygiene Press, p. 68.
Cheng, Siwei (2008), 'Institutional Innovation is the Core of the Reform', *2008 China Reform Forum*, Foshan, China, 8 April.
Cheng, Siyan (ed.) (1999), *General Cultural Economics*, Shanghai: Shanghai Finance and Economics Press, p. 61.
China Economic Times (2001), 4 December.
China's National Health Statistics Yearbook 2009 (2009).
China Statistical Yearbook (1991).
China Statistical Yearbook (2009).
Coase, R., Alchain, A. and North, D. (2000), *Property Rights and Institutional Changes*, Shanghai: SDX Joint Publishing Co./Shanghai People's Publishing House, p. 329.
Coleman, James (1988), 'Social Capital in the Creation of Human Capital', *American Journal of Sociology*.
The Comprehensive Dictionary of Chinese Language (1989), Shanghai: Shanghai Lexicographical Publishing House, pp. 210; 627; 1731.
Concise Encyclopaedia Britannica (1985a), 1, Beijing, Shanghai: Encyclopaedia of China Publishing House, p. 256.
Concise Encyclopaedia Britannica (1985b), 3, Beijing, Shanghai: Encyclopaedia of China Publishing House, pp. 422; 524.
Concise Encyclopaedia Britannica (1985c), 4, Beijing, Shanghai: Encyclopaedia of China Publishing House, p. 233.
Concise Encyclopaedia Britannica (1986a), 7, Beijing, Shanghai: Encyclopaedia of China Publishing House, pp. 641–644.
Concise Encyclopaedia Britannica (1986b), 9, Beijing, Shanghai: Encyclopaedia of China Publishing House, p. 541.
Cui, Jiehe (comp. and ed.) (1988), *An Introduction to Physical Distribution*, Beijing: China Commercial Publishing House, pp. 1–3.
Davidow, William H. and Malone, Michael S. (1992), *The Virtual Corporation: Structuring and Revitalizing the Corporation for the 21st century*, New York: Harper Collins.
Deiaco, Enrico (2000), 'Technology Foresight in Sweden', *International Conference on Technology Foresight*, Tokyo, Japan, March.
Deng, Biquan (2003), 'The Development of Cultural Industry Is Imperative', *Xinhua net*, 29 September.
Dixon, J. et al. ([1997] trans. 1998), 'Expanding the Measure of Wealth: Indicators of Environmentally Sustainable Development', *Environmentally Sustainable Development Studies and Monographs Series*, 17, Washington D.C: Environmental Protection Bureau of World Bank, (trans. Kunmin Zhang et. al.), Beijing: China Environmental Science Press, p. 160.

Don, Bruce (2000), 'Changes in the US Approach to Technology Foresight and Critical Technology Assessment', *International Conference on Technology Foresight*, Tokyo, Japan, March.

Dong, Guangbi (1996), 'On Strategy of Chinese Culture', *Guangming Daily*, 13 January.

Dong Guangbi (1998), 'The History and Prospect of Chinese Science Modernization', *Historical Reference and Implications of the Success of Science & Technology Development*, Beijing: Science Press, p. 59.

Dong, Guangbi (1989), 'On Social Technology', *Dialectics of Nature Newspaper*, 269.

Dukas, Helen and Hoffman, Banesh (eds.) ([1981] trans. 1984), *Albert Einstein: The Human Side*, New Jersey: Princeton University Press, (trans. Zhikai Gao), Beijing: World Knowledge Publishing House, p. 78.

Duplaga, Mariusz, Radziszowski, Dominik et. al. (unknown date), 'Technical and Non-Technical Factors Influencing the Process of Teleconsultation Services Development Carried Out in Krakow Centre of Telemedicine', http://portal.ics.agh.edu.pl:8001/papers/TR-03-4.doc.

Drucker, Peter (1964), *The Creative Manager*.

E-Square Inc. ([2007] trans. 2008), 'Sustainable Agriculture Survey', Japan (trans. Center for the Future of Beijing Academy of Soft Technology), Beijing: Xinhua Publishing House.

The Economist (2001), 30 January.

Edsall, Thomas B. (2006), 'Lobbyists Emergence Reflects Shift in Capital Culture', *The Washington Post*, 12 January. Reprinted by *Reference News*, 16 January 2006.

Encyclopaedia China – Economy, Finance, Agriculture (1994), Beijing, Shanghai: Encyclopaedia of China Publishing House, pp. 1476–1495.

Encyclopaedia China – Finance, Revenue, Banking and Price (1993), Beijing, Shanghai: Encyclopaedia of China Publishing House, p. 8.

Encyclopaedia China – Sociology, (1991), Beijing, Shanghai: Encyclopaedia of China Publishing House, pp. 409–418.

Encyclopaedia Britannica (1999), 8, Beijing, Shanghai: Encyclopaedia of China Publishing House, p. 389.

An Encyclopaedia of Finance (1990), Beijing: China Development Press, p. 2097.

Etymological Dictionary of Chinese Characters (1998), Beijing: The Commercial Press, p. 658.

Fang, Xianming (comp.) (1989), *Insurance Guide*, Shanghai: Shanghai Science and Technology Press, p. 39.

Fukumitu, Hirosi (1986), *Securities market in the age of financial deregulation*, Tokyo: Nihon Keizai Hyoronsha.

'The Future of Entertainment Technology' (2001), *Nikkei Science*, February.

The Future of High-Tech – Positive and Negative Sides (2007), Beijing: China Science and Technology Press.

Gan, Shijun et al. (1989), *Soft Science in China*, Wuhan: Huazhong University of Science and Engineering Press.

Gan, Tairoku and Hiromasa, Naohiko (1986), *On Technology*, Tokyo: Tokai University Press.

Gan, Zangchun (2001), 'Market Economy and Reshaping Government Role', *China Economic Times*, 20 June.

Gao, Lianghua (1996), *Technology in the View of Humanities*, Beijing: China Social Science Press.

Gao, Shangquan (2000), 'Institutional innovation is the Key', *China Business Times*, 24 October.

Gao, Weijiang (ed.) (1998), *A Course of Public Relations*, Suzhou: Suzhou University Press, pp. 27–29.

Garten, Jeffrey E. (2001), 'Intellectual Property: New Answers to New Problems', *BusinessWeek*, 2 April.

Georghiou, Luke (2000), 'Third Generation Foresight – Integrating the Socio-economic Dimension', *International Conference on Technology Foresight*, Tokyo, Japan, March.

'GEP – Chinese made the rule of game' (2001), *Science and Technology Daily*, 10 December.
Giridharadas, Anand (2006), 'For many Indians, higher education does more harm than good', *International Herald Tribune*, 26 November.
Gloss, N. and Port, A. (1998), 'Next Upsurge', *American Business Weekly*, 31 August.
Grayson, David, Jin, Zhouying, Lemon, Mark, Slaughter, Sarah, Rodriguez, Miguel A. and Tay, Simon (2008), 'A New Mindset for Corporate Sustainability', *White Paper sponsored by BT and Cisco*.
Gross, Daniel (2005), 'What Makes a Nation More Productive? It's Not Just Technology', *The New York Times*, 25 December.
Guan, Xin (2000), 'The Statistics of American R & D investment', *World Economic Outlook*, 10.
Gui, Shou (2000), *Chinese Macro-economy Information Web*, 3 November, www.macrochina.com.cn/info.shtml.
Hasegawa, Yosaku (2001), '30 Years of Future Technology Forecasting', *Technology and Economics Magazine*, January.
He, Zhiyong (ed.) (1997), *Stock System Innovation*, Chengdu: Southwestern University of Finance and Economics Press, pp. 2–5.
He, Zhonghua (2002), 'The Spirit of the University and its Paradoxes in the Modern Context', *Journal of Literature, History and Philosophy*, 1.
Helmer, Olaf, Brown, Bernice and Gordon, Theodore (1966), *Social Technology*, New York: Basic Books.
Henderson, Hazel (2002), 'Beyond Globalization: Building A Win-Win World', *Closing Plenary Session of the 2002 Forum of World Future Society*, Philadelphia, USA, 21 July.
'High-tech Industries and Coffee Shops Propel a Revolution' (1999), *Financial Times*, 2 March.
Hirasawa, Ryo (1999), 'Innovation Competitiveness in the Era of Knowledge Economy: Lessons from Japanese Enterprises', *International Forum on Knowledge Economy and Industrialization of High Technology*, October.
Homer (trans. 1997), *Homeric Hymns*, (trans. Huansheng Wang), Beijing: People's Literature Publishing House.
Honda, Syuro (1975), *Technological Anthropology*, Tokyo: Asakura Bookstore.
Howkins, John (2006), *The Creative Economy*, (trans. Qingfu Hong, Weiwei Sun and Maoling Liu), Shanghai: SDX Joint Publishing Co.
IBM Global Service (2008), *Soft Manufacturing*, Beijing: Oriental Press.
'IBM's report submitted to the U.S. Securities and Exchange Commission (SEC)' (2009), *Netease Science and Technology News*, 8 September.
IMD (2000), *The World Competitiveness Yearbook 2000*.
'Incorporating Military Tactics into Military scientific system' (2001), *Liberation Army Daily*, 18 December.
The Institute for Future Technology (1973a), 'Basic Design of Japan-Type Science and Technology Development System – The Status Quo and Trend of Soft Science in the United States', 4: 1.
The Institute for Future Technology (1973b), 'Basic Design of Japan-Type Science and Technology Development System – The Status Quo and Trend of Soft Science in the United States', 6: 1.
'Interview of Milton Friedman, the Nobel Prize Winner' (2001), *Times Weekly* (Germany), 21 June. Reprinted by *Reference News*, 10 July 2001.
Ishiguro, Kazunori (2000), 'Globalization and Law', *Mainichi Shinbunsha* (Japan). Reprinted in *The Economist*, 30 January 2001.
Ito, Shuntaro et. al. (eds.) ([1983] trans. 1984), *Chronological Table of Brief World History of Science and Technology*, Tokyo: KOBUNDO, (trans. Zhenhuan Jiang et. al.), Harbin: Harbin Industry University Press.

Jantsch, Erich ([1967] trans. 1968), *Technological Forecasting in Perspective*, Paris: Organization For Economic Co-operation and Development – OECD, (trans. Japanese Business Management Centre), pp. 15; 290.

Japan External Trade Organization (JETRO) (1995), 'Institutions Regarding Science and Technology Activities: Investigation of Europe and America', *Japanese Academy of Industrial Technology*.

Japan Nomura Research Institute (1990), *The Strategy for Technology Toward 2000*, November.

Jewkes, J., Sawers, D. and Stillerman R. (1968), *The Sources of Invention*, Second Edition, London: Macmillan.

Ji, Sha (1999), *The Thinking for The Future Medicine*, Beijing: China Medicine Science Publishing Company.

Jin, Wulun (1997), 'The Purpose of Science is to seek knowledge and truth', *Guangming Daily*, 12 April.

Jin, Wulun (2000), 'It is Necessary to Make a Clear Distinction between Science and Technology', *Science and Technology Daily*, 15 December.

Jin, Zhouying (1997), 'Technology Driving Force – The Principle of Harmony and Balance', *I3UPDATE*, http://www.skyrme.com/updates.

Jin, Zhouying (1998a), 'Knowledge Economy Practice is the Key' in Graduate School of Chinese Academy of Social Sciences and Graduate School of Chinese Academy of Sciences (ed.), *Knowledge Economy and the National Innovation System*, Beijing: Economic Management Press.

Jin, Zhouying (1998b), 'Organizational Innovation and Virtual Institute', *Chinese Soft Science*, 4.

Jin, Zhouying (1998c), 'To View Knowledge Economy from Another Angle', *Science and Technology Daily*, 14 November.

Jin, Zhouying (1999a), *World Forum 1999*, San Francisco, USA, October.

Jin, Zhouying (1999b), 'High-tech in Knowledge Economy Age', *Journal of Quantitative & Technical Economics*, August.

Jin, Zhouying (1999c), *S863 Project Soft Science Research – Strategy Research Report*, June.

Jin, Zhouying (1999d), *Internet World*, 2.

Jin, Zhouying (1999e), 'Knowledge Economy and Soft Technology', *International Conference on 'Knowledge Economy and China'*, Beijing, China, 3 November.

Jin, Zhouying (1999f), 'Knowledge Economy and Soft Technology', *China Entrepreneur Newspaper*, 26 November.

Jin, Zhouying (1999g), *Journal of Knowledge Management*, MCB University Press, 3: 1.

Jin, Zhouying (2000a), 'Soft-tech Industry/High-tech Industry – A Thinking on Economic Development of Beijing', *Centre for Technology Innovation and Strategy Studies, Chinese Academy of Social Sciences*.

Jin, Zhouying (2000b), 'Soft Technology', *GIST Academic Journal Series*, Yokohama: GIST.

Jin, Zhouying (2000c), 'Soft Technology', *PAF Conference*, Kobe, Japan, February.

Jin, Zhouying (2001), 'Technology Driving Force – The Principle of Harmony and Balance', *AI & Society MS 193*.

Jin, Zhouying (2005), *Global Technological Change – From Hard Technology to Soft Technology*, London: Intellect Books.

Jin, Zhouying (2007), 'Towards a Quadruple Bottom Line: Achieving Sustainability in China', *A New Mindset for Corporate Sustainability*, UK.

Jin, Zhouying (2008), 'From National Soft Power to Corporate Soft Power', *China Soft Science*, 212: 8.

Jin, Zhouying and Bai, Ying (2009), 'Beyond the financial crisis, look forward for the future – The building 4th Industrial Revolution', *AI & Society*, 24: 4.

Jin, Zhouying, Jiang, Jinhe and Gong, Feihong (2006), *Long-term Strategy-System Integration and Sustainable Development*, Beijing: China Social Sciences Academic Press.

Jin, Zhouying and Ren, Lin (2004), *Service Innovation and Social Resource*, Beijing: China Financial and Economic Publishing House.

Johansen, Bob (2007), *Getting There Early*, San Francisco: Berrett-Koeher Publishers.

Jones, Charles (1999), 'Was an Industrial Revolution Inevitable? Economic Growth Over the Very Long Run', in *Stanford University, Department of Economics, Working Papers*, October 1999. Reprinted by *Science and Technology Daily*, 18 January 2000.

Jorgenson, Dale W. (2001), 'IT and American Economy', *American Economic Review*, March.

Journal of Taiwan Economy Forum (2009), 7: 8.

Joy, Bill (2000), 'Why the Future Doesn't Need Us', *Wired*, April.

Juan, Juan (2007), 'A Comparative Study of the Logistics Cost Between the USA and China', *Journal of Northern Economy*, 10.

Kasper, Wolfgang and Streit, Manfred E. (1998), *Institutional Economics: Social Order and Public Policy*, Cheltenham, UK & Northampton, MA: Edward Elgar Publishing, pp. 32–142.

Kato, Hidetoshi (1983), *Sociology of Technology*, PHP Institute, pp. 248; 262.

Kenney, Martin (2000), 'Institution for New Firm Formation in Silicon Valley', *International Seminar on Technological Innovation*, Beijing, China, 5–7 September.

Kishimoto, Yoshiyuki (2000), 'Management', *Hitotsubashi Business Review*, WIN.

Kiuchi, Tachi and Shireman, Bill (2002), *What we learned in the rainforest*, San Francisco: Berrett-Koehler Publishers.

Kondo, Takao (1997), 'To Fabricate Value in Value Management', *Value Development Speeches Collection*, Institute of Science and Technology Policy of Japan STA.

Kurihara, Siro (1987), *Technology Philosophy for the Future*, Tokyo: Ohmsha.

Kusaka, Kimindo (1978), *The New Culture Industry*, Tokyo: Toyo Keizai Shinposha.

Kusaka, Kimindo (comp.) (1980), *New Map of Culture Industry*, Tokyo: Nikkei Publishing.

Kusaka, Kimindo (1985), 'Japanese Technology or American Wisdom?', *Ushio Magazine*, July.

Kuwahara, Terutaka (2001), 'Technology Forecasting – Past, Today and Future', Seminar at the National Institute of Science and Technology Policy, Tokyo, Japan, 8 February.

Landes, David S. ([1999] trans. 2001), *The Wealth And Poverty of Nations: Why Some are so Rich and Some are so Poor*, New York: W.W. Norton & Company, (trans. Honghua Men et. al.), Beijing: Xinhua Publishing House, p. 278.

Le, Housheng (2000), *Golden Industry of the 21st Century – Economic Upsurge of Cultural Industry*, Beijing: Chinese Society Press.

Leadbeater, Charles ([1997] trans. 2006), 'The Rise of the Social Entrepreneur', *Demos*, (trans. Fan Li et. al.), Global Links Initiative.

Lee, Kaifu (2006), 'Microsoft's way to success', *A Walk into the Future*, Beijing: People's Publishing House.

Lesser, Eric L. (ed.) (2000), *Knowledge and Social Capital: Foundations and Applications*, Boston, MA: Butterworth Heinemann, p. 97.

Li, Jianliang (comp. and ed.) (1999), *Operational Guide of Venture Capital*, Beijing: China Industry & Commerce Associated Press, pp. 16–21.

Li Jiqun (comp. and ed.) (1999), *An Introduction to Patentology*, Beijing: China Textile & Apparel Press, pp. 11–13.

Li, Rong (2001), *Science and Technology Daily*, 3 June.

Li, Yifei (1999), 'New Concept in Public Relations: Cognitive Management', *Science and Technology Daily*, 24 January.

Li, Yining (1990), 'Economic Culture', *Journal of Cultural Economics*, 4.

Lietaer, Bernard A. (1999), *The Future of Money: A New Way to Create Wealth, Work and a Wiser World*, London: Random House.

Lin, Dafeng and Liu, Meizhu (2003), 'Exploration on the Meaning and Development of Somatics', *Sports Journals of Taitung University*, 1, pp. 249–272.

Liu, Changli (2003), 'Toyota's Real-Time Logistics Management Strategy', *Website of China's Logistics Network*.

Liu, Manhong (ed.) (1998), *Venture Capital: Innovation and Finance*, Beijing: China Renming University Press, pp. 1–51.

Liu, Tianjun (1994), *On concrete thinking – Thought Operation in the Process of Keeping Still and In Deep Meditation*, Beijing: People's Physical Education Press.

Liu, Tianjun (1995), 'Concrete Thinking is the fundamental thinking model of traditional Chinese medical', *Chinese Journal of Basic Medicine in Traditional Chinese Medicine*, 1.

Liu, Tianjun (1996a), 'Experimental Science and Experience Science – Contrast between Traditional Chinese Medicine and Western Medicine', *Chinese Journal of Basic Medicine in Traditional Chinese Medicine*, 2: 1.

Liu, Tianjun (1996b), 'Experimental Science and LPFE Science – A Comparison of the Methodology of TCM and West Medicine', *Chinese Journal of Basic Medicine in Traditional Chinese Medicine*, 2: 1.

Liu, Tianjun (1996c), 'The Methodological Framework of LPFE Science', *Chinese Journal of Basic Medicine in Traditional Chinese Medicine*, 2: 3.

Liu, Tianjun (1999), *Traditional Chinese Medicine – QiGong*, Beijing: People's Hygiene Press.

Liu, Tonggui (2009), 'Guaranteed Chicken', *Dazhong Daily*, 5 February.

'Logistics Revolution' (2000), *Handelsblatt* (Germany), 25 September.

Luo, Zhihua (2007), 'Introduction of Tactics Technology', *Strategy Weekly of China Strategy Science Network China Strategy Web of Science*, 5.

Ma, Kai (2008), 'To Promote the Better and Faster Development of MBA Education in China', *China Education Daily*, 14 November.

Machlup, Fritz ([1958] trans. 1975), *An Economic Review of the Patent System*, Washington D.C: US Government Printing Office, (trans. Teruo Doi), Tokyo: Nikkei Publishing.

Machlup, Fritz ([1962] trans. 1969), 'The Production and Distribution of knowledge in the United States', Princeton, NJ: Princeton University Press, (trans. Tatuo Takahasi and Hirosi Kita), Tokyo: Sangyo Nohritsu Tanki Daigaku.

Maddison, Angus (1995), *Monitoring the World Economy 1820–1992*, Paris, 4 September, Development Centre of the Organisation for Economic Co-operation and Development.

Maeda, Kazuhisa (1986), 'An Exploration for the Key for Softwarized Age', *Technology and Economics Monthly*, November. Reprinted by *World Economy Science*, 3 March 1987.

Makino, Noboru (1983), *A Misjudgement for Future Industry?*, Tokyo: Toyo Keizai Shinposha.

Malinowski, Bronislaw ([1944] trans. 1946), *A Scientific Theory of Culture*, Chapel Hill: The University of North Carolina Press, (trans. Xiaotong Fei et. al.), Beijing: The Commercial Press.

Malone, Michael and Davidow, Bill (1994), 'Welcome to the Age of Virtual Corporations', *Computer Currents*, 12: 1.

Martin, Ben R. and Irvine, John (1989), *Research Foresight: Priority-Setting in Science*, London & New York: Pinter, pp. 22–24; 109.

Meadows, Donella H. and Meadows, Dennis L. et. al. ([1972] trans. 1997), *The Limits To Growth: A Report for the Club of Rome's Project on the Predicament of Mankind*, New York: Universe Books, (trans. Baoheng Li), Changchun: Jilin Renmin Press.

Meng, Xisheng (ed.) (1997), *An Introduction to Modern Logistics*, Beijing: The Great Wall Publishing House, pp. 1–15.

Min, Qingfei and Tang, Keyue (2003), 'Pay More Attention on Non-Technological Factors of the Informatization in Chinese Enterprises', *Science and Technology Management Research*, 23: 1.

Misumi, Jyuji (1955), *Introduction of Social Technology*, Tokyo: HAKUA Press.

Modern Chinese Dictionary (1991), Beijing: The Commercial Press, p. 533.

Morris, James H. (2001), 'An Update: US Patents on Business Methods', *Licensing Executive Society Annual Conference*, South Africa, 1 May.

Muramatsu, Shinobu (1988), *Merger, Acquisition and Enterprise Evaluation*, Tokyo: Dobunkan.

Naisbitt, John, Naisbitt Nana and Philips, Douglas ([1999] trans. 2000), *High Tech, High Touch: Technology and Our Search for Meaning*, New York: Broadway Books 1999, (trans. Ping Yin), Kuala Lumpur: Mentor.

Nakayama, Yichiro et al.(eds.) (1971), *Yuhikaku Dictionary of Economics*, Tokyo: Yuhikaku Publishing, p. 427.

Nihon, Keizai Shimbun (ed.) (1987), *Money Guide*, Tokyo: Nikkei Publishing.

NISTEP (National Institute of Science and Technology Policy of Japanese) (1988), 'A Survey on Present Situation and Future Trend of R&D of Series of Soft Science and Technology', March.

NISTEP (National Institute of Science and Technology Policy of Japanese) (1989), 'A Survey on Present Situation and Future Trend of R&D of Series of Soft Science and Technology', March.

Nonaka, Ikujiro and Takeuchi, Hirotaka (1997), *To Seek Success through Innovation – Knowledge Creates Companies*, (trans. Zijang Yang and Meiyin Wang et. al.), Taiwan: YuanLiu Press.

North, Douglass C. (1981), *Structure and Change in Economic History*, First Edition, New York: W.W. Norton & Company, (trans. Yu Chen, Huaping Luo et. al.), Shanghai: SDX Joint Publishing Co.; Shanghai People's Publishing House, pp. 132–325.

North, Douglass C. (1994), *Institution, Institutional Changes and Economic Performance*, Shanghai: SDX Joint Publishing Co.

Nye, Joseph S. (1990), *Bound to Lead: The Changing Nature of American Power*, New York: Basic Books.

OECD (2000), 'The Service Economy', *Business and Industry Policy Forum Series*.

OECD (2001), 'Innovation and Productivity in Services', *OECD Proceedings*.

Petersen, Craig and Lewis, Cris (2000), *Managerial Economics*, Beijing: China Renmin University Press, p. 431.

Pine, J. and Gilmore, J. ([1999] trans. 2008), *The Experience Economy*, Boston: Harvard Business School Press, (trans. Yeliang Xia et. al.), Beijing: China Machine Press.

Popenoe, David ([1995] trans. 1999), *Sociology*, Tenth Edition, Englewood Cliffs, NJ: Prentice Hall, (trans. Qiang Li et. al.), Beijing: China Renmin University Press.

Porat, Marc U. and Rubin, Michael R. ([1977] trans. 1987), 'The Information Economy', *Office of Telecommunications Special Publication 77-12 (US Department of Commerce, Washington D.C.)*, (trans. Junshi Yuan et. al.), Beijing: China Zhanwang Press.

Porter, Michael (1980), *Competitive Strategy: Techniques for Analyzing Industries and Competitors*, New York: The Free Press, 1980.

Porter, Michael (1985), *Competitive Advantage: Creating and Sustaining Superior Performance*, New York: The Free Press.

Porter, Michael (1990), *The Competitive Advantage of Nations*, New York: The Free Press.

Portes, Alejandro (2000), 'Social Capital: Its Origins and Applications in Modern Sociology', in: Eric L. Lesser (ed.), *Knowledge and Social Capital*, Boston: Butterworth Heinemann.

Prime Minister of Japan (1993), 'On R&D Basic Project of Soft Series of Science and Technology', *Recommendation No. 19 of the Council for Science and Technology*, 11 January.

Project Team of Shanghai Municipal Government Research Office, Publicity Dept. CPC Shanghai Committee (comp. and ed.) (1998), 'Research on Developing Culture Industry in Shanghai', Shanghai: Shanghai Education Press, p. 82.
Putnam, Robert D. (1993), 'The Prosperous Community', *American Prospect*.
Putnam, Robert D. (1995), 'Bowling Alone', *Journal of Democracy*.
Qian, Xueshen (2001), 'Qian Xueshen and Metasynthetic Wisdom', *People's Daily Overseas Edition*, 24 February.
Qin, Lichun and Zhou, Zhihong (comp. and eds.) (1995), *Public Relations*, Changsha: Hunan Science and Technology Press, pp. 49–54.
Qiu, Hongzhong (1993), *Medicine and Human Culture*, Changsha: Hunan Science & Technology Press.
Qiu, Hongzhong (2001), 'On the Scientific Spirit and the Humanistic Methods of Chinese Traditional Medicine', *Medicine and Philosophy*, September.
Rader, Randall (2007), *Speech in the International Forum on 'WTO: IPR Issues in Standardization'*.
Reference News (1999), 6 April.
Reference News (2001), 5 May.
Reference News (2008), 21 January.
Ren, Tianyuan (comp. and ed.) (2000), *Operation and Evaluation of Venture Capital*, Beijing: China Economic Press, pp. 20–58.
Rivette, Kevin G. and Kline, David (2000), *Rembrandts in the Attics: Unlocking the Hidden Value of Patents*, Boston: Harvard Business School Press, p. 16.
Roan, Shari (2003), 'The mind's role comes into focus', *The Los Angeles Times*, 20 January.
Ru, Xin (2009), 'A Deep Understanding of the Nature of Contemporary Capitalism', *Journal of Dynamics of World Socialism*, 9.
'S863 Soft Science Research Strategy Report' (1999), March.
Salamon, Lester M. and Anheier, Helmut K. (1992), 'In Search of the Non-profit Sector II: The problem of Classification', *Voluntas* (USA), 3: 3, p. 5.
Savage, Charles M. (1990), *Fifth Generation Management: Integrating Enterprises through Human Networking*, Bedford, MA: Digital Press.
Schumpeter, Joseph A. ([1912] trans. 1934), *The Theory of Economic Development: An Inquiry into Profits, Capital, Credit, Interest, and the Business Cycle*, (trans. Opie Redvers), Cambridge MA: Harvard University Press, p.66.
Science & Technology Agency of Japan (ed.) (no date), *White Paper on Science and Technology*, 1971–1987 Edition.
Science & Technology Agency of Japan (ed.) (no date), *White Paper on Science and Technology*, 1988–1998 Edition.
Science and Technology Daily (1999), 10 January.
Science and Technology Daily (2000), 4 June.
Science and Technology Policy Bureau of State Science and Technology Commission (ed.) (1988), *The Rise of Soft Science*, Beijing: The Earthquake Press.
Shen, Kang (2002), 'Non-Technological Factors Affecting the Hospital Network Security', *Proceedings of CMIA'02*, Beijing, China.
Sheridan, Tatsuno (2000), *The 2000 Annual Conference of Institute of Innovation, Creation & Capital*, Austin, TX, USA.
Shi, Zhihua (2000), *A Fresh Interpretation of Sun Tzu's Art of War*, Shanghai: Xuelin Press.
Siwek, Stephen E. (2009), 'Copyright Industries in the US Economy: The 2003–2007 Report', in The International Intellectual Property Alliance (IIPA), *Economists Incorporated*.

Skyrme, David (1997), 'Virtual Teaming and Virtual Organizations – 25 Principles of Proven Practice', *I3UPDATE*, 11.
Smith, Adam ([1880] trans. 1997), *An Inquiry into the Nature and Caused of the Wealth of Nations*, (trans. Dali Guo and Yanan Wang), Beijing: Commercial Press, pp. 1–26.
Song, Jian (ed.) (1994), *Essentials of Modern Sciences*. Beijing: Science Press.
State Council Information Office (2006), *Environmental Protection in China (1996–2005)*, White Paper, June.
State Science and Technology Commission (no date), *Chinese Science and Technology Index*, pp. 246–268.
Steiner, George A. (1969), *Top Management Planning*.
Strategy Research Group on National Innovation System (2004), *The Report on Development Strategy of Samsung Group*, 19 February.
Sue, Roger (1999), 'Le Social – Principale Ressource Economique/ Social Life – Main Economic Resource', *Le Monde*, 2 March.
Sun, Qixiang (2000), *Insurance*, Beijing: Peking University Press, pp. 20–36.
'A Survey for New Master of Event' (2000), *Thursday Event Weekly* (France). Reprinted by *Reference News*, 5 April 2000.
Swiss Re-insurance Company (2004a), *Sigma*, 3.
Swiss Re-insurance Company (2004b), *Sigma*, 8.
Takeya, Mitsuo (1968), *Collected Works of Takaya Mitsuo I, Dialectic Issues*, Tokyo: Keisoshobo, p. 139.
Tang, Hao (2007a), 'Weathering the storm', *China Dialogue*, 10 May.
Tang, Hao (2007b), 'An environmental approach to technology', *China Dialogue*, 21 November.
Tang, Lu (2001), 'New Economy, New Life', *Reference News*, 8 March.
Tanouchi, Koichi and Murata, Shoji (eds.) (1985), *Basic Theory of Modern Marketing*, Tokyo: Dobunkan.
Tao, Deyan (2000), 'Second Market: The Engine of New Economics', *Reference News*, 7 June.
Tegard, Greg (2000), 'Foresight Studies in Australia', *International Conference on Technology Foresight*, Tokyo, Japan, March.
Tu, Songbai (2008), 'Soft Manufacturing: Innovation Increases the Value of Manufacturing', *Seminar of The Study on Transnational Corporations*, December.
Tu, Zhengge and Xiao, Geng (2005), 'China's Industrial Productivity Revolution', *Economic Research Journal*, February.
Uchida, Moriya (1987), *Knowledge Capital – 21st Century Creative Technology Strategy*, Tokyo: Nikkan Kogyo Publication.
UNCTAD (2008), 'Creative Economy Report 2008', NCTAD/DITC.
United Kingdom Department of Culture, Media and Sport (DCMS) (1998), 'The Creative Industries Mapping Document (CIMD)', London: HMSO.
United Kingdom Department of Culture, Media and Sport (DCMS) (2001), 'The Creative Industries Mapping Document (CIMD)', London: HMSO.
USA Today (2001), October. Reprinted by *Reference News*, 28 October 2001.
van den Boom, Marien (2009), 'Intellectual Capital, Intellectual Property and Tourism: An Empirical Study in Southeast Asia', *Regional Symposium on Management of Intellectual Capital, Intellectual Assets and Intellectual Property*, Hong Kong SAR, 29–30 October, World Intellectual Property Organization.
The volume of historical economy statistical data for Soviet Union and the main capitalism countries (1800–1982) (1989), Beijing: People Press.

von Braun, Christoph-Friedrich ([1997] trans. 1999), *The Innovation War*, New Jersey: Prentice-Hall PTR, (trans. Ministry of Science and Technology International Cooperation of China), Beijing: Machine Industry Press.
Wang, Changya (ed.) (1985), *Fundamental Patentology*, Wuhan: Hubei Science and Technology Press, pp. 2–5.
Wang, Chiwei (2000), *Science and Technology Daily*, 28 May.
Wang, Dingding (2001), 'The newest economy', *Tianjin Daily*, 24 September.
Wang, Guohua (2002), 'Zhang Ruimin on Innovation Strategy', *China Economic Weekly*, 20 May.
Wang, Hongsheng (2001), *A History of World's Science and Technology*, Beijing: China Renmin University Press.
Wang, Hui and Zhang, Jin (comp. and eds.) (1989), *Public Relations*, Beijing: Guangming Daily Press, pp. 10–11.
Wang, Ling and Luo, Jetao et al. (comp. and eds.) (no date), *Modern Enterprise Logistics*, Beijing: Economic Science Press, pp. 1–7.
Wang, Liusheng and Huang, Xinxian (comp.) (1990), *Chronology of Big Events in the Chinese and Foreign History of Education*, Changchun: Jilin University Press.
Wang, Yuying and Zhen, Xiaohua (eds.) (1994), *Culture Industry*, China Peasant's Press.
Wang, Zhitai (1995), *Modern Logistics*, Beijing: China Logistics Publishing House, pp. 26–59.
Wei, Jinghong, Zhou, Yanfan and Qi, Jinbo (2000), *Management Accounting*, Beijing: China Agricultural Technology Press, pp. 3–8.
Wen, Tian (2001), 'The Hot Professions In Four Years', *Science and Technology Daily*, 16 March.
Williams, Trevor I. (ed.) ([1979] trans. 1989), *A History of Technology*, 6, New York: Oxford University Press. Chinese version published by Changsha: Central-South University of Technology Press, p. 37; 147.
Willoughby, Kelvin (1990), *Technology Choice: A Critique of the Appropriate Technology Movement*, Boulder and San Francisco: Westview Press.
Wood, John (2000), 'Current Foresight Activities in the UK', *International Conference on Technology Foresight*, Tokyo, Japan, March.
World Executive (2007), http://ceo.icxo.com/htmlnews/2007/12/18. Accessed 18 December.
Wu, Dongxi (1999), 'Misunderstanding of Knowledge Economy', *Tsinghua University R&D Newsletter*, 8.
Xenophon (trans. 1983), *Anabasis*, (trans. Jinrong Cui), Beijing: The Commercial Press.
Xiao, Yongliang (2009), 'US Copyright Industries Challenge the World', http://blog.cbice.com/xiaoyongliang/2009/09/24.
XinHua net (2010), www.news.cn. Accessed 12 March.
Xu, Guozhi, Gu, Jifa and Che, Hongan (comp.) (2000), *A Study on Systems Science and Engineering*, Shanghai: Shanghai Science and Technology Education Press.
Xu, Ling and Ma, Le (2007), 'Non-Technical Factors in Tasks of Using Science and Technology to Defeat Terrorism', *Studies in Dialectics of Nature*, 23: 9.
Yamada, Toshihiro (2000), 'Base of the Internet Revolution', *Toyokeizai* (Japan), 15 January.
Yan, Jianjun (1993), *The Rise of the Fourth Industry*, Kunming, China: Yunnan People's Press.
Yang, Ronggang (1987), *Modern Advertising*, Beijing: Economic Science Press.
Yang, Wenzhi (2003), 'Several deep-rooted problems in the current association reform', *Journal of Research on Association Reform*.
Yang, Xinhui (2005), *Modern Psychological Technology Studies*, Shanghai: Shanghai Education Press.
Yao, Haiming and Kun, Duan (comp. and eds.) (1999), *Insurance*, Shanghai: Fudan University Press, pp. 35–37.

Yi, Zhibin and Ma, Xiaoming (2009), 'Why do local environment regulations seize up', *Journal of Chinese Social Sciences*, August.

Yu, Jiwei (no date), *An Introduction to Economics of Share Holding System*, Shanghai: Fudan University Press, pp. 62–248.

Yuhikaku Dictionary of Economics (1979), Tokyo: Yuhikaku Publishing, p. 88.

Zhan, Dexiong (2008), 'The Hot Topic on "China Model" in Foreign Countries and its Implications', *Reference News*, 27 March.

Zhang, Gangzhu (comp. and ed.) (1988), *Stock Company – Establishment, Organization and Management*, Beijing: China Economic Press, pp. 13–19.

Zhang, Qinsheng (2002), 'The Analysis on Management Factors Affecting the Power of Technological Innovation and Technological Diffusion', *Missiles and Space Vehicles*, 1.

Zhang, Wei (2002), 'Did Arthur Andersen Lose Faith?', *China Economic Times*, 25 January.

Zhao, Jingxia (1998), 'Transnational Merger – Mainstream of International Direct Investment', *Transnational Enterprise Studies*, 6.

Zhao, Mulan (1999), 'The Road of Zhongguancun Science Park', *High and New-tech Industry Development and Programme Compendium of Zhongguancun Science Park*, Zhongguancun Science Park Management Commission, 21 July.

Zhao, Tao (1997), *Stock System – An Important Form of Modern Enterprises*, Beijing: Economic Science Press, pp. 5–6.

Zhao, Yuji (comp. and ed.) (1987), *Modern Advertising*, Beijing: China Business Press, pp. 6–7.

'Zhongguancun is on the rise' (2009), *China High-Tech Industry Herald*, 20 April.

Zhou, Qihong (2009), 'Agricultural Service Industry: An Entry Point for Modern Agriculture', *Hubei Daily*, 8 February.

Zhou, Shuqun, Zhang, Guowei and Yang, Fengming (comp. and eds.) (1999), *100 Techniques of Management Tactics*, Zhengzhou: Henan People's Press, p.10.

Zhou, Zhenhua (1999), *System Reform and Economic Growth*, Shanghai: SDX Joint Publishing Co. & Shanghai People's Publishing House.